Metal Fatigue: Theory and Design

The Authors

HOWARD L. LEVE	Head, Dynamic Ground Loads Group, Structural Mechanics Section, Douglas Aircraft Division, McDonnell Douglas Company, Long Beach, California.
KERRY S. HAVNER	Section Chief, Solid Mechanics Research, Missile and Space Systems Division, McDonnell Douglas Company, Santa Monica, California.
PRITCHARD H. WHITE	Member of Senior Staff, Measurement Analysis Corporation, Santa Monica, California.
CHARLES C. S. YEN	Research Projects Engineer, Structures Analysis and Test, Hughes Tool Company, Aircraft Division, Culver City, California.
ANGEL F. MADAYAG	Coordinator, Advanced Technical Courses, Institute of Aerospace Safety and Management, University of Southern California, Los Angeles.
HENRY G. SMITH	Chief, Structures Analysis and Test, Hughes Tool Company, Aircraft Division, Culver City, California.
WILLIAM H. RASER, JR.	Project Engineer, Datatron Instruments Company, Reading, Pennsylvania.
MITCHELL H. WEISMAN	Specialist-Research, Materials and Processes, Rocketdyne, Division of North American Rockwell, Canoga Park, California.
LUIS YOUNG	Design Specialist, Lockheed California Company, Van Nuys, California.
JAMES A. SCOTT	Group Leader, Materials and Process Engineering, Douglas Aircraft Division, McDonnell Douglas Company, Long Beach, California.

Metal Fatigue: Theory and Design

Edited by

ANGEL F. MADAYAG

University of Southern California, Los Angeles

John Wiley & Sons, Inc. New York · London · Sydney · Toronto

Library of Congress Catalog Card Number: 68-24798
SBN 471 56315 3
Printed in the United States of America

Preface

The continued search for structural materials with high static strength to achieve minimum weight may not, because of fatigue, necessarily ensure an improved performance. For this reason recent designs do not specify static strength alone as a primary design criterion but also include fatigue analysis. The demand for improved aircraft performance indicated by higher speeds and maneuverability, new fabrication processes, the constant need for changes in design concepts and procedures, the low constraint typical of minimum weight designs, and FAA certification requirements on fail-safe configurations makes fatigue analysis as routine as those analyses associated with stress and deflection formulations. In addition, it is likely that some nondestructive testing methods that may enhance fatigue crack propagation are presently available.

The emphasis in this book, which is proportionally balanced between theory and design concepts, is on current practical design applications. It is lamentable to show design applications and fatigue-life prediction methods in their general form when the methods or procedures are proprietory; the method or procedure presented, however, can be readily extended without much difficulty. In addition, it serves to bridge the gap between theoretical knowledge and practical understanding of the many typical fatigue problems encountered in the field. The book starts by introducing the basic concepts of fatigue and cumulative damage theories. The latter part focuses on design applications. It is hoped that the information provided will contribute to a safer utilization of the increasingly complex equipment in the field of aerospace operations.

I should like to express my sincere thanks for the support provided by the Boeing Company, Wichita, Kansas, the Lockheed California Company,

Burbank, California, the McDonnell Douglas Corporation, Missile and Space Systems Division, Santa Monica, and Aircraft Division, Long Beach, California, the Hughes Tool Company, Aircraft Division, Culver City, California, the Measurement Analysis Corporation, Santa Monica, California, the Rocketdyne Division of North American Rockwell, Inc., Canoga Park, California, and the Datatron Instruments Company, Reading, Pennsylvania, for providing materials and personnel time in the preparation and editing of the manuscripts that were used by the authors as lecture notes. In addition, I should also like to express my appreciation for the cooperation of all my coauthors.

To all those organizations and individuals, too numerous to mention, who gave us their permission to use part of their works, publications, illustrations, and reports in this book and also to the many Federal aviation engineering personnel and aerospace engineers from the various industries who attended the course and gave us feedback to improve the course contents, in behalf of all the authors I wish to express my appreciation and most sincere thanks.

ANGEL F. MADAYAG
Editor

Los Angeles, California
June 1968

Contents

Metal Fatigue: Theory and Design

1

Causes and Recognition of Fatigue Failures

ANGEL F. MADAYAG

1.1 GENERAL CAUSES OF MATERIAL FAILURES

There can be numerous reasons why a structural member subjected to various types of load will fail. Some primary causes of material failure can be due to any one or combined effects of the following general set of conditions: design deficiencies, manufacturing deficiencies, improper and insufficient maintenance, operational overstressing, environmental factors such as heat, corrosion, and secondary stresses not considered in the normal operating conditions set at the preliminary design stages, and fatigue failures [1,3].

Continuous studies are being performed to minimize the abovementioned causes. Improper and insufficient maintenance seems to be one of the most contributing factors influenced by some improper designs such as areas that are hard to inspect and maintain and the need for better maintenance procedures. In many circumstances the true loads are hard to predict. It is then possible to subject the structure beyond its normal carrying capability and structural limitations. When a structure is subjected to cyclic loads, areas that are critical to a fatigue type failure must, if possible, be accurately identified. This is quite hard to analyze in a highly composite structure for which analysis has a high degree of uncertainty; thus, in general, structural fatigue testing is frequently resorted to in such cases.

In general, we define "failure" to mean actual rupture of a material or member and may also refer to a condition when the member has just attained its maximum or limit load. In addition, failure may refer also to a condition when a structure ceases to carry the applied loads.

1.2 DEFINITION AND CRITERION

What is fatigue? Generally speaking, fatigue is a progressive failure of a part under repeated, cyclic, or fluctuating loads. It should be noted that if we provide proper precautions against creep and corrosion a structure subjected to a steady, static load less than the limit strength of the metal should theoretically last forever. On the other hand, if a structure is subjected to a cyclic, repeated, or fluctuating load, it may fracture at a stress level less than that required to cause failure under static conditions. This phenomena is known as fatigue which is a common source of primary failures of metals in service [10].

Fatigue failures can be classified into two main groups, namely, simple and compound. Simple results when a fatigue failure starts from a single crack and propagates until ultimate failure occurs. A compound fatigue failure results when the origin of the fatigue crack originates from two or more locations and propagates; the joint effects cause total failure [3]. The sequences by which failure occurs are in three parts; the initial damage occurs in a submicroscopic scale, the crack initiates and propagates, and the final rupture identifies the third stage.

The *criterion* for fatigue failure is the simultaneous action of *cyclic stress*, *tensile stress*, and *plastic strain*. If any one of these three items is eliminated, fatigue is also eliminated. The cyclic stress is readily visualized. The tensile stress is apparent to provide crack formation, but we should also bear in mind that though *compressive stresses will not cause fatigue compression loads will.*

In some instances fatigue fractures are often erroneously blamed on crystallization of the metal. Structural metals, however, are crystalline from the time they solidify from the molten state so that the term crystallization in connection with fatigue is meaningless and should be avoided [10].

1.3 RECOGNITION

In investigating a fracture surface due to fatigue, two zones are evident, namely, a fatigue zone and a rupture zone. The fatigue zone is the area of the crack propagation and the area of final failure is called the rupture or instantaneous zone. In investigating a failed specimen, the instantaneous zone provides the following: ductility of the material, type of loading, and direction of loading. The distortion and damage pattern will be apparent to clarify the type and direction of loading. In addition, the relative size of the instantaneous zone compared with the fatigue zone relates the degree of overstress applied to the structure. The degree of overstressing can be mentioned as follows: highly overstressed if the area of the fatigue zone is very

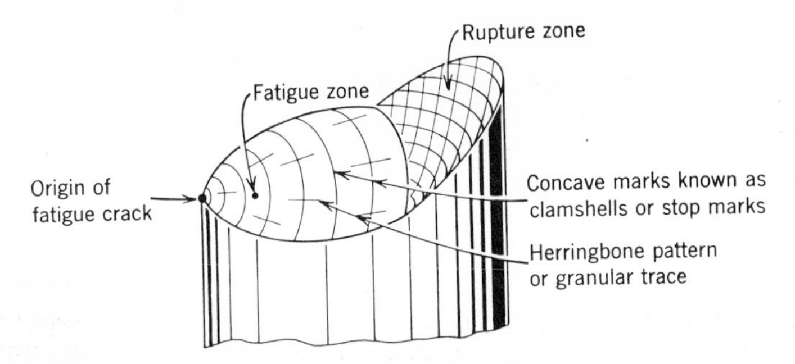

Figure 1.1 A typical fatigue section showing the identifying marks.

small compared with the area of the rupture zone; medium overstress if the size or area of both zones are nearly equal, and low overstress if the area of the instantaneous zone is very small.

The features associated with the fatigue zone are the following: it has a smooth, rubbed, and velvety appearance, the presence of waves known as "clam-shells" or "oyster-shells," "stop marks" and "beach marks," and the herringbone pattern or granular trace which shows the origin of the crack. Most clam-shell marks are concave with respect to the origin of the crack but can also be convex depending on the brittleness of the material, degree of overstressing and the influence of stress concentrations [1]. In general, these stop marks indicate the variations in the rate of crack propagation due to variations in stress amplitude in a cyclic application varying with time. There are some aluminum alloys that may not exhibit these waves but instead show a smooth type appearance. (See Figure 1.1.)

When considering ductile failures under static overload, evidence of necking-down exists. A brittle material under static overload will not show any evidence of necking-down portion. A fatigue fracture, whether the material is ductile or brittle, follows that of a brittle fracture. Not all brittle failures are fatigue failures, however. The most recognizable features of a fatigue failure are lack of deformation pattern and the existence of a singular plane of fracture usually a 90-degree cross section [3].

Beach marks are formed when a fatigue fracture originates from two or more locations and the intersection of these cracks forms a "slip shelf" commonly referred to as beach marks. Some probable causes of beach marks are material nonhomogeneity, which tends to induce multiple fracture areas, stress concentrations on areas not coincident with a plane perpendicular to the principal tensile stress, and a high-overstress loading condition. A typical beach mark or slip shelf is shown in Figure 1.2.

Most of the fatigue cracks discussed were caused by tension loads, tension strains, and tension stress. Typical fracture appearances of fatigue failures in

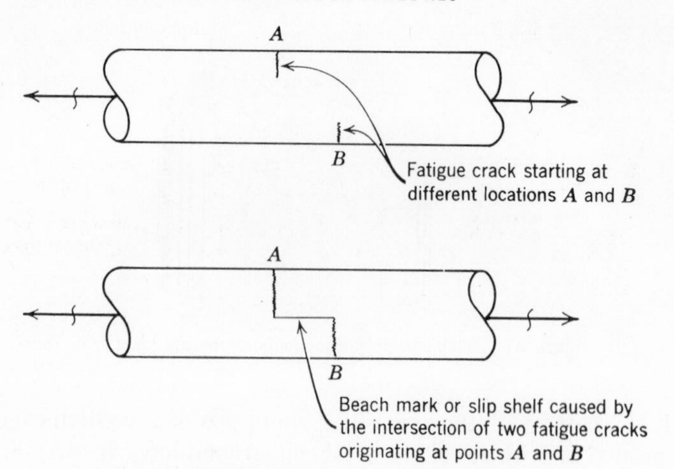

Figure 1.2 Typical illustration of a beach mark.

bending and torsion are shown in Figures 1.3 and 1.4. Bending fatigue failure can be divided into the three classifications according to the type of bending load, namely, one-way, two-way, and rotary. The fatigue crack formations associated with the type of bending load are shown in Figure 1.3. The type of torsional fatigue failures occur in two modes [1] which are longitudinal or transverse along planes of maximum shear helical at 45 degrees to the axis of the shaft along planes of maximum tension. Transverse fractures are commonly associated with a smooth surface due to the rubbing of both sides, a characteristic that can be used to identify this type of fracture [3].

1.4 OTHER DESIGN CONSIDERATIONS

Fatigue considerations in the design of aerospace vehicles produced a broad range of activity to cope with the problems introduced by these phenomena. In the past most vehicles were designed primarily on strength criteria. Studies are constantly seeking better and more efficient ones.

Even if careful attention to good design practices is constantly the aim of design engineers, fatigue problems are sometimes introduced into the structure. Fatigue failures are often the result of geometrical or strain discontinuities, poor workmanship or improper shop techniques, material defects, and the introduction of residual stresses that may add to existing service stresses.

Some typical factors affecting fatigue strength are the following: STRESS RAISERS. This can be in the form of a notch or inclusion. Most fatigue fractures can be attributed to notch effects and those by inclusions are seldom encountered because of good quality control practices. Fatigue specimens originating from an inclusion are extremely rare. It should be noted that subsurface defects are frequently undetectable when inspected by the usually

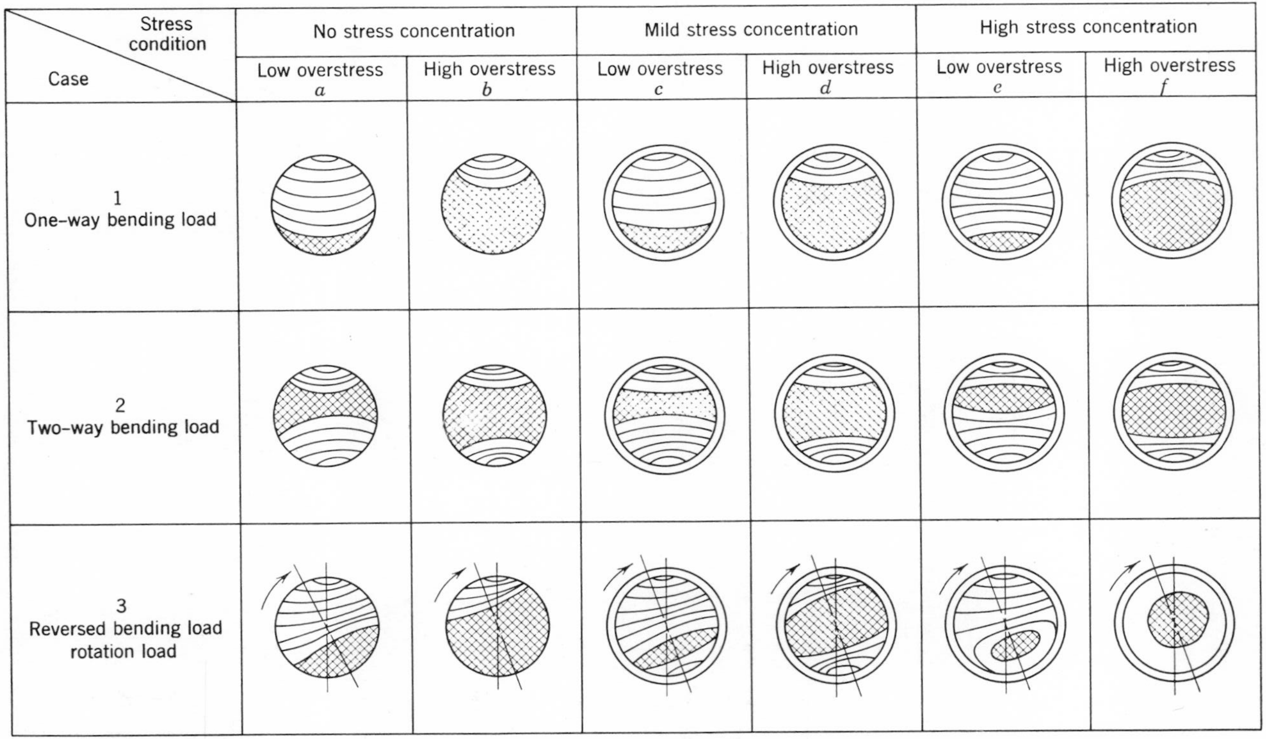

Stress condition / Case	No stress concentration		Mild stress concentration		High stress concentration	
	Low overstress *a*	High overstress *b*	Low overstress *c*	High overstress *d*	Low overstress *e*	High overstress *f*
1 One–way bending load						
2 Two–way bending load						
3 Reversed bending load rotation load						

Figure 1.3 Fracture appearances of fatigue failures in Bending by Dr. Charles Lipson, "Why Machine Parts Fail," Machine Design, Penton, Cleveland 13, Ohio.

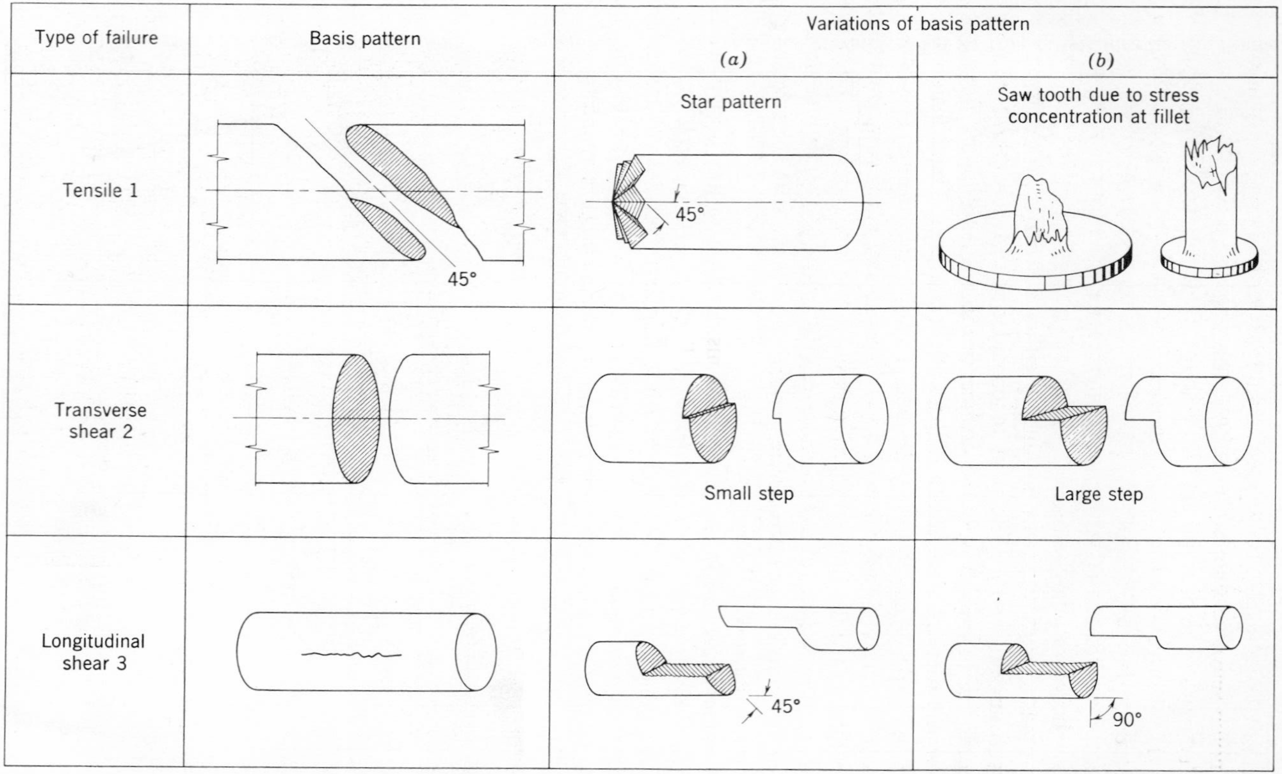

Type of failure	Basis pattern	Variations of basis pattern (a)	Variations of basis pattern (b)
Tensile 1	45°	Star pattern 45°	Saw tooth due to stress concentration at fillet
Transverse shear 2		Small step	Large step
Longitudinal shear 3		45°	90°

Figure 1.4 Fracture appearances of fatigue failures in Torsion by Dr. Charles Lipson, "Why Machine Parts Fail," Machine, Penton, Cleveland 13, Ohio.

magnetic-particle inspection process. Materials with high strength are much more notch-sensitive than softer alloys. The next one is CORROSION. The corroded parts form pits and act like notches. It also reduces the amount of material thus increasing the actual stresses. This type of failure is referred to as "corrosion-fatigue." Another type of corrosion fatigue occurs when two parts are press-fitted or clamped together under vibratory loadings. This type is called "fretting-corrosion." Another contributor stems from PRESS-FIT OR TIGHT CLAMP ASSEMBLY producing either a stress concentration or a change in material size. The next to consider is DECARBURIZATION which defines the loss of carbon from the surface, producing a soft skin. Because of bending and torsion, stresses are highest at the surface. This consideration should be emphasized in spring design. The last to consider is RESIDUAL STRESSES. When a residual stress adds to the design stress, the combined effect may readily exceed the limit stress as imposed in the initial design. There are various ways in which residual stresses can be introduced into a structure; thus careful examinations should be performed to make sure the stresses are acting with you rather than against you.

In summary, fatigue is an important consideration in any design phase. The fact is that fatigue problems will very likely always be with us. Every effort should be expended to reduce the failure rate further. Appropriate management action should be constantly pointed toward improved design techniques and manufacturing procedures. The general principle followed is that failure analysis is highest when it is not available [10,11].

Figure 1.5 The fatigue crack shown by the circled area started at the upper longeron. Courtesy of Boeing Company, Structural Test Program, Wichita, Kansas.

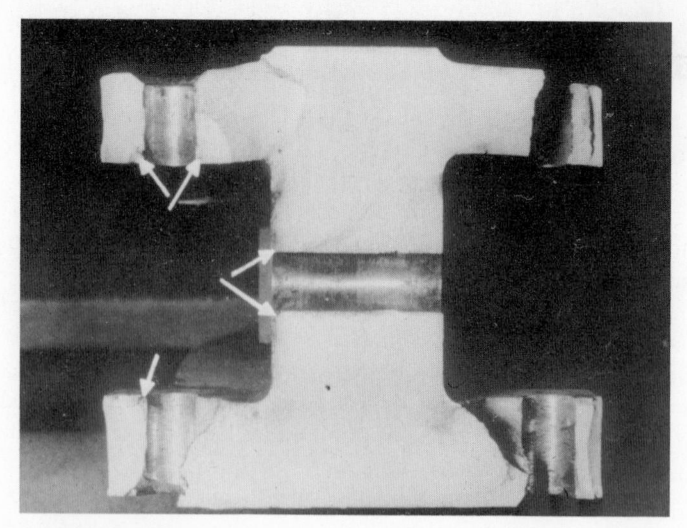

Figure 1.6 Fatigue crack originating at the five bolt holes. Courtesy of Boeing Company, Structural Test Program, Wichita, Kansas.

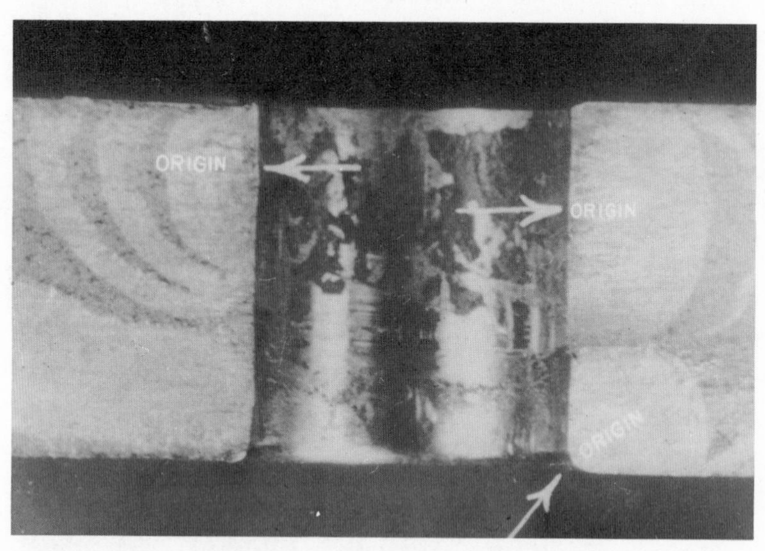

Figure 1.7 Fatigue crack magnified nine times. Fastener had been placed two and half years under indoor, ambient temperature and humidity conditions. Courtesy of Boeing Company, Structural Test Program, Wichita, Kansas.

Figure 1.8 Fatigue crack originating at the countersunk portion of a fastener hole taken from a laboratory component structural member. Courtesy of Boeing Company, Wichita, Kansas.

Figure 1.9 Fatigue fractures originating in the center region of the hole from a laboratory component fatigue test panel. Courtesy of Boeing Company, Structural Test Program, Wichita, Kansas.

Figure 1.10 Fatigue fracture from a pinion gear. Arrow indicates the nucleating region in the form of subsurface inclusions. Courtesy of Douglas Aircraft Division, McDonnell Douglas Company, Long Beach, California.

Figure 1.11 A tension fatigue failure of a helicopter rotor blade flapping link. Fatigue crack originated at arrow B, propagated to arrows C. From "Metal Fatigue and Its Recognition," Civil Aeronautics Board, Bureau of Safety, Bulletin No. 63-1, April 1963, by Frank R. Stone, Jr.

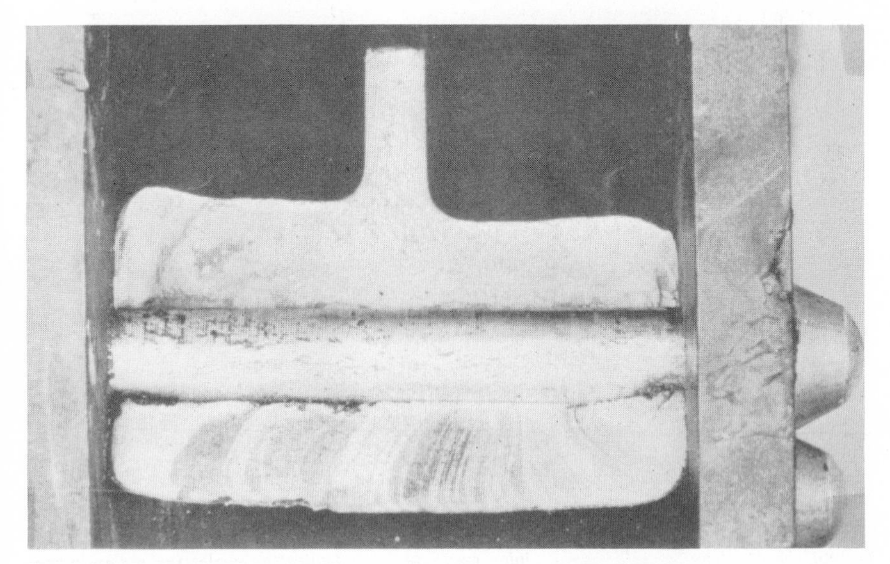

Figure 1.12 An enlarged view of the lower spar cap. From "Metal Fatigue and Its Recognition," Civil Aeronautics Board, Bureau of Safety, Bulletin No. 63-1, April 1963, by Frank R. Stone, Jr.

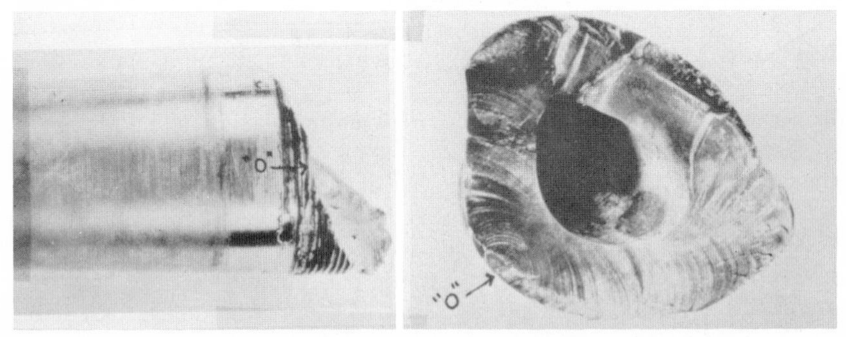

Figure 1.13 Torsional fatigue failure of a crankshaft. Fatigue crack originated at arrow "O". Right side shows enlarged view of fracture surface. From "Metal Fatigue and Its Recognition," Civil Aeronautics Board, Bureau of Safety, Bulletin No. 63-1, April 1963, by Frank R. Stone, Jr.

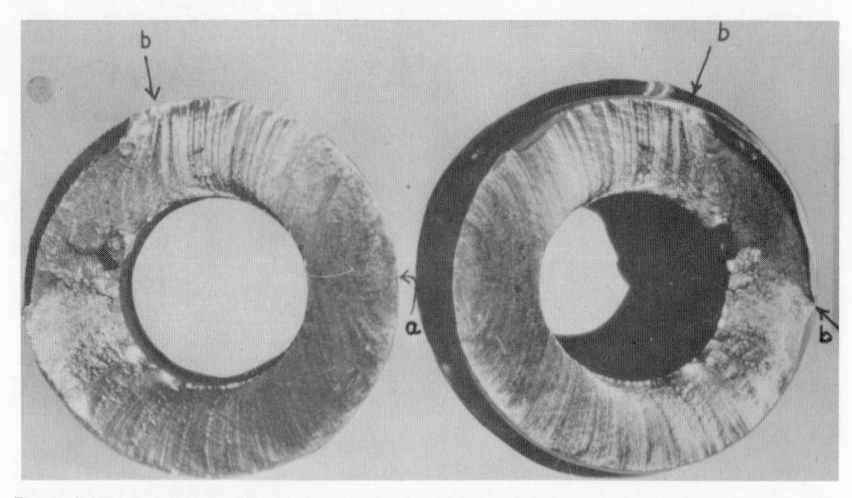

Figure 1.14 One-way bending fatigue fracture of a nose landing-gear axle. Fatigue crack originated at arrow "a" and penetrated to arrows "b" before final failure. From "Metal Fatigue and Its Recognition," Civil Aeronautics Board, Bureau of Safety, Bulletin No. 63-1, April 1963, by Frank R. Stone, Jr.

1.5 OTHER DEFINITIONS

The definitions included here are elementary. They are introduced for convenience.

Nominal stress Obtained from the simple theory in tension, bending, and torsion neglecting geometric discontinuities.

Maximum stress The largest or highest algebraic value of a stress in a stress cycle. Positive for tension.

Minimum stress The smallest or lowest algebraic value of a stress in a stress cycle. Positive for tension.

Mean stress The algebraic mean of the maximum and minimum stress in one cycle.

Stress range The algebraic difference between the maximum and minimum stresses in one cycle.

Stress amplitude Half the value of the algebraic difference between the maximum and minimum stresses in one cycle or half the value of the stress range.

Stress ratio The ratio of minimum stress to maximum stress.

Fatigue life The number of stress cycles which can be sustained for a given test condition.

Fatigue strength The highest or greatest stress which can be sustained by a member for a given number of stress cycles without fracture.

REFERENCES

[1] Charles Lipson, *Why Machine Parts Fail*, Penton, Cleveland 13, Ohio, 1950.

[2] E. P. Polushkin, *Defects and Failure of Metals*, Elsevier, Berlin, 1956; Distributors, Van Nostrand, Princeton, N.J.

[3] Frank R. Stone, Jr., "Metal Fatigue and Its Recognition," Engineering Division, Bulletin No. 63-1, Bureau of Safety, Civil Aeronautics Board, April 1963.

[4] G. Sines and J. Waisman, *Metal Fatigue*, McGraw-Hill, New York, 1959.

[5] A. M. Freudenthal, *Fatigue in Aircraft Structures*, Academic, New York, 1950.

[6] John A. Bennett and G. Williard Quiclo, "Mechanical Failures of Metals in Service," National Bureau of Standards Circular 550, United States Department of Commerce, September 27, 1954.

[7] E. P. Polushkin, *Defects and Failures of Metals*, Elsevier, Berlin, 1956; Distributors, Van Nostrand, Princeton, N.J.

[8] B. F. Langer, "Fatigue Failures From Stress Cycles of Varying Amplitude," *ASME J. Appl. Mech.*, **59,** A-160 (1937).

[9] R. W. Hess and R. W. Fralich and H. H. Hubbard, "Studies of Structural Failure Due to Acoustic Leading," NACA Tech. Note 4050, July 1957.

[10] Slide-tape presentation on fatigue, by Frederick K. Fox, Senior Group Engineer, Structures Staff, Strength Group Supervisor, Boeing Company, Wichita, Kansas and A. F. Madayag, University of Southern California.

[11] Slide-tape presentation on failures in service, by Richard M. Ehlenft, Process Engineer, Materials and Research Process Engineering, Douglas Aircraft Company, Long Beach, California.

[12] H. W. Liu and H. T. Corten, "Fatigue Damage Under Varying Stress Amplitudes," NASA Tech. Note D-647, November 1960.

[13] H. W. Hayden, W. G. Moffatt, and J. Wulff, *The Structure and Properties of Materials*, Vol. III, Wiley, New York, 1965.

[14] "Strength of Metal Aircraft Elements," Military Handbook—5, March 1961, Armed Forces Supply Support Center, Washington, D.C., FSC 1500.

2

Mathematical Theories of Material Behavior

KERRY S. HAVNER

The discipline known as mechanics (or strength) of materials, including the applied mathematical sciences of elasticity and plasticity, assumes a fundamental role in the study of fatigue and failure in metals. Every failure criterion is predicated on a fairly detailed knowledge of the distribution of both stress and mechanical strain in a structure, and a determination of these quantities is made possible only through a reasonable mathematical description of the material response to complex loading. In this chapter stress and strain are discussed in detail as separate macroscopic concepts and then related through mathematical laws of material behavior for both the elastic and plastic ranges. Both initial yielding and theories of strain-hardening are studied, and the chapter closes with a brief exposition of continuum theories of static fracture.

2.1 STATE OF STRESS

The mathematical description of the state of macroscopic stress at a point in a solid is certainly a necessary prelude to any strength or failure analysis of a structural or machine part. Principal stress, maximum shearing stress, and octahedral shearing stress each play an important part in one or more of the various theories of the strength of metals. Thus consideration of the basic transformation equations for determining normal and shearing stress components on an arbitrary plane is the place where one logically should begin.

In working out the algebraic details of stress transformation, it is both convenient and helpful to utilize tensor subscript notation and the tensor concept of stress. Tensor notation will also prove valuable in later considerations of strain-displacement relations and the basic stress-strain laws for elastic and plastic theories of material behavior. In the following section the notation is described and the necessary vector and tensor concepts are presented as we proceed through the development of the transformation equations.

2.1.1 *Transformation Matrix and the Tensor Concept of Stress*

The state of stress at a point in a three-dimensional body is a second-order tensor (by definition) due to the way the stress components transform under rotation of the coordinate axes. Before establishing these equations of stress transformation, it is necessary first to establish the equations of vector (e.g., force) transformation in cartesian coordinates and to define the transformation matrix. Consider two sets of rectilinear orthogonal coordinate axes x_1, x_2, x_3, and x_1', x_2', x_3' (Figure 2.1) and define $a_{ij} = \cos(x_i', x_j)$, that is, the cosine of the angle between the x_i' and x_j axes. Choose an arbitrary vector **A** having components A_1, A_2, A_3 in the x_1, x_2, x_3 directions, respectively. The components of the vector in the primed coordinate system, denoted as A_1', A_2', A_3', can be determined from the unprimed components through the transformation equations

$$
\begin{aligned}
A_1' &= a_{11}A_1 + a_{12}A_2 + a_{13}A_3, \\
A_2' &= a_{21}A_1 + a_{22}A_2 + a_{23}A_3, \\
A_3' &= a_{31}A_1 + a_{32}A_2 + a_{33}A_3.
\end{aligned} \tag{2.1}
$$

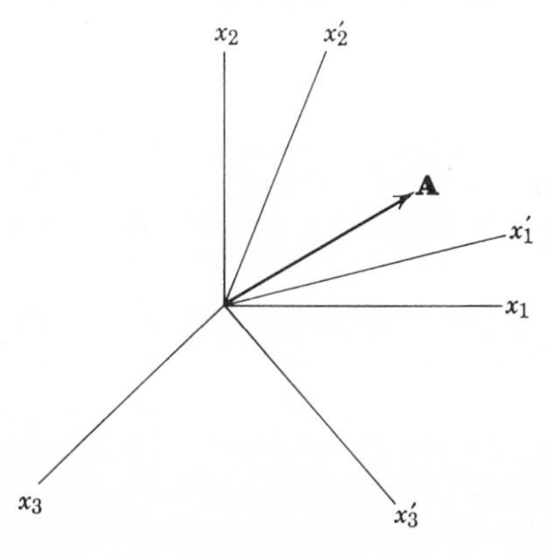

Figure 2.1

By introducing the summation convention of cartesian tensor analysis, we can write the set of equations (2.1) in compact form as

$$A'_i = a_{ij}A_j \qquad i = 1, 2, 3. \tag{2.2}$$

The repeated index j indicates summation over the integers 1, 2, 3. Index i, appearing singly on each side of the equation, is a free index that ranges over these same integers to yield the three separate equations. Similarly, the reverse transformation can be written

$$A_i = a_{ji}A'_j. \tag{2.3}$$

The set of nine coefficients a_{ij} constitutes a 3×3 matrix called the transformation matrix. It is readily shown that these coefficients satisfy the relations

$$a_{ik}a_{jk} = \delta_{ij}, \tag{2.4}$$

where δ_{ij} is the Kronecker delta or unit tensor, defined as

$$\delta_{ij} = \begin{cases} 1, & i = j, \\ 0, & i \neq j. \end{cases}$$

Expanding (2.4), six independent conditions are obtained on the elements of the transformation matrix: the normality conditions

$$\begin{aligned} a_{11}^2 + a_{12}^2 + a_{13}^2 &= 1, \\ a_{21}^2 + a_{22}^2 + a_{23}^2 &= 1, \\ a_{31}^2 + a_{32}^2 + a_{33}^2 &= 1, \end{aligned} \tag{2.5}$$

and the orthogonality conditions

$$\begin{aligned} a_{11}a_{21} + a_{12}a_{22} + a_{13}a_{23} &= 0, \\ a_{11}a_{31} + a_{12}a_{32} + a_{13}a_{33} &= 0, \\ a_{21}a_{31} + a_{22}a_{32} + a_{23}a_{33} &= 0. \end{aligned} \tag{2.6}$$

The direction cosines of the primed coordinate system must satisfy these equations in order for the transformation to be rectilinear and orthogonal.

We are now in a position to look at the basic equations of stress transformation at a point in a solid. Consider an arbitrary body (Figure 2.2) cut by a plane whose normal is given as the axis n. The stress resultant at a point P of the cut surface is defined as

$$\boldsymbol{\sigma}_n = \lim_{\Delta A \to 0} \frac{\Delta \mathbf{F}}{\Delta A} = \frac{d\mathbf{F}}{dA}, \tag{2.7}$$

where $\Delta\mathbf{F}$ is the resultant force acting upon the incremental area ΔA. Since a different stress resultant is obtained if a differently oriented plane is passed

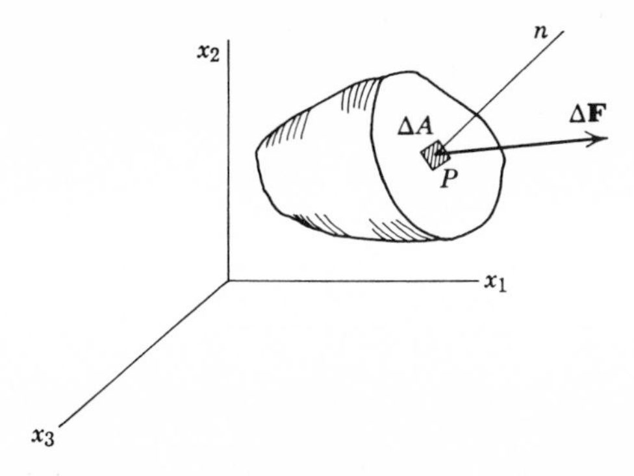

Figure 2.2

through point P, $\boldsymbol{\sigma}_n$ is called the vector that the stress tensor (state of stress) associates with the direction n.

Superimposing a cartesian coordinate system on the body at P and considering the equilibrium of an elementary tetrahedron (Figure 2.3), a stress resultant vector $\boldsymbol{\sigma}_i$ with components σ_{ij} can be associated with each of the axes x_i. (The stress component σ_{ij} is in the direction x_j acting on the face normal to direction x_i.) Thus the state of stress at point P, denoted as σ, can be represented conceptually in terms of either its vector or tensor components:

$$\sigma = \begin{Bmatrix} \boldsymbol{\sigma}_1 \\ \boldsymbol{\sigma}_2 \\ \boldsymbol{\sigma}_3 \end{Bmatrix} = \begin{bmatrix} \sigma_{11} & \sigma_{12} & \sigma_{13} \\ \sigma_{21} & \sigma_{22} & \sigma_{23} \\ \sigma_{31} & \sigma_{32} & \sigma_{33} \end{bmatrix}. \tag{2.8}$$

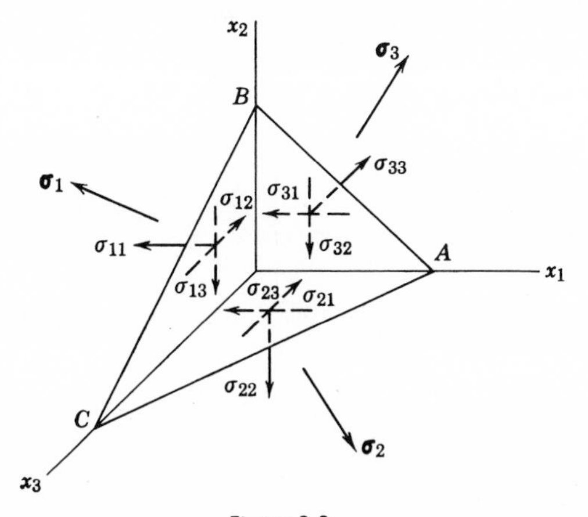

Figure 2.3

The stress resultants $\boldsymbol{\sigma}_i$ are

$$\begin{aligned}\boldsymbol{\sigma}_1 &= \sigma_{11}\mathbf{i}_1 + \sigma_{12}\mathbf{i}_2 + \sigma_{13}\mathbf{i}_3,\\ \boldsymbol{\sigma}_2 &= \sigma_{21}\mathbf{i}_1 + \sigma_{22}\mathbf{i}_2 + \sigma_{23}\mathbf{i}_3,\\ \boldsymbol{\sigma}_3 &= \sigma_{31}\mathbf{i}_1 + \sigma_{32}\mathbf{i}_2 + \sigma_{33}\mathbf{i}_3,\end{aligned}$$

or, in summation notation,

$$\boldsymbol{\sigma}_i = \sigma_{ij}\mathbf{i}_j, \tag{2.9}$$

where $\mathbf{i}_j$ is a unit vector in the direction of the coordinate axis x_j. Denoting dA_i as the area of the face normal to x_i (projection of the area dA of face ABC on the plane normal to x_i), a vector equilibrium equation for the differential force $d\mathbf{F}$, in terms of the tensor components σ_{ij}, can be written

$$\begin{aligned}d\mathbf{F} &= \boldsymbol{\sigma}_1\,dA_1 + \boldsymbol{\sigma}_2\,dA_2 + \boldsymbol{\sigma}_3\,dA_3 = \boldsymbol{\sigma}_i\,dA_i\\ &= \sigma_{ij}\mathbf{i}_j\,dA_i.\end{aligned}$$

Prescribing a rotated coordinate system x'_l with the axis x'_k chosen to coincide with the normal to ABC (Figure 2.4), it is evident that $dA_i = a_{ki}\,dA$. Thus

$$d\mathbf{F} = \sigma_{ij}\mathbf{i}_j a_{ki}\,dA, \tag{2.10}$$

which can be expanded as

$$d\mathbf{F} = \begin{matrix} +\sigma_{11}\mathbf{i}_1 a_{k1}\,dA & + & \sigma_{21}\mathbf{i}_1 a_{k2}\,dA & + & \sigma_{31}\mathbf{i}_1 a_{k3}\,dA \\ +\sigma_{12}\mathbf{i}_2 a_{k1}\,dA & + & \sigma_{22}\mathbf{i}_2 a_{k2}\,dA & + & \sigma_{32}\mathbf{i}_2 a_{k3}\,dA. \\ \underbrace{+\sigma_{13}\mathbf{i}_3 a_{k1}\,dA}_{\boldsymbol{\sigma}_1\,dA_1} & + & \underbrace{\sigma_{23}\mathbf{i}_3 a_{k2}\,dA}_{\boldsymbol{\sigma}_2\,dA_2} & + & \underbrace{\sigma_{33}\mathbf{i}_3 a_{k3}\,dA}_{\boldsymbol{\sigma}_3\,dA_3} \end{matrix}$$

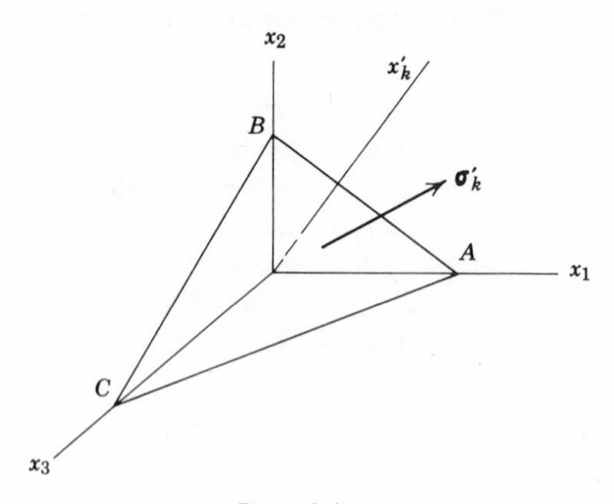

Figure 2.4

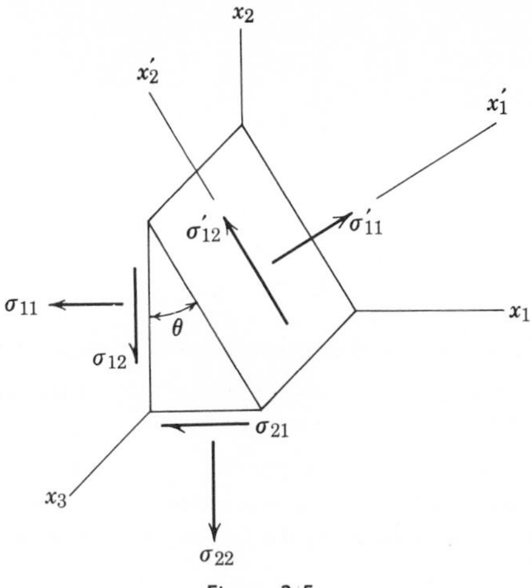

Figure 2.5

Denoting by $\mathbf{i}_l'$ the unit vectors in the directions of the primed coordinate axes x_l' and using the vector transformation equations (2.3),

$$d\mathbf{F} = a_{ki}a_{lj}\sigma_{ij}\, dA\mathbf{i}_l'. \tag{2.11}$$

In terms of the stress resultant $\boldsymbol{\sigma}_k'$ on the face ABC, $d\mathbf{F}$ can be expressed

$$\begin{aligned} d\mathbf{F} &= \boldsymbol{\sigma}_k'\, dA = \sigma_{k1}'\mathbf{i}_1'\, dA + \sigma_{k2}'\mathbf{i}_2'\, dA + \sigma_{k3}'\mathbf{i}_3'\, dA \\ &= \sigma_{kl}'\, dA\mathbf{i}_l'. \end{aligned} \tag{2.12}$$

If these two equations for $d\mathbf{F}$ are set equal, we obtain

$$(\sigma_{kl}' - a_{ki}a_{lj}\sigma_{ij})\, dA\mathbf{i}_l' = 0,$$

and the components of the stress tensor are seen to transform according to the law

$$\sigma_{kl}' = a_{ki}a_{lj}\sigma_{ij}, \tag{2.13}$$

which is precisely the transformation law in cartesian coordinates of a tensor of order 2. The two summation indices i and j yield a total of nine terms in each equation. The two free indices k and l result in nine different equations, corresponding to the nine components of the stress tensor. As a simple illustration of the use of (2.13), consider the transformation of a two-dimensional state of stress (Figure 2.5). The transformation matrix is

$$\begin{bmatrix} \cos\theta & \sin\theta & 0 \\ -\sin\theta & \cos\theta & 0 \\ 0 & 0 & 1 \end{bmatrix}.$$

The complete equation for the stress component σ'_{12} (the shearing stress on the inclined plane) is

$$\sigma'_{12} = a_{11}a_{21}\sigma_{11} + a_{11}a_{22}\sigma_{12} + a_{11}a_{23}\sigma_{13} + a_{12}a_{21}\sigma_{21} + a_{12}a_{22}\sigma_{22} + a_{12}a_{23}\sigma_{23} + a_{13}a_{21}\sigma_{31} + a_{13}a_{22}\sigma_{32} + a_{13}a_{23}\sigma_{33}.$$

From moment equilibrium considerations of a differential particle, the stress tensor is symmetric ($\sigma_{ji} = \sigma_{ij}$). Thus, substituting the elements of the transformation matrix,

$$\sigma'_{12} = -\sigma_{11}\sin\theta\cos\theta + \sigma_{12}(\cos^2\theta - \sin^2\theta) + \sigma_{22}\sin\theta\cos\theta$$

or

$$\sigma'_{12} = \sigma_{12}\cos 2\theta - \frac{\sigma_{11} - \sigma_{22}}{2}\sin 2\theta. \tag{2.14}$$

Similarly, the equation for the stress component σ'_{11} (the normal stress) is

$$\sigma'_{11} = \sigma_{11}\cos^2\theta + 2\theta_{12}\sin\theta\cos\theta + \sigma_{22}\sin^2\theta$$

or

$$\sigma'_{11} = \frac{\sigma_{11} + \sigma_{22}}{2} + \frac{\sigma_{11} - \sigma_{22}}{2}\cos 2\theta + \sigma_{12}\sin 2\theta. \tag{2.15}$$

Equations 2.14 and 2.15 are the well-known equations of coplanar stress transformation associated with Mohr's circle of stress.

2.1.2 *Principal Stresses and Principal Directions*

The normal to a plane on which no shearing stresses are acting is called a principal axis (direction) of stress, and the corresponding normal stress (which may be zero) is called a principal stress. Hence we can make the following definition: the direction specified by the unit vector $\boldsymbol{\mu}$ is a principal direction of the symmetric stress tensor σ_{ij} if the stress vector $\boldsymbol{\sigma}_\mu$ associated with this direction is parallel to $\boldsymbol{\mu}$. The unit vector $\boldsymbol{\mu}$ can be written in terms of the vectors $\mathbf{i}_j$ as (Figure 2.6)

$$\begin{aligned}\boldsymbol{\mu} &= \mu_j\mathbf{i}_j = a_{\mu j}\mathbf{i}_j\\ &= a_{\mu 1}\mathbf{i}_1 + a_{\mu 2}\mathbf{i}_2 + a_{\mu 3}\mathbf{i}_3.\end{aligned} \tag{2.16}$$

From (2.10), the stress vector is

$$\begin{aligned}\boldsymbol{\sigma}_\mu &= \frac{d\mathbf{F}}{dA} = \sigma_{ij}\mathbf{i}_j a_{\mu i}\\ &= \sigma_{ij}\mu_i\mathbf{i}_j,\end{aligned} \tag{2.17}$$

or

$$\begin{aligned}\boldsymbol{\sigma}_\mu = (\sigma_{11}\mu_1 + \sigma_{21}\mu_2 + \sigma_{31}\mu_3)\mathbf{i}_1\\ +(\sigma_{12}\mu_1 + \sigma_{22}\mu_2 + \sigma_{32}\mu_3)\mathbf{i}_2\\ +(\sigma_{13}\mu_1 + \sigma_{23}\mu_2 + \sigma_{33}\mu_3)\mathbf{i}_3.\end{aligned}$$

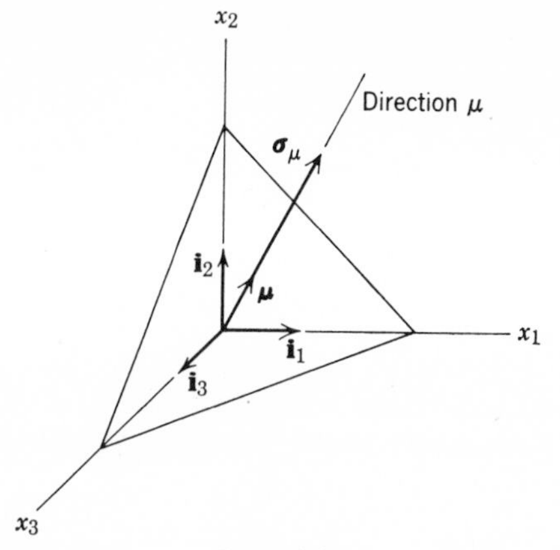

Figure 2.6

As $\boldsymbol{\sigma}_\mu$ and $\boldsymbol{\mu}$ are to be parallel, $\boldsymbol{\sigma}_\mu$ must equal $\lambda\boldsymbol{\mu}$, where λ, a scalar, is the magnitude of the principal stress (since $\boldsymbol{\mu}$ is a unit vector). Thus

$$\sigma_{ij}\mu_i\mathbf{i}_j = \lambda\mu_j\mathbf{i}_j = \lambda\delta_{ij}\mu_i\mathbf{i}_j,$$

from which

$$(\sigma_{ij} - \lambda\delta_{ij})\mu_i = 0. \tag{2.18}$$

Equation 2.18 represents a set of three simultaneous equations, homogeneous in the unknown direction numbers μ_i:

$$\begin{aligned}
(\sigma_{11} - \lambda)\mu_1 + \sigma_{12}\mu_2 + \sigma_{13}\mu_3 &= 0,\\
\sigma_{12}\mu_1 + (\sigma_{22} - \lambda)\mu_2 + \sigma_{23}\mu_3 &= 0,\\
\sigma_{13}\mu_1 + \sigma_{23}\mu_2 + (\sigma_{33} - \lambda)\mu_3 &= 0.
\end{aligned}$$

For this system to have a nontrivial solution the determinant of the matrix of coefficients must be zero:

$$D(\lambda) = |\sigma_{ij} - \lambda\delta_{ij}| = 0, \tag{2.19}$$

or

$$D(\lambda) = \begin{vmatrix} \sigma_{11} - \lambda & \sigma_{12} & \sigma_{13} \\ \sigma_{12} & \sigma_{22} - \lambda & \sigma_{23} \\ \sigma_{13} & \sigma_{23} & \sigma_{33} - \lambda \end{vmatrix} = 0.$$

Expanding this determinant, we obtain a cubic equation in λ which can be written

$$\lambda^3 - C_1\lambda^2 + C_2\lambda - C_3 = 0. \tag{2.20}$$

The quantities C_1, C_2, C_3 are given as

$$\begin{aligned} C_1 &= \sigma_{kk} \equiv \sigma_{11} + \sigma_{22} + \sigma_{33}, \\ C_2 &= \tfrac{1}{2}(\sigma_{kk}{}^2 - \sigma_{ij}\sigma_{ij}) \\ &\equiv \sigma_{11}\sigma_{22} + \sigma_{11}\sigma_{33} + \sigma_{22}\sigma_{33} - \sigma_{12}{}^2 - \sigma_{13}{}^2 - \sigma_{23}{}^2, \\ C_3 &= D(\sigma_{ij}) \\ &\equiv \sigma_{11}\sigma_{22}\sigma_{33} + 2\sigma_{12}\sigma_{13}\sigma_{23} - \sigma_{11}\sigma_{23}{}^2 - \sigma_{22}\sigma_{13}{}^2 - \sigma_{33}\sigma_{12}{}^2. \end{aligned} \tag{2.21}$$

[The expression $D(\sigma_{ij})$ indicates the determinant of the stress tensor of (2.8).]

Equation 2.20 will yield three roots, corresponding to the three principal values of stress at the point P in the body. Since these values are independent of the particular coordinate system x_i or x'_j, etc., chosen to describe the components of the stress tensor, the quantities C_1, C_2, C_3 must be invariant with respect to transformation from one coordinate system to another. For this reason, the expressions given in (2.21) are called the stress invariants.

Denoting the three distinct (in general), real roots of (2.20) by λ_{I}, λ_{II}, λ_{III} and the corresponding principal directions as μ_{I}, μ_{II}, μ_{III}, the direction cosines $a_{\mu i}$ of the principal unit vector for a particular principal stress are determined from the set of equations

$$(\sigma_{ij} - \lambda\delta_{ij})a_{\mu i} = 0$$

and

$$a_{\mu i}a_{\mu i} = 1.$$

To show that the principal directions are orthogonal for distinct principal stresses consider (2.18) written for each of the two roots λ_{I} and λ_{II}:

$$\begin{aligned} (\sigma_{ij} - \lambda^{\mathrm{I}}\delta_{ij})\mu_i{}^{\mathrm{I}} &= 0, \\ (\sigma_{ij} - \lambda^{\mathrm{II}}\delta_{ij})\mu_i{}^{\mathrm{II}} &= 0. \end{aligned}$$

Taking the scalar product of the first equation with respect to $\mu_j{}^{\mathrm{II}}$ and the second with respect to $\mu_j{}^{\mathrm{I}}$ (that is, multiplying the equations by these quantities and then summing on the repeated index j to obtain single scalar values):

$$\begin{aligned} (\sigma_{ij} - \lambda_{\mathrm{I}}\delta_{ij})\mu_i{}^{\mathrm{I}}\mu_j{}^{\mathrm{II}} &= \sigma_{ij}\mu_i{}^{\mathrm{I}}\mu_j{}^{\mathrm{II}} - \lambda_{\mathrm{I}}\mu_i{}^{\mathrm{I}}\mu_i{}^{\mathrm{II}}, \\ (\sigma_{ij} - \lambda_{\mathrm{II}}\delta_{ij})\mu_i{}^{\mathrm{II}}\mu_j{}^{\mathrm{I}} &= \sigma_{ji}\mu_i{}^{\mathrm{I}}\mu_j{}^{\mathrm{II}} - \lambda_{\mathrm{II}}\mu_i{}^{\mathrm{I}}\mu_i{}^{\mathrm{II}}. \end{aligned}$$

Subtracting the first equation from the second and taking advantage of the symmetry of the stress tensor, we obtain

$$(\lambda_{\mathrm{I}} - \lambda_{\mathrm{II}})\mu_i{}^{\mathrm{I}}\mu_i{}^{\mathrm{II}} = 0. \tag{2.22}$$

Since $\lambda_{\mathrm{I}} \neq \lambda_{\mathrm{II}}$,

$$\mu_i{}^{\mathrm{I}}\mu_i{}^{\mathrm{II}} = 0, \tag{2.23}$$

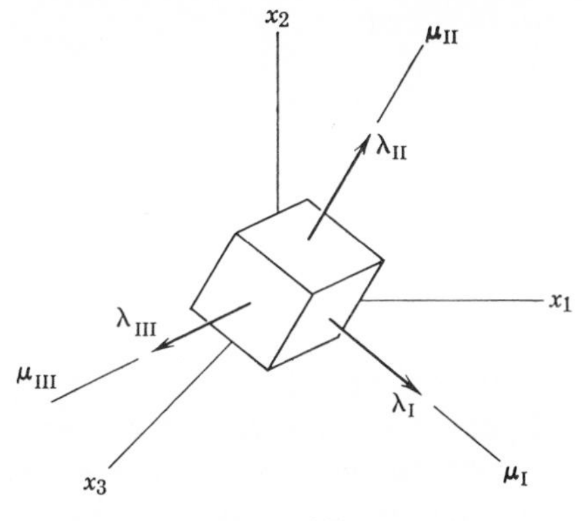

Figure 2.7

which is precisely the orthogonality condition on the unit vectors $\boldsymbol{\mu}_I$ and $\boldsymbol{\mu}_{II}$. Similarly, if $\lambda_I \not\equiv \lambda_{III}$ and $\lambda_{II} \not\equiv \lambda_{III}$,

$$\mu_i^{\mathrm{I}}\mu_i^{\mathrm{III}} = \mu_i^{\mathrm{II}}\mu_i^{\mathrm{III}} = 0, \tag{2.24}$$

from which the principal stress directions are orthogonal. Thus for the general state of stress σ_{ij} at point P we can find an orthogonal transformation that permits the stress tensor to be defined solely by its principal values λ_I, λ_{II}, λ_{III} (Figure 2.7).

As an example, consider the state of stress shown in Figure 2.8. The reduced algebraic values for the stress invariants are

$$C_1 = \sigma_{11} + \sigma_{22} + \sigma_{33},$$
$$C_2 = \sigma_{11}\sigma_{22} + \sigma_{11}\sigma_{33} + \sigma_{22}\sigma_{33} - \sigma_{12}^2,$$
$$C_3 = \sigma_{11}\sigma_{22}\sigma_{33} - \sigma_{33}\sigma_{12}^2,$$

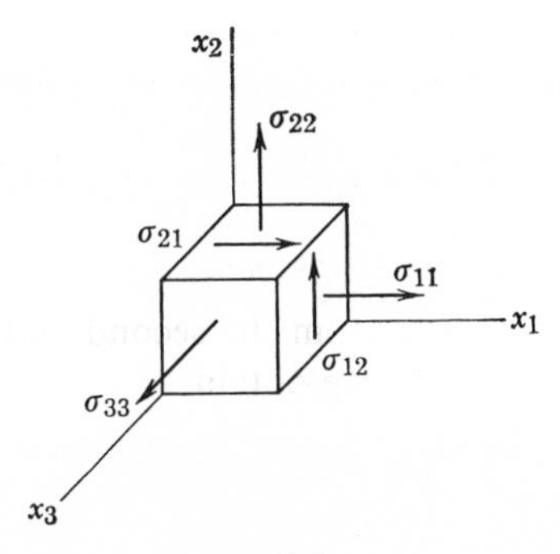

Figure 2.8

and (2.20) for the principal stress values becomes

$$\lambda^3 - (\sigma_{11} + \sigma_{22} + \sigma_{33})\lambda^2 + [\sigma_{11}\sigma_{22} - \sigma_{12}{}^2 + \sigma_{33}(\sigma_{11} + \sigma_{22})]\lambda - \sigma_{33}(\sigma_{11}\sigma_{22} - \sigma_{12}{}^2) = 0,$$

which can be factored as

$$(\lambda - \sigma_{33})[\lambda^2 - (\sigma_{11} + \sigma_{22})\lambda + (\sigma_{11}\sigma_{22} - \sigma_{12}{}^2)] = 0. \tag{2.25}$$

Thus

$$\lambda - \sigma_{33} = 0$$

and

$$\lambda^2 - (\sigma_{11} + \sigma_{22})\lambda + (\sigma_{11}\sigma_{22} - \sigma_{12}{}^2) = 0.$$

From the first of these equations $\lambda_{\text{III}} = \sigma_{33}$, which must be the case, since x_3 is a principal direction by definition for the given state of stress.

From the second equation

$$\begin{aligned} \lambda_{\text{I}} &= \frac{\sigma_{11} + \sigma_{22}}{2} + \left[\left(\frac{\sigma_{11} - \sigma_{22}}{2}\right)^2 + \sigma_{12}{}^2\right]^{1/2}, \\ \lambda_{\text{II}} &= \frac{\sigma_{11} + \sigma_{22}}{2} - \left[\left(\frac{\sigma_{11} - \sigma_{22}}{2}\right)^2 + \sigma_{12}{}^2\right]^{1/2}. \end{aligned} \tag{2.26}$$

Equations 2.26 are seen to be identical to the expressions obtainable from Mohr's circle of stress for coplanar stress transformations.

2.1.3 Principal Shearing Stresses and Octahedral Shearing Stresses

Principal shearing stresses and octahedral shearing stresses are of considerable significance in the theories of initial and continued yielding of ductile metals, and it is important that we develop the general equations for these stresses. Consider an elementary tetrahedron with coordinate axes x_1', x_2', x_3' oriented in the directions of the principal stresses λ_{I}, λ_{II}, λ_{III}, respectively (Figure 2.9). Denoting the direction cosines of the normal to the plane *EFG* as ν_i, the equation for the normal component of stress σ_N is

$$\sigma_N = \lambda_{\text{I}}\nu_1{}^2 + \lambda_{\text{II}}\nu_2{}^2 + \lambda_{\text{III}}\nu_3{}^2. \tag{2.27}$$

The equation for the resultant stress vector $\boldsymbol{\sigma}_\nu$ is [from (2.17)]

$$\boldsymbol{\sigma}_\nu = \sigma_{ij}'\nu_i\mathbf{i}_j' = \lambda_{\text{I}}\nu_1\mathbf{i}_1' + \lambda_{\text{II}}\nu_2\mathbf{i}_2' + \lambda_{\text{III}}\nu_3\mathbf{i}_3'. \tag{2.28}$$

Thus the equation for the magnitude of the resultant shearing σ_S on the plane *EFG* is

$$\sigma_S{}^2 = \boldsymbol{\sigma}_\nu \cdot \boldsymbol{\sigma}_\nu - \sigma_N{}^2 = \lambda_{\text{I}}{}^2\nu_2{}^2 + \lambda_{\text{II}}{}^2\nu_2{}^2 + \lambda_{\text{III}}^2\nu_3{}^2 - \sigma_N{}^2,$$

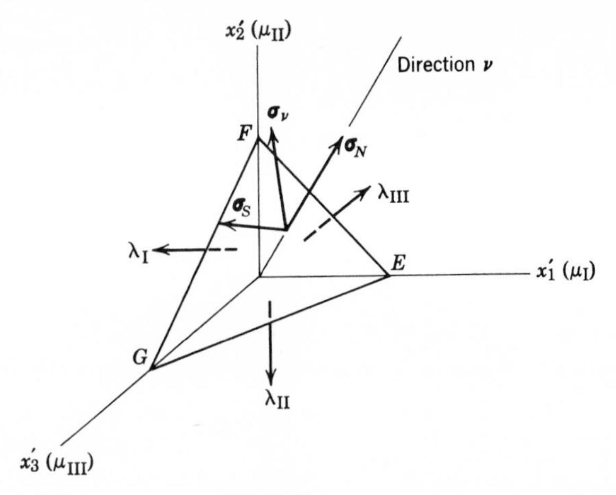

Figure 2.9

where $\boldsymbol{\sigma}_\nu \cdot \boldsymbol{\sigma}_\nu$ denotes the scalar product of the stress resultant vector with itself (i.e., the square of the magnitude of $\boldsymbol{\sigma}_\nu$). The final equation for σ_S becomes

$$\sigma_S{}^2 = \lambda_I{}^2\nu_1{}^2 + \lambda_{II}{}^2\nu_2{}^2 + \lambda_{III}^2\nu_3{}^2 - (\lambda_I\nu_1{}^2 + \lambda_{II}\nu_2{}^2 + \lambda_{III}\nu_3{}^2)^2. \quad (2.29)$$

Since the direction cosines ν_i must satisfy the normality condition $\nu_i\nu_i = 1$, we can substitute the relation

$$\nu_3{}^2 = 1 - \nu_1{}^2 - \nu_2{}^2 \quad (2.30)$$

into (2.29) to obtain

$$\sigma_S{}^2 = (\lambda_I{}^2 - \lambda_{III}^2)\nu_1{}^2 + (\lambda_{II}{}^2 - \lambda_{III}^2)\nu_2{}^2 + \lambda_{III}^2 - [(\lambda_I - \lambda_{III})\nu_1{}^2 + (\lambda_{II} - \lambda_{III})\nu_2{}^2 + \lambda_{III}]^2. \quad (2.31)$$

To establish the directions of the principal shearing stresses, we determine the values of ν_1, ν_2 that make the rates of change of σ_S with respect to these direction numbers vanish:

$$2\sigma_S \frac{\partial \sigma_S}{\partial \nu_1} = 0 = 2(\lambda_I{}^2 - \lambda_{III}^2)\nu_1 - 4(\lambda_I - \lambda_{III}) \cdot [(\lambda_I - \lambda_{III})\nu_1{}^2 + (\lambda_{II} - \lambda_{III})\nu_2{}^2 + \lambda_{III}]\nu_1,$$

$$2\sigma_S \frac{\partial \sigma_S}{\partial \nu_2} = 0 = 2(\lambda_{II}{}^2 - \lambda_{III}^2)\nu_2 - 4(\lambda_{II} - \lambda_{III}) \cdot [(\lambda_I - \lambda_{III})\nu_1{}^2 + (\lambda_{II} - \lambda_{III})\nu_2{}^2 + \lambda_{III}]\nu_2,$$

from which, for distinct principal stress values,

$$\begin{aligned} \nu_1[(\lambda_{\mathrm{I}} - \lambda_{\mathrm{III}})(1 - 2\nu_1^2) - 2(\lambda_{\mathrm{II}} - \lambda_{\mathrm{III}})\nu_2^2] &= 0, \\ \nu_2[(\lambda_{\mathrm{II}} - \lambda_{\mathrm{III}})(1 - 2\nu_2^2) - 2(\lambda_{\mathrm{I}} - \lambda_{\mathrm{III}})\nu_1^2] &= 0. \end{aligned} \tag{2.32}$$

The three sets of roots of (2.30) and (2.32) and of a similar set obtained by eliminating ν_2 are

$$\begin{aligned} &\nu_1^2 = \tfrac{1}{2}, \quad &\nu_2^2 = \tfrac{1}{2}, \quad &\nu_3 = 0, \\ &\nu_1^2 = \tfrac{1}{2}, \quad &\nu_2 = 0, \quad &\nu_3^2 = \tfrac{1}{2}, \\ &\nu_1 = 0, \quad &\nu_2^2 = \tfrac{1}{2}, \quad &\nu_3^2 = \tfrac{1}{2}, \end{aligned} \tag{2.33}$$

with the corresponding principal shearing stress values

$$\sigma_S = \tfrac{1}{2}\,|\lambda_{\mathrm{I}} - \lambda_{\mathrm{II}}|, \qquad \tfrac{1}{2}\,|\lambda_{\mathrm{I}} - \lambda_{\mathrm{III}}|, \qquad \tfrac{1}{2}\,|\lambda_{\mathrm{II}} - \lambda_{\mathrm{III}}|, \tag{2.34}$$

respectively. Thus the principal shearing stresses act on planes oriented at 45 degrees with respect to the planes of principal (normal) stresses.

The octahedral planes are the eight planes whose normals make equal angles with the principal stress directions. Thus the direction cosines of the octahedral planes are given as $\nu_1^2 = \nu_2^2 = \nu_3^2 = \frac{1}{3}$ and the equation for the octahedral shearing stress is

$$(\overset{\circ}{\sigma}_S)^2 = \tfrac{1}{3}(\lambda_{\mathrm{I}}^2 + \lambda_{\mathrm{II}}^2 + \lambda_{\mathrm{III}}^2) - \tfrac{1}{9}(\lambda_1 + \lambda_{\mathrm{II}} + \lambda_{\mathrm{III}})^2.$$

Introducing the deviatoric stress tensor

$$s_{ij} = \sigma_{ij} - \tfrac{1}{3}\sigma_{kk}\delta_{ij}, \tag{2.35}$$

with principal values (in the principal stress directions)

$$\begin{aligned} S_{\mathrm{I}} &= \lambda_{\mathrm{I}} - \tfrac{1}{3}\sigma, \\ S_{\mathrm{II}} &= \lambda_{\mathrm{II}} - \tfrac{1}{3}\sigma, \\ S_{\mathrm{III}} &= \lambda_{\mathrm{III}} - \tfrac{1}{3}\sigma \end{aligned} \tag{2.36}$$

(where $\sigma = \lambda_{\mathrm{I}} + \lambda_{\mathrm{II}} + \lambda_{\mathrm{III}} = \sigma_{kk}$ is the first stress invariant) and noting that $S_{\mathrm{I}} + S_{\mathrm{II}} + S_{\mathrm{III}} \equiv 0$, the equation for the shearing stress can be written

$$(\overset{\circ}{\sigma}_S)^2 = \tfrac{1}{3}(S_{\mathrm{I}}^2 + S_{\mathrm{II}}^2 + S_{\mathrm{III}}^2). \tag{2.37}$$

Denoting

$$J_2 = \tfrac{1}{2}s_{ij}s_{ij} = \tfrac{1}{2}(S_{\mathrm{I}}^2 + S_{\mathrm{II}}^2 + S_{\mathrm{III}}^2), \tag{2.38}$$

the octahedral shearing stress becomes

$$\overset{\circ}{\sigma}_S = (\tfrac{2}{3}J_2)^{\frac{1}{2}}. \tag{2.39}$$

The quantity J_2, called the second deviatoric stress invariant, also can be expressed in terms of the stress invariants C_1 and C_2 [from (2.21)]:

$$J_2 = \tfrac{1}{3}C_1^2 - C_2. \tag{2.40}$$

The normal stress on the octahedral planes is determined from (2.27) as

$$\sigma_N^\circ = \tfrac{1}{3}\sigma = \tfrac{1}{3}(\lambda_{\text{I}} + \lambda_{\text{II}} + \lambda_{\text{III}}), \tag{2.41}$$

which corresponds to a state of equal triaxial stress.

2.2 STATE OF STRAIN

The concept of strain in a solid body arises in a natural and straightforward way from a consideration of the changes in displacements between two points a vanishingly small distance apart. By limiting our investigation to small displacements the components of the state-of-strain tensor can be physically interpreted in terms of extensions and distortions and linearly related to the displacement rates of change.

The development of the equations for the strain components from an analysis of the displaced configuration of the solid is a generalization of the "geometry of deformation" step in elementary mechanics of materials theory. Here, however, we have not prefaced this step by a specified relationship between stress and strain. Thus, although stress and strain both transform linearly and the strain-displacement equations given below are linear, the stress-strain relationship itself need not be linear and remains to be defined by a specific theory of elasticity or plasticity in each problem treated.

2.2.1 *Displacement Vector and Strain Tensor*

Consider two points P and P_1 a distance $d\mathbf{x}$ apart in an initially undeformed body (Figure 2.10). After some arbitrary deformation the corresponding material points of the body have been displaced by the amounts $\mathbf{u}^0$ and $\mathbf{u}$, respectively. From the vector diagram we can write

$$d\mathbf{x} + \mathbf{u} = \mathbf{u}^0 + d\mathbf{x} + d\mathbf{u},$$

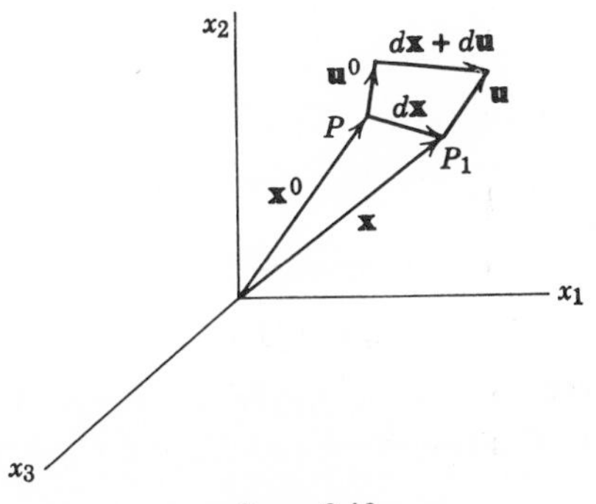

Figure 2.10

from which, in component form

$$u_i = u_i^\circ + du_i. \tag{2.42}$$

By restricting the displacements in size, a Taylor's series expansion about point P can be made for each component of displacement:

$$u_i = u_i^\circ + \frac{\partial u_i^\circ}{\partial x_j} dx_j + 0(|d\mathbf{x}|^2). \tag{2.43}$$

Neglecting the terms of second and higher order in the vector $d\mathbf{x}$ and substituting into (2.42),

$$du_i = u_{i,j}\, dx_j, \tag{2.44}$$

where the subscript notation $u_{i,j} \equiv \partial u_i^\circ/\partial x_j$ is introduced for the sake of brevity. Since there are three components u_i° of the displacement vector and three components x_j of the position vector, the displacement rate of change $u_{i,j}$ has nine components and correspondingly is a second-order tensor quantity. Adding and subtracting terms, we can write

$$du_i = \tfrac{1}{2}(u_{i,j} - u_{j,i})\, dx_j + \tfrac{1}{2}(u_{i,j} + u_{j,i})\, dx_j. \tag{2.45}$$

By denoting

$$\begin{aligned} \xi_{ij} &= \tfrac{1}{2}(u_{i,j} - u_{j,i}), \\ \epsilon_{ij} &= \tfrac{1}{2}(u_{i,j} + u_{j,i}), \end{aligned} \tag{2.46}$$

it becomes evident that the first quantity ξ_{ij} is an antisymmetric tensor

$$\xi = \tfrac{1}{2}\begin{bmatrix} 0 & u_{1,2} - u_{2,1} & u_{1,3} - u_{3,1} \\ u_{2,1} - u_{1,2} & 0 & u_{2,3} - u_{3,2} \\ u_{3,1} - u_{1,3} & u_{3,2} - u_{2,3} & 0 \end{bmatrix}, \tag{2.47}$$

whereas the second quantity ϵ_{ij} is symmetric:

$$\epsilon = \tfrac{1}{2}\begin{bmatrix} 2u_{1,1} & u_{1,2} + u_{2,1} & u_{1,3} + u_{3,1} \\ u_{2,1} + u_{1,2} & 2u_{2,2} & u_{2,3} + u_{3,2} \\ u_{3,1} + u_{1,3} & u_{3,2} + u_{2,3} & 2u_{3,3} \end{bmatrix}. \tag{2.48}$$

Denoting the first term in (2.45) by du_i' and the second by du_i'', we have

$$\begin{aligned} du_i' &= \xi_{ij}\, dx_j, \\ du_i'' &= \epsilon_{ij}\, dx_j. \end{aligned} \tag{2.49}$$

It can be shown that the vector $d\mathbf{u}'$ corresponds to a small rigid body rotation of the magnitude ω about point P:

$$d\mathbf{u}' = (\boldsymbol{\omega}) \times (d\mathbf{x}), \tag{2.50}$$

where $\times$ represents the vector or cross product of $\boldsymbol{\omega}$ and $d\mathbf{x}$. Thus it is the incremental vector $d\mathbf{u}''$ which represents that portion of the differential displacement due to the deformation (i.e., distortion and volume change) of the body, and the symmetric tensor ϵ correspondingly is called the strain tensor of small displacements.

2.2.2 *Physical Interpretation of Strains*

A general equation that relates physically defined extensional and shearing strains (in arbitrary directions) to the components of the strain tensor ϵ_{ij} can be developed in the following manner. Consider the three points P, P_1, P_2 (Figure 2.11) whose relative positions with respect to one another are defined by the direction numbers μ_i, ν_i and distances ds and δs:

$$\begin{aligned} dx_i &= \mu_i\, ds, \\ \delta x_i &= \nu_i\, \delta s. \end{aligned} \tag{2.51}$$

The displacements of these points after deformation of the body (shown greatly exaggerated in the figure) are $\mathbf{u}^\circ$, $\mathbf{u}^\circ + d\mathbf{u}$, and $\mathbf{u}^\circ + \delta\mathbf{u}$, respectively.

To obtain the desired equation, we first form the scalar product of the relative position vectors $d\mathbf{x}$ and $\delta\mathbf{x}$:

$$d\mathbf{x} \cdot \delta\mathbf{x} = ds\, \delta s \cos\theta = dx_i\, \delta x_i, \tag{2.52}$$

where θ is the angle between the direction lines μ and ν. Taking the total differential of the right side of this equation, and noting that, from Figure 2.11, $\Delta(dx_i) = du_i$ and $\Delta(\delta x_i) = \delta u_i$ (where Δ denotes the differential).

$$\Delta(dx_i\, \delta x_i) = du_i\, \delta x_i + dx_i\, \delta u_i, \tag{2.53}$$

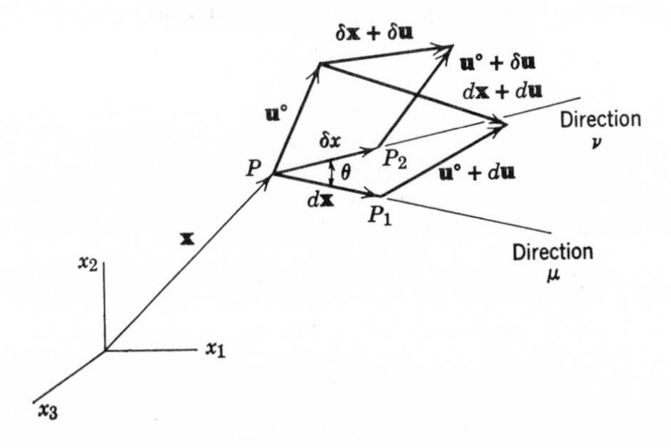

Figure 2.11

which (from 2.44) can be written

$$\Delta(dx_i\,\delta x_i) = u_{i,j}\,dx_j\,\delta x_i + u_{i,j}\,\delta x_j\,dx_i = (u_{j,i} + u_{i,j})\,dx_i\,\delta x_j = 2\epsilon_{ij}\,dx_i\,\delta x_j. \tag{2.54}$$

Thus, combining with (2.51) and (2.52),

$$\Delta(ds\,\delta s\cos\theta) = 2\epsilon_{ij}\mu_i\nu_j\,ds\,\delta s. \tag{2.55}$$

Performing the differential operation on the left-hand side and dividing through by $ds\,\delta s$, we obtain the final equation

$$\left(\frac{\Delta(ds)}{ds} + \frac{\Delta(\delta s)}{\delta s}\right)\cos\theta - \Delta\theta\sin\theta = 2\epsilon_{ij}\mu_i\nu_j. \tag{2.56}$$

Consider now the physical strain terms appearing on the left-hand side. The two terms within the parentheses are clearly the extensional strains in the μ and ν directions, denoted as $\epsilon_{\mu\mu}$ and $\epsilon_{\nu\nu}$:

$$\begin{aligned}\frac{\Delta(ds)}{ds} &= \epsilon_{\mu\mu},\\ \frac{\Delta(\delta s)}{\delta s} &= \epsilon_{\nu\nu}.\end{aligned} \tag{2.57}$$

(Note that, for Greek subscripts, repeated indices do not imply summation.) Thus, taking $\theta = 0$ in (2.56), the general equation for extensional (normal) strain in an arbitrary direction μ in terms of the tensor components ϵ_{ij} becomes

$$\epsilon_{\mu\mu} = \epsilon_{ij}\mu_i\mu_j,$$

which can be expanded as

$$\epsilon_{\mu\mu} = \epsilon_{11}\mu_1^2 + \epsilon_{22}\mu_2^2 + \epsilon_{33}\mu_3^2 + 2\epsilon_{12}\mu_1\mu_2 + 2\epsilon_{13}\mu_1\mu_3 + 2\epsilon_{23}\mu_2\mu_3. \tag{2.58}$$

For the case $\theta = 90°$ the shearing strain $\gamma_{\mu\nu}$ is defined as the decrease in the right angle between the direction lines μ and ν:

$$\gamma_{\mu\nu} = -\Delta\theta = 2\epsilon_{ij}\mu_i\nu_j, \tag{2.59}$$

with

$$\mu_i\nu_i = 0,$$

from the orthogonality of μ and ν. Letting direction μ coincide in turn with each of the coordinate axes x_i, we see that the diagonal elements of the strain tensor are precisely the extensional strains in the corresponding coordinate directions. The shearing strains, with $\theta = 90°$, are

$$\begin{aligned}\gamma_{12} &= -\Delta\theta_{12} = 2\epsilon_{12},\\ \gamma_{13} &= -\Delta\theta_{13} = 2\epsilon_{13},\\ \gamma_{23} &= -\Delta\theta_{23} = 2\epsilon_{23},\end{aligned} \tag{2.60}$$

or, in general,

$$\gamma_{ij} = 2\epsilon_{ij} = u_{i,j} + u_{j,i}, \qquad i \neq j. \tag{2.61}$$

The off-diagonal element ϵ_{ij} of the strain tensor matrix is therefore equal to one-half the decrease in the right angle formed by the corresponding coordinate directions x_i and x_j.

Since the state of strain is a second order symmetric tensor, it transforms in exactly the same way as the stress tensor. Hence all the equations for principal directions, principal normal stresses, principal shearing stresses, etc., can be repeated for strains simply by replacing σ_{ij} by ϵ_{ij} in all terms.

2.3 ELASTIC STRESS, STRAIN, ENERGY RELATIONS

Most metals behave elastically over the greater part of their stress range in a tensile test. Consequently, in the strength analysis of a structural element one first is concerned with elasticity concepts and relationships. Even in the investigation of plastic yielding or fracture in metals, the elastic components of strain and elastic strain energy are quantities requiring evaluation. Thus we proceed to a study of the basic equations of elasticity.

Within general elasticity theory are included materials with nonlinear yet elastic stress-strain laws and orthotropic materials with pronounced differences in properties in different directions. The large majority of structural metals, however, are both isotropic (in a statistical sense) at the macroscopic level as well as reasonably linear up to the yield stress. Thus restricting attention within the following sections to this class of elastic materials will not significantly limit the range of application of the basic equations presented.

2.3.1 Generalized Hooke's Law

The most general linear relationship between the stress and strain tensors σ and ϵ for an arbitrary anisotropic solid can be expressed by the set of six equations:

$$\begin{aligned}
\sigma_{11} &= C_{1111}\epsilon_{11} + C_{1112}\gamma_{12} + C_{1113}\gamma_{13} + C_{1122}\epsilon_{22} \\
&\quad + C_{1123}\gamma_{23} + C_{1133}\epsilon_{33}, \\
\sigma_{12} &= C_{1211}\epsilon_{11} + C_{1212}\gamma_{12} + C_{1213}\gamma_{13} + C_{1222}\epsilon_{22} \\
&\quad + C_{1223}\gamma_{23} + C_{1233}\epsilon_{33}, \\
&\cdots\cdots\cdots\cdots\cdots\cdots \\
\sigma_{33} &= C_{3311}\epsilon_{11} + C_{3312}\gamma_{12} + C_{3313}\gamma_{13} + C_{3322}\epsilon_{22} \\
&\quad + C_{3323}\gamma_{23} + C_{3333}\epsilon_{33},
\end{aligned} \tag{2.62}$$

with 36 elastic constants C_{ijkl}. In tensor notation

$$\sigma_{ij} = C_{ijkl}\epsilon_{kl}, \tag{2.63}$$

where

$$C_{ijkl} = C_{jikl} = C_{ijlk} \tag{2.64}$$

from the symmetry of the stress tensor, thereby reducing the 81 constants that would arise from (2.63) to the 36 independent constants given in (2.62). In the simplest case of a statistically isotropic elastic material this number can be shown to reduce to two and the corresponding equations may be written

$$\sigma_{ij} = \lambda\epsilon_{kk}\delta_{ij} + 2\mu\epsilon_{ij}, \tag{2.65}$$

in which λ and μ are called the Lamé constants. The six independent equations represented by (2.65) are

$$\begin{aligned} \sigma_{11} &= (\lambda + 2\mu)\epsilon_{11} + \lambda(\epsilon_{22} + \epsilon_{33}), \\ \sigma_{22} &= (\lambda + 2\mu)\epsilon_{22} + \lambda(\epsilon_{11} + \epsilon_{33}), \\ \sigma_{33} &= (\lambda + 2\mu)\epsilon_{33} + \lambda(\epsilon_{11} + \epsilon_{22}), \end{aligned} \tag{2.66}$$

the equations of normal stress, and

$$\begin{aligned} \sigma_{12} &= 2\mu\epsilon_{12} = \mu\gamma_{12}, \\ \sigma_{13} &= 2\mu\epsilon_{13} = \mu\gamma_{13}, \\ \sigma_{23} &= 2\mu\epsilon_{23} = \mu\gamma_{23}, \end{aligned} \tag{2.67}$$

the equations of shearing stress. Since μ equals the ratio of shearing stress to shearing strain, it is identically the shear modulus (modulus of rigidity) G of the material and is replaced by G in all subsequent equations.

To obtain the inverse relationships (the general strain-stress law), we begin by contracting the indices in (2.65) [i.e., summing (2.66)] to obtain

$$\sigma_{kk} = (3\lambda + 2G)\epsilon_{kk}. \tag{2.68}$$

As $\sigma_{kk} = \sigma_{11} + \sigma_{22} + \sigma_{33}$ is the first stress invariant (2.21), $\sigma_{av} = \frac{1}{3}\sigma_{kk}$ is the average of the principal stresses at the point. The quantity ϵ_{kk} is (to a first-order approximation) the volumetric strain ϵ_V, which can be demonstrated as follows. Since ϵ_{kk} is the first strain invariant, we have

$$\epsilon_{kk} = \epsilon_{I} + \epsilon_{II} + \epsilon_{III}, \tag{2.69}$$

where ϵ_{I}, ϵ_{II}, ϵ_{III} are the principal strains. Consider a unit cube of the material (Figure 2.12) whose faces are oriented normal to the principal directions. The change in volume (volumetric strain) under load is

$$\Delta V \equiv \epsilon_V = (1 + \epsilon_{I})(1 + \epsilon_{II})(1 + \epsilon_{III}) - 1 = \epsilon_{I} + \epsilon_{II} + \epsilon_{III} + \text{strain terms of higher order.}$$

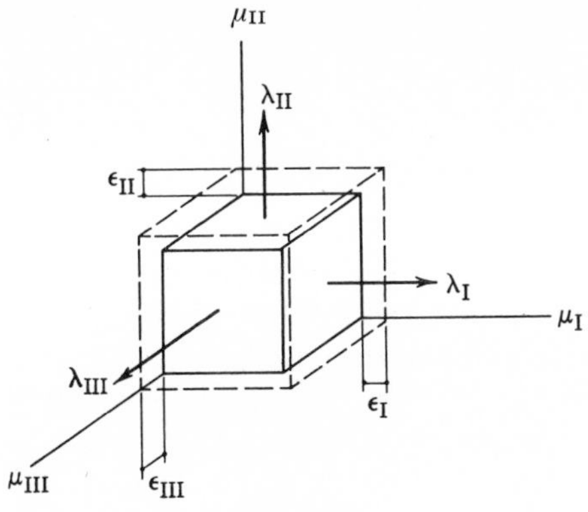

Figure 2.12

Neglecting these products of strains,

$$\epsilon_V = \epsilon_{\text{I}} + \epsilon_{\text{II}} + \epsilon_{\text{III}} \equiv \epsilon_{kk}. \tag{2.70}$$

Thus (2.68) can be written

$$\sigma_{\text{av}} = \tfrac{1}{3}(3\lambda + 2G)\epsilon_V. \tag{2.71}$$

The quantity $\frac{1}{3}(3\lambda + 2G)$, the ratio of the average normal stress to the volumetric strain, is called the volumetric or bulk modulus of elasticity E_V:

$$\sigma_{kk} = 3E_V\epsilon_{kk}. \tag{2.72}$$

Returning to (2.65) and dividing by $2G$ (i.e., 2μ), we can write, with the aid of (2.72),

$$\epsilon_{ij} = \frac{1}{2G}\left(\sigma_{ij} - \frac{\lambda}{3E_V}\,\sigma_{kk}\delta_{ij}\right) \tag{2.73}$$

or

$$\epsilon_{ij} = \frac{1}{2G}\left(\sigma_{ij} - \frac{\lambda}{E_V}\,\sigma_{\text{av}}\delta_{ij}\right).$$

Equation 2.73 is one form of the strain-stress law for the isotropic, elastic material. An alternate form of this law can be determined by first defining a deviatoric strain tensor e_{ij} analogous to the deviatoric stress tensor s_{ij} of (2.35):

$$e_{ij} = \epsilon_{ij} - \tfrac{1}{3}\epsilon_{kk}\delta_{ij}. \tag{2.74}$$

Substituting this equation and the equation for s_{ij} into (2.65),

$$s_{ij} + \sigma_{\text{av}}\delta_{ij} = 2Ge_{ij} + \tfrac{1}{3}(3\lambda + 2G)\epsilon_{kk}\delta_{ij},$$

from which, combining with (2.71)

$$s_{ij} = 2Ge_{ij}. \tag{2.75}$$

Therefore, from (2.72), (2.74), and (2.75), the equation for the strain can be written

$$\epsilon_{ij} = \frac{1}{2G} s_{ij} + \frac{1}{3E_V} \sigma_{av}\delta_{ij}. \tag{2.76}$$

Contracting the subscripts in this equation and noting that $s_{kk} \equiv 0$, we obtain $\epsilon_{kk} = \sigma_{av}/E_V$ as required by (2.70) through (2.72).

It is important to relate the constants G and λ to the Young's modulus E and Poisson's ratio ν obtained from the tensile test. Consider the uniaxial state of stress σ_{11}, all other components being zero. From (2.73)

$$\epsilon_{11} = \frac{1}{2G}\left(\sigma_{11} - \frac{\lambda}{3E_V}\sigma_{11}\right) = \frac{\lambda + G}{3\lambda + 2G}\frac{\sigma_{11}}{G}$$

$$\epsilon_{22} = \epsilon_{33} = -\frac{\lambda}{3\lambda + 2G}\frac{\sigma_{11}}{2G}.$$

Since E is defined as the ratio of σ_{11} to ϵ_{11} and ν is the ratio of the lateral contraction $-\epsilon_{22}$ (or $-\epsilon_{33}$) to ϵ_{11},

$$E = \frac{(3\lambda + 2G)G}{(\lambda + G)}$$

and

$$\nu = \frac{\lambda}{2(\lambda + G)}. \tag{2.77}$$

Solving for λ and G in terms of the experimentally determined values E and ν,

$$\lambda = \frac{\nu E}{(1 + \nu)(1 - 2\nu)},$$

$$G = \frac{E}{2(1 + \nu)}, \tag{2.78}$$

and the bulk modulus is expressed

$$E_V = \tfrac{1}{3}(3\lambda + 2G) = \frac{E}{3(1 - 2\nu)}. \tag{2.79}$$

A value of $\nu = \frac{1}{2}$ yields an infinite value of E_V, corresponding to a perfectly incompressible material. For most metals ν is one third or less.

The final form of the strain-stress law is obtained by substituting (2.78) and (2.79) into (2.73):

$$\epsilon_{ij} = \frac{1 + \nu}{E}\sigma_{ij} - \frac{\nu}{E}\sigma_{kk}\delta_{ij} \tag{2.80}$$

or

$$\epsilon_{11} = \frac{1}{E}[\sigma_{11} - \nu(\sigma_{22} + \sigma_{33})],$$

$$\epsilon_{22} = \frac{1}{E}[\sigma_{22} - \nu(\sigma_{11} + \sigma_{33})], \qquad (2.81a)$$

$$\epsilon_{33} = \frac{1}{E}[\sigma_{33} - \nu(\sigma_{11} + \sigma_{22})]$$

for the normal (extensional) strains and

$$\gamma_{12} = 2\epsilon_{12} = \frac{1}{G}\sigma_{12},$$

$$\gamma_{13} = 2\epsilon_{13} = \frac{1}{G}\sigma_{13}, \qquad (2.81b)$$

$$\gamma_{23} = 2\epsilon_{23} = \frac{1}{G}\sigma_{23}$$

for the shearing strains.

2.3.2 *Strain Energy*

The internal elastic strain energy in a deformed solid under load can be resolved into two separate parts, strain energy of volume change and strain energy of distortion, corresponding to the two distinct characteristics of deformation. To derive the basic equations of strain energy we begin by expressing the work done (per unit volume) by a slowly applied force F acting normal to an area A in producing an elastic displacement δ over a length L:

$$Wk = \frac{\frac{1}{2}F\delta}{AL} = \frac{1}{2}\sigma\epsilon,$$

where σ and ϵ are the uniaxial stress and associated strain in the unit volume. Generalizing to the case of a three-dimensional state of stress σ_{ij} and denoting the total strain energy per unit volume by U, we have

$$U = \tfrac{1}{2}\sigma_{ij}\epsilon_{ij}, \qquad (2.82)$$

or

$$U = \tfrac{1}{2}(\sigma_{11}\epsilon_{11} + \sigma_{22}\epsilon_{22} + \sigma_{33}\epsilon_{33} + \sigma_{12}\gamma_{12} + \sigma_{13}\gamma_{13} + \sigma_{23}\gamma_{23}).$$

To separate U into its two components of volume change and distortion, the stress and strain tensors must be resolved into their corresponding components of average and deviatoric stress and strain, respectively. Thus, substituting from (2.35) and (2.74),

$$U = \tfrac{1}{2}(s_{ij} + \sigma_{\text{av}}\delta_{ij})(e_{ij} + \epsilon_{\text{av}}\delta_{ij})$$

$$= \tfrac{1}{2}(s_{ij}e_{ij} + \sigma_{\text{av}}e_{kk} + s_{kk}\epsilon_{\text{av}} + 3\sigma_{\text{av}}\epsilon_{\text{av}}),$$

where $\epsilon_{\text{av}} \equiv \frac{1}{3}\epsilon_{kk} = \frac{1}{3}\epsilon_V$. Noting that s_{kk} and e_{kk} are identically zero.

$$U = \tfrac{1}{2}\sigma_{\text{av}}\epsilon_V + \tfrac{1}{2}s_{ij}e_{ij}. \tag{2.83}$$

The first term in this equation is the work done by the average stress on the volumetric strain and represents the contribution of the change in volume to the energy. The second term, the work done by the deviatoric stress on the deviatoric strain, corresponds to zero volume change and is the contribution of the distortion of the body. This separation of stress and deformation states is depicted in Figure 2.13 for a unit volume of material whose axes are aligned with the principal directions of stress. Denoting the energy of volume change as U_V and the distortional strain energy as U_D, we have [utilizing (2.71), (2.41), (2.75), (2.38) and (2.39)]

$$U_V = \tfrac{1}{2}\sigma_{\text{av}}\epsilon_V = \frac{1}{2E_V}\,{\sigma_{\text{av}}}^2 = \frac{1}{2E_V}\,(\sigma_N^{\circ})^2, \tag{2.84}$$

and

$$U_D = \tfrac{1}{2}s_{ij}e_{ij} = \frac{1}{2G}\,s_{ij}s_{ij} = \frac{J_2}{G} = \frac{3}{2G}\,(\sigma_S^{\circ})^2. \tag{2.85}$$

Thus we see that the octahedral normal and shearing stresses are related to the energies of volume change and distortion, respectively. It is through this relationship with the distortional component of strain energy that the octahedral shear stress plays such an important part in plasticity laws, since plastic (permanent) deformation is primarily distortional in nature.

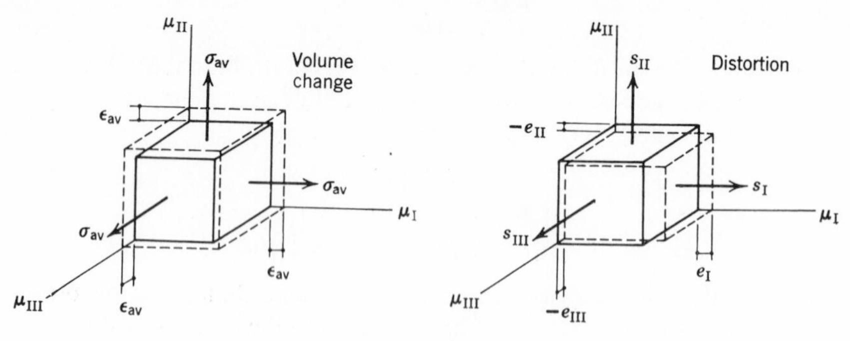

Figure 2.13

Returning to the expression for the total strain energy U, we find from (2.84) and (2.85) that this energy can be written

$$U = \frac{1-2\nu}{E}(\sigma_{kk})^2 + \frac{1+\nu}{E}(s_{ij}s_{ij}). \tag{2.86}$$

Alternatively, U can be expressed in terms of the strains alone as

$$U = \frac{E}{3(1-2\nu)}(\epsilon_{kk})^2 + \frac{E}{1+\nu}(e_{ij}e_{ij}). \tag{2.87}$$

The use of the total strain energy function will be required in the investigation of static fracture phenomena in metals.

2.4 THEORIES OF INITIAL AND CONTINUED YIELDING

The theory of plasticity—that branch of mechanics which is concerned with the inelastic behavior of materials—finds its physical basis in the crystal structure of metals and the nature of atomic slip (the source of permanent deformation in metals). These subjects are the domain of dislocation theory, whereas plasticity theory itself is restricted to the phenomenological study of initial and continued yielding of metals (under combined states of stress) and the development of mathematical laws to predict observed macroscopic behavior.

Before we begin our brief survey of some of the fundamental concepts and general results that have been obtained in the theory of plasticity, consider the simple case of uniaxial tension and the true stress-true strain diagram of a typical material in Figure 2.14. As the specimen is loaded from the virgin state (the state of the material before it has experienced permanent deformation), the strain at any level of stress is given by the corresponding point on

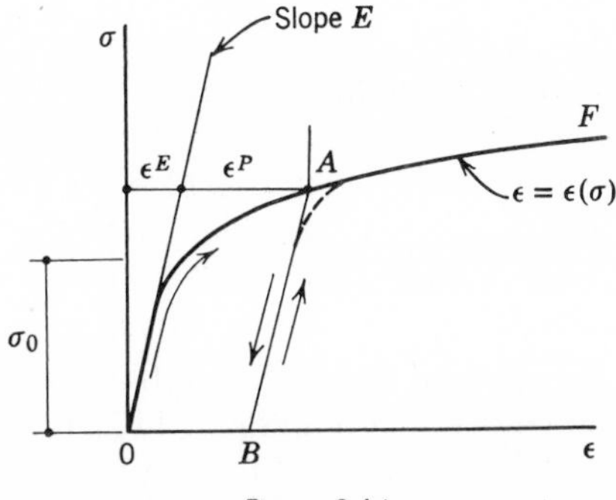

Figure 2.14

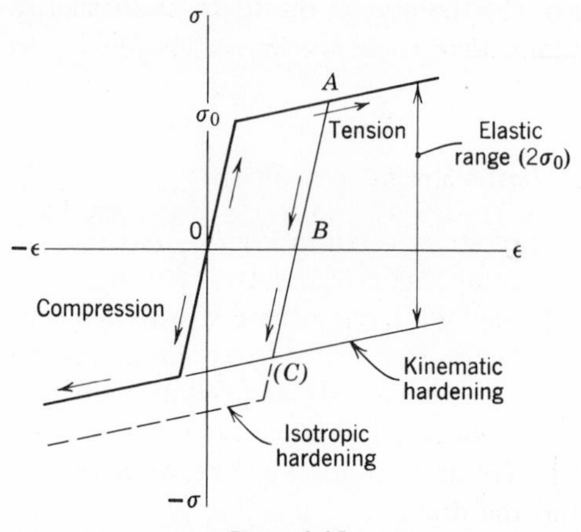

Figure 2.15

the curve $\epsilon = \epsilon(\sigma)$. If the specimen is unloaded (as depicted by the arrow) from some arbitrary point A, the response will be linearly elastic and follow the straight line AB with a slope equal to the initial slope E of the stress-strain curve. Thus the strain is composed of an elastic (recoverable) part ϵ^E and a plastic or permanent strain ϵ^P, as shown. On reloading, the response remains elastic until the stress level once more reaches point A (approximately) after which, with continued loading, the curve $\epsilon = \epsilon(\sigma)$ again is traced. Now consider an idealization of this curve—the elastic, linear strain-hardening curve shown in Figure 2.15, with an abrupt yield point at the stress σ_0. If the material is isotropic in the virgin state, the yield stress in compression will in general equal the value in tension, as shown. If we hypothesize a loading to point A and then a load reversal, following the elastic unloading path ABC, the question immediately is, at what stress does the material yield in compression? Should the material remain isotropic, the stress of course would be equal in magnitude to the tensile stress at A. However, macroscopic plastic deformation is a result of an accumulation of microscopic slips along crystallographic planes oriented in the general direction of planes of maximum shearing stress. Thus at the crystalline level, at least, we are not dealing with an isotropic phenomenon. We therefore expect the material to yield at a compressive stress less than σ_A. If the material actually yields at a compressive stress less than its initial compressive yield stress σ_0, the phenomenon is referred to as the Bauschinger effect. The ideal Bauschinger effect corresponds to the case in which the material retains its elastic range $(2\sigma_0)$ and yields at a compressive stress lowered by an amount equal to the increase in the tensile yield stress. We can expect all metals of interest to lie between the extremes of isotropic hardening and the ideal Bauschinger effect (also called kinematic hardening). Two of the theoretical results to be developed in the following sections are the extension of these concepts of hardening to the general state of stress σ_{ij}.

2.4.1 *Fundamental Plastic Strain-Stress Relations*

Consider the state of stress given in Figure 2.5. Since there are but three independent nonzero components of the stress tensor, σ_{11}, σ_{22}, and $\sigma_{12} = \sigma_{21}$, a stress space can be defined having these stress components as coordinate axes (Figure 2.16). The position vector to a point in this space is the vectorial representation of the stress tensor σ_{ij} and is called the stress vector $\boldsymbol{\sigma}$. It is assumed that a loading surface (or function) exists in this stress space; that is, a surface that serves as the boundary between the domain (region) of elastic response and the domain of possible plastic response. For a virgin material this surface is called the yield surface. In an isotropic material the yield surface intersects the axes σ_{11} and σ_{22} at the points $\sigma_0, 0, 0$, $-\sigma_0, 0, 0$, $0, \sigma_0, 0$, and $0, -\sigma_0, 0$, corresponding to uniaxial tension or compression in the x_1 or x_2 direction (Figure 2.5), respectively. The intersections of the yield surface with the σ_{12} axis correspond to the yield stress in pure shear from a torsion test, denoted as τ_0. As plastic deformation of a work-hardening material takes place, the surface will, in general, change both its shape and position in stress space. Extending the concepts introduced in the preceding section, we find that the uniform expansion of the yield surface corresponds to isotropic hardening, whereas a loading surface that retains the size of its initial elastic domain while translating in stress space corresponds to kinematic hardening.

Consider now the generalization of Figure 2.16 to a nine-dimensional stress space whose coordinates are the nine components of the stress tensor σ_{ij}

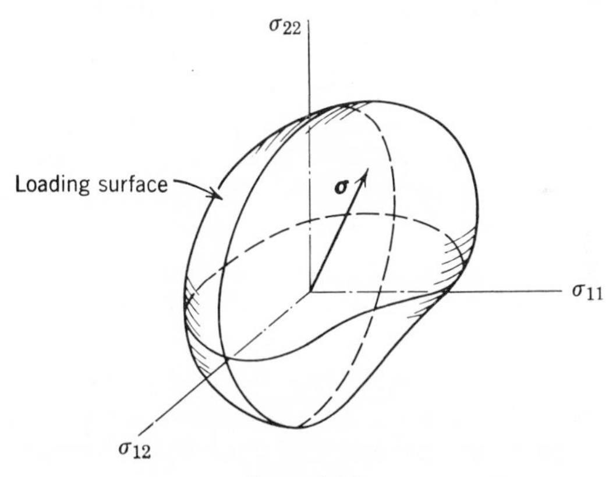

Figure 2.16

(again referred to as the stress vector). It is possible to establish certain fundamental properties of both the loading surface in this general stress space and the basic stress-strain relations of plasticity through use of what has come to be known as Drucker's postulate of material stability. This postulate, given in a paper [5] that has become something of a landmark in the modern theory of plasticity, is essentially a quasi-thermodynamic definition of a stable material, expressed in terms of the (isothermal) work done by an external agency which slowly applies and then removes an additional set of stresses to an initially stressed element in equilibrium. The postulate is stated as follows:

1. The element in a work-hardening material remains in equilibrium during the cycle of application and removal of stresses.
2. The work done by the external agency during the application of additional stresses is positive.
3. The work done by the external agency during the complete cycle is nonnegative and equal to zero if and only if elastic changes in strain alone are produced.

Expressed differently, work hardening means that useful net energy cannot be removed from the system consisting of the material body and the forces acting on it in a closed cycle of loading and unloading, and energy must be put into the system for plastic deformation to occur.

To see the consequences of Drucker's postulate consider an initial state of stress σ_{ij}^* at a point in a stressed material element. The corresponding vector $\boldsymbol{\sigma}^*$ in nine-dimensional stress space will be on or within the loading surface $f(\sigma_{ij}) = 0$ (represented conceptually in Figure 2.17). Consistent with the statement of the postulate, an external agency is chosen to apply additional stresses along an elastic path until a state of stress σ_{ij} on the loading surface is attained. (An elastic path is a path that lies wholly within the elastic domain.) At this stage only elastic changes in strain have taken place.

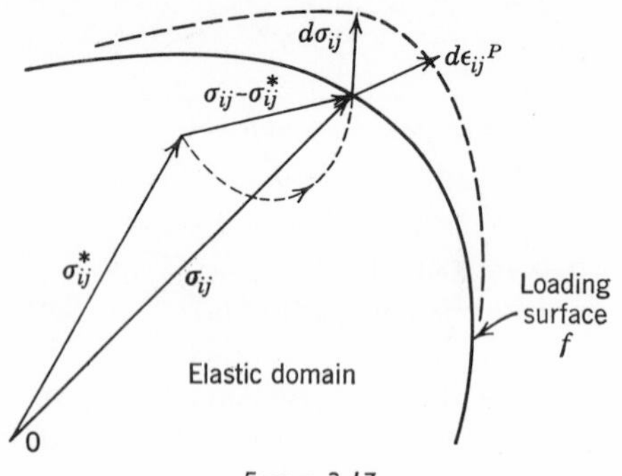

Figure 2.17

Consequently, the state of strain at σ_{ij} is path-independent. If now the external agency adds an increment in stress $d\sigma_{ij}$ such that $\sigma_{ij} + d\sigma_{ij}$ is in the plastic domain (thereby changing the surface f in the vicinity of σ_{ij} as shown), a plastic strain increment $d\epsilon_{ij}{}^P$ will be produced in addition to the elastic increment $d\epsilon_{ij}{}^E$. By associating a nine-dimensional plastic strain vector with each point in stress space, we can represent this plastic strain increment on the diagram. On removing the increment in stress and returning the element to its initial state of stress σ_{ij}^* along an elastic path, all the elastic energy will be recovered and only the work done by the stresses on the incremental plastic strains will be nonzero. The equivalent mathematical statement of Drucker's postulate is therefore

$$\delta Wk = (\sigma_{ij} - \sigma_{ij}^*)\, d\epsilon_{ij}{}^P + d\sigma_{ij}\, d\epsilon_{ij}{}^P > 0. \tag{2.88}$$

The first term in this equation is the work done per unit volume by the external agency $(\sigma_{ij} - \sigma_{ij}^*)$, corresponding to the state of stress σ_{ij}, on the increment in plastic strain. The second term is the work done by the increment in the external agency $(d\sigma_{ij})$ on this incremental strain. Geometrically, the terms can be interpreted as scalar products of the stress vectors $\boldsymbol{\sigma} - \boldsymbol{\sigma}^*$ and $d\boldsymbol{\sigma}$, respectively, with the plastic strain vector $d\boldsymbol{\epsilon}^P$ in the nine-dimensional space. As the initial state of stress σ_{ij}^* is arbitrary, it can be chosen in the loading surface f, in which case σ_{ij}^* will be identically σ_{ij}. Therefore the second term in (2.88) must be positive:

$$d\sigma_{ij}\, d\epsilon_{ij}{}^P > 0. \tag{2.89}$$

Equation 2.88 also must hold for any state of stress in the elastic domain and a finite distance from the loading surface (such that $|\sigma_{ij} - \sigma_{ij}^*| \gg |d\sigma_{ij}|$). Thus the first term in (2.88) is non-negative:

$$(\sigma_{ij} - \sigma_{ij}^*)\, d\epsilon_{ij}{}^P \geq 0. \tag{2.90}$$

Equations 2.89 and 2.90, the fundamental consequences of Drucker's postulate, have the following geometrical interpretation. From the non-negative value of the scalar product in (2.90), the incremental plastic strain vector $d\boldsymbol{\epsilon}^P$ must be perpendicular to or make an acute angle with the stress vector $\boldsymbol{\sigma} - \boldsymbol{\sigma}^*$ for every stress point σ_{ij}^* in the elastic domain. Thus all points σ_{ij}^* lie on one side of a hyperplane (the generalization of a plane in higher dimensional space) passing through σ_{ij} and perpendicular to $d\boldsymbol{\epsilon}^P$, and therefore the loading surface is convex, as shown in Figure 2.18.

The inequality (2.89) requires that the incremental plastic strain vector $d\boldsymbol{\epsilon}^P$ make an acute angle with the incremental stress vector $d\boldsymbol{\sigma}$. From the convexity of the loading surface, this requirement leads to the conclusion that $d\boldsymbol{\epsilon}^P$ must be normal to the surface at a regular (smooth) point (as in

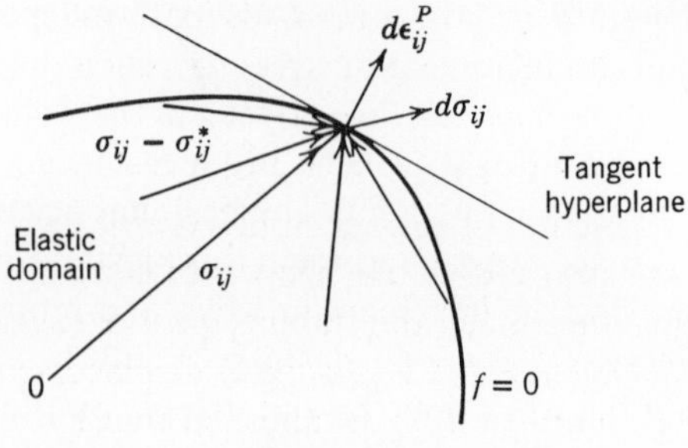

Figure 2.18

Figure 2.18) and lie within the outward cone of limiting normals to the surface at a singular point (corner), as shown by Figure 2.19. At a singular point nothing further can be concluded about the incremental plastic strain vector. Thus the fundamental postulate alone cannot uniquely determine the direction of this vector but can only impose a restriction (the forward cone of normals) on the possible choice of directions. We can hypothesize that the direction of the incremental plastic strain vector depends in some way on the direction of the incremental stress vector, and there is considerable experimental evidence in support of this view. It is emphasized again, however, that Drucker's postulate can say nothing whatsoever about this, and some additional basis must be provided for a determination of the relationship at a singular point.

For a regular point the situation is altogether different, as the direction of the incremental plastic strain vector is both known and completely independent of the incremental stress vector (Figure 2.18). Since $d\boldsymbol{\epsilon}^P$ is normal to the loading surface, it must coincide with the vector gradient of the scalar

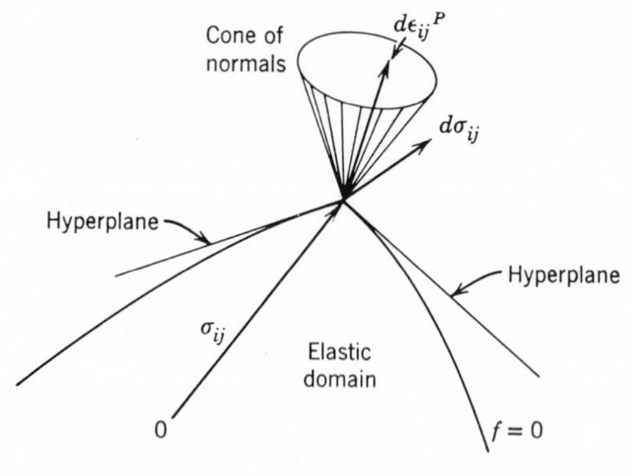

Figure 2.19

function f. Thus the mathematical statement of normality is

$$d\epsilon_{ij}{}^{P} = \lambda \frac{\partial f}{\partial \sigma_{ij}}, \tag{2.91}$$

where λ is a scalar function of proportionality which may depend on stress, increment in stress, strain, and the history of loading. In a more recent paper [6] Drucker has extended the stability postulate to state that the work done by the external agency on the *change* in displacements it produces must be non-negative. From this he has established that linearity, as well as convexity and normality, is a necessary consequence of material stability for a smooth loading surface. By linearity is meant that the incremental plastic strain $d\epsilon_{ij}{}^{P}$ depends only on the component of the incremental stress vector normal to the loading surface f (i.e., collinear with $d\boldsymbol{\epsilon}^{P}$). Since this component is proportional to the scalar product of $d\boldsymbol{\sigma}$ with the vector gradient of f (2.91) becomes

$$d\epsilon_{ij}{}^{P} = G \frac{\partial f}{\partial \sigma_{ij}} \frac{\partial f}{\partial \sigma_{kl}} d\sigma_{kl}, \tag{2.92}$$

where G as well as f is a scalar function that may depend on stress, strain, and history of loading but is independent of the increment in stress $d\sigma_{ij}$. Equation 2.92 is the general stress-strain law of the linear incremental theory of small plastic deformation. It has no built-in restrictions (other than the requirement of a smooth loading surface), and it is sufficiently general to include both initially isotropic and initially anisotropic materials that exhibit the Bauschinger effect and possess loading functions that change size, shape, and position as plastic straining takes place. Thus this equation is the plastic counterpart of the elastic generalized Hooke's law (2.63), and the equation for the total incremental strain can be written

$$d\epsilon_{ij} = C_{ijkl}\, d\sigma_{kl} + G \frac{\partial f}{\partial \sigma_{ij}} \frac{\partial f}{\partial \sigma_{kl}} d\sigma_{kl}. \tag{2.93}$$

The mathematical statements of loading, unloading, and neutral loading follow from (2.92) for an incremental change in stress at a point on the loading surface $f = 0$:

$$\begin{aligned} &\text{Unloading} && \frac{\partial f}{\partial \sigma_{kl}} d\sigma_{kl} < 0 \\ &\text{Neutral loading} && \frac{\partial f}{\partial \sigma_{kl}} d\sigma_{kl} = 0 \\ &\text{Loading} && \frac{\partial f}{\partial \sigma_{kl}} d\sigma_{kl} > 0. \end{aligned} \tag{2.94}$$

Correspondingly, the incremental stress vector is directed inward, outward, or tangent to the convex loading surface for unloading, loading, or neutral loading, respectively.

2.4.2 *Phenomenological Stress-Strain Laws*

In formulating specific stress-strain laws for the correlation of experimental data of initial and continued yielding, three basic assumptions commonly are made:

1. The material is statistically isotropic in the virgin state.
2. Plastic deformation is independent of hydrostatic stress (of an order of magnitude of the yield stress).
3. There is no plastic (irrecoverable) volume change.

Using these assumptions as limitations on the scope of this review, the first serves to exclude from consideration any structural metal which exhibits preferred orientation in the elastic state (consistent with Section 2.3). The second assumption implies that plastic strain depends only on the components of the deviatoric stress tensor s_{ij} (2.35) and not on the average principal stress, from which

$$\frac{\partial f}{\partial \sigma_{kk}} = 0. \tag{2.95}$$

The third assumption, which requires that volume change be a purely elastic phenomenon, can be expressed mathematically as

$$d\epsilon_{kk}{}^P = 0. \tag{2.96}$$

These assumptions are justified to within a reasonable first approximation by the experimental work of Bridgman [7] and others. Based on these hypotheses, a simple representation of the loading surface f is now possible in three-dimensional, principal stress space. Consider first the initial yield surface f_0 of the virgin material. Since the material is initially isotropic, the yield surface must be independent of the direction of the applied stress and depend only on the stress invariants (2.21). Furthermore, from (2.95) f_0 is independent of the first invariant σ_{kk}; hence f_0 can be expressed solely in terms of the second and third invariants J_2 and J_3 of the deviatoric stress tensor s_{ij}, with J_2 given by (2.38) and

$$J_3 = D(s_{ij}) = \tfrac{1}{3}s_{ij}s_{jk}s_{ki}. \tag{2.97}$$

If, in addition, it is assumed that there is no Bauschinger effect at initial yield, $f_0(s_{ij})$ must equal $f_0(-s_{ij})$ and, since $J_3(s_{ij}) = -J_3(-s_{ij})$, f_0 must be an even function of J_3:

$$f_0(J_2, J_3{}^2) = 0. \tag{2.98}$$

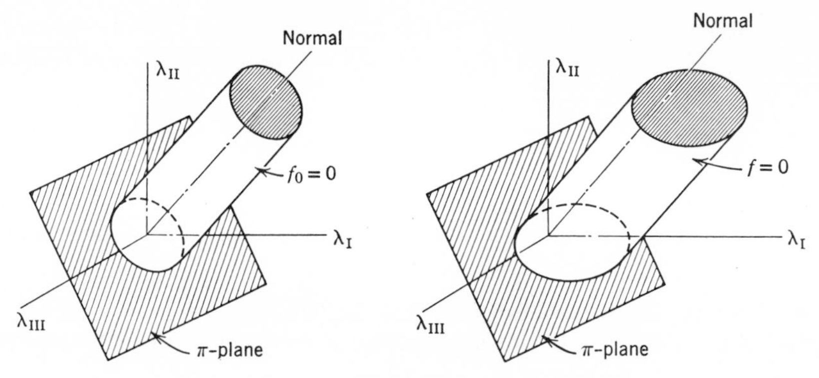

Figure 2.20

After plastic straining takes place a Bauschinger effect will be observed (in general), and the work hardening will be anisotropic. The loading surface remains independent of σ_{kk}, however, from (2.95), and the functional dependent of f can be expressed as

$$f(J_2, J_3, \epsilon_{ij}{}^P) = 0. \tag{2.99}$$

The geometric interpretation of (2.98) and (2.99) in principal stress space is shown in Figure 2.20. The initial and subsequent loading surfaces are represented by infinite, convex right cylinders parallel to the line of hydrostatic (equal triaxial) stress which passes through the origin and makes equal angles with the principal stress axes. This representation corresponds to the assumption that plastic deformation is not influenced by hydrostatic stress $\lambda_{\mathrm{I}} = \lambda_{\mathrm{II}} = \lambda_{\mathrm{III}}$. The π-plane is perpendicular to this line and to the cylinders and has the equation

$$\lambda_{\mathrm{I}} + \lambda_{\mathrm{II}} + \lambda_{\mathrm{III}} = 0. \tag{2.100}$$

The initial yield cylinder f_0, assumed to be isotropic and without a Bauschinger effect, must have six symmetries in the π-plane (Figure 2.21) in addition to being convex. Isotropy requires f_0 to be symmetric with respect to the projections of the stress coordinate axes on this plane. The absence of the Bauschinger effect requires f_0 to be symmetric with respect to the intersections of the stress coordinate planes with the π-plane, represented by the dotted lines in Figure 2.21. The points A, A', B, B', C, C' correspond to the yield points in tension and compression in the three coordinate directions. The range of all possible convex (hence permissible) yield surfaces that pass through these points is shown by the shaded area, and a typical yield curve is represented by the curved line. Thus Drucker's postulate, together with the basic assumptions stated at the beginning of this section, enables us to bound

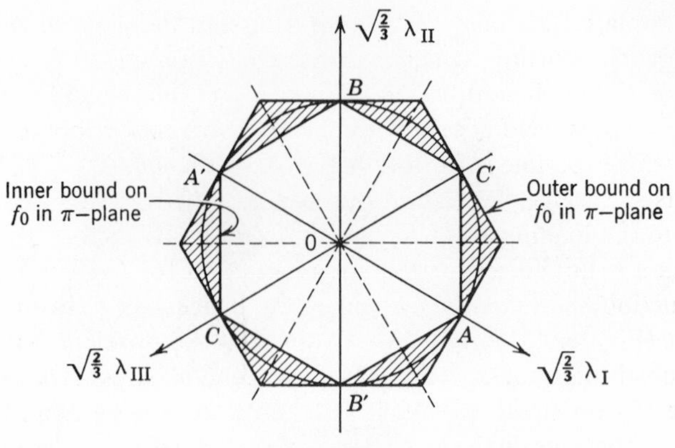

Figure 2.21

the yield surface of a stable isotropic material within quite narrow limits. Unfortunately, such is not the case for subsequent loading surfaces, since these need only be convex, there being no requirements of symmetry in (2.99). Thus the intersection of f with the π-plane in Figure 2.20 can be a convex closed curve of any shape, depending on the nature of the work hardening and the loading history of the material considered. It need not bound the origin of principal stress space and is not required to be smooth.

Returning to a consideration of initial yield criteria, the inner bound on f_0 in Figure 2.21 is called the Tresca criterion and corresponds to a material that yields when the maximum shear stress reaches the value of the maximum shear stress at yielding ($\frac{1}{2}\sigma_0$) in a tensile test. From (2.34) the Tresca criterion can be expressed

$$\max |\lambda_{I} - \lambda_{II}|, |\lambda_{I} - \lambda_{III}|, |\lambda_{II} - \lambda_{III}| = \sigma_0. \tag{2.101}$$

The outer bound on f_0 is not an established yield criterion and serves merely as a mathematical limitation on the criteria that could be considered. The geometrically simplest criterion is represented by a circle (a circular cylinder in principal stress space) of radius $\sqrt{\frac{2}{3}}\sigma_0$ passing through the yield points A, A', etc. This criterion is due to von Mises [8] and is expressible by the equation

$$f_0 = J_2 - \tfrac{1}{3}\sigma_0^{\,2} = 0, \tag{2.102}$$

which can be written

$$s_{kl}s_{kl} = \tfrac{2}{3}\sigma_0^{\,2}. \tag{2.103}$$

From (2.39) the von Mises criterion is seen to be related to the octahedral shearing stress σ_S°. Thus a material obeying this criterion yields at a value of octahedral shear stress equal to that attained at yielding ($\sqrt{\frac{2}{9}}\sigma_0$) in a tensile test. An alternate statement of this criterion can be given in terms of the elastic energy of distortion, which is expressed as a function of the octahedral shear stress in (2.85). In addition to having a simple mathematical representation, the von Mises yield criterion has been shown repeatedly to be in close agreement with experiment for a number of ductile metals. For this reason, as well as its simplicity, we restrict our attention here to this yield law.

Turning to the loading surface f, we find that a satisfactory mathematical representation is not so easy to obtain as was that for f_0. The fact that the loading function in nine-dimensional stress space has a two-dimensional counterpart (Figure 2.21) does little to simplify the problem, for it permits only the substitution of s_{ij} for σ_{ij} in the general stress-strain law. The dependence of f on stress, plastic strain, and path history does not in itself impose any restrictions on the size, shape, or position of the curve representing this function in the π-plane. In order to correlate experimental data in the plastic range and perform mathematical stress analyses of elastic-plastic problems, it is necessary to formulate phenomenological hardening laws. The remainder of this section is devoted to three of the more widely used laws.

The simplest and undoubtedly the best known incremental theory of hardening considers the loading surface f to be an isotropic, homogeneous function of stress. In this theory, named isotropic hardening, successive loading surfaces are uniform expansions of the initial yield surface (as previously mentioned). For the special case of the von Mises yield criterion the equation for the loading function is

$$f = J_2 - \tfrac{1}{3}\sigma_{\text{eq}}^{\,2} = 0. \tag{2.104}$$

or

$$s_{kl}s_{kl} = \tfrac{2}{3}\sigma_{\text{eq}}^{\,2}, \tag{2.105}$$

where σ_{eq} is an equivalent stress which increases with increasing strain and is equal to σ_0 at initial yield [see (2.102) and (2.103)]. The stress-strain law corresponding to this loading function is determined as follows. From (2.104) and (2.38)

$$\frac{\partial f}{\partial \sigma_{ij}} = \frac{\partial J_2}{\partial \sigma_{ij}} = s_{ij} \tag{2.106}$$

and

$$\frac{\partial f}{\partial \sigma_{kl}}\,d\sigma_{kl} = \frac{\partial J_2}{\partial \sigma_{kl}}\,d\sigma_{kl} = s_{kl}\,ds_{kl} = dJ_2. \tag{2.107}$$

Substituting these relations into (2.92), we obtain the desired stress-strain law:

$$d\epsilon_{ij}^{\,P} = Gs_{ij}\,dJ_2 = \tfrac{2}{3}Gs_{ij}\sigma_{\text{eq}}\,d\sigma_{\text{eq}}. \tag{2.108}$$

The function G is determined from the requirement that the correct value of incremental plastic strain from a tensile test be given by (2.108). For a

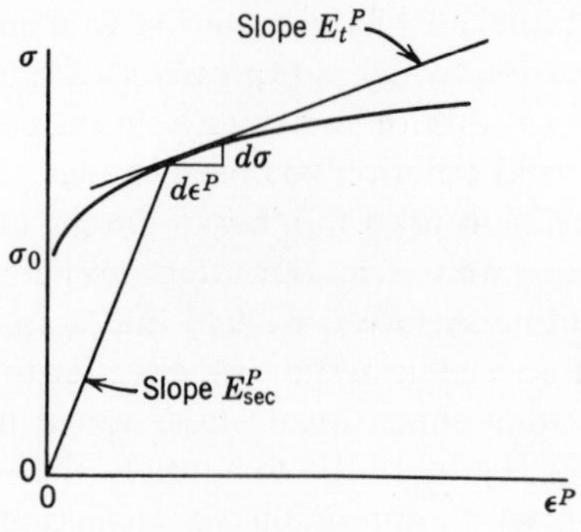

Figure 2.22

tensile stress σ_{11}: $s_{11} = \frac{2}{3}\sigma_{11}$, $s_{22} = s_{33} = -\frac{1}{3}\sigma_{11}$, all other components are zero and

$$dJ_2 = s_{kl}\, ds_{kl} = \tfrac{2}{3}\sigma_{11}\, d\sigma_{11},$$
$$d\epsilon_{11}{}^P = (\tfrac{2}{3}\sigma_{11})\, G\, (\tfrac{2}{3}\sigma_{11}\, d\sigma_{11}) = (\tfrac{4}{9}\sigma_{11}{}^2\, d\sigma_{11})\, G. \tag{2.109}$$

From a plot of stress vs. plastic strain in the tensile test (Figure 2.22),

$$d\epsilon_{11}{}^P = \frac{d\sigma_{11}}{E_t{}^P}, \tag{2.110}$$

where $E_t{}^P$ is the tangent modulus of the stress-plastic strain curve. Thus, equating these two expressions for $d\epsilon_{11}{}^P$,

$$G = \frac{9}{4}\frac{1}{\sigma_{11}{}^2 E_t{}^P} = \frac{3}{4}\frac{1}{J_2 E_t{}^P}. \tag{2.111}$$

Substituting (2.111) into (2.108), we obtain the final stress-strain law of uniform isotropic hardening from the Mises yield surface:

$$d\epsilon_{ij}{}^P = \frac{3}{4E_t{}^P}\frac{dJ_2}{J_2}\, s_{ij} \tag{2.112}$$

or, from (2.104),

$$d\epsilon_{ij}{}^P = \frac{3}{2}\frac{1}{E_t{}^P}\frac{d\sigma_{\text{eq}}}{\sigma_{\text{eq}}}\, s_{ij}. \tag{2.113}$$

This equation can also be expressed in terms of the tangent modulus E_t of the stress-total strain curve (Figure 2.14) and the initial elastic modulus E by using the identity

$$\frac{d\epsilon_{11}}{d\sigma_{11}} \equiv \frac{d\epsilon_{11}{}^E}{d\sigma_{11}} + \frac{d\epsilon_{11}{}^P}{d\sigma_{11}}.$$

Thus

$$d\epsilon_{ij}{}^P = \frac{3}{4}\left(\frac{1}{E_t} - \frac{1}{E}\right)\frac{dJ_2}{J_2}\, s_{ij}. \tag{2.114}$$

Equation 2.114 has been called the Laning hardening law [9] and is an extension of the equations for an elastic—perfectly plastic material (known as the Prandtl-Reuss equations) to include the effects of strain-hardening. The Laning law is considered adequate to represent the behavior of many metals in the plastic range as long as there is but one loading, followed, perhaps, by unloading. It is not appropriate for cases of reversed loading, as previously pointed out for uniaxial tension (Figure 2.15), since isotropic hardening predicts a negative Bauschinger effect (increased rather than decreased yield stress) after reloading in the opposite direction from an initial plastically strained state.

In collating experimental data with the predictions of the Laning stress-strain law, it is desirable to introduce a single strain parameter, called the equivalent plastic strain, as a basis for comparison. From the tensile test and (2.110)

$$\epsilon_{11}{}^P = \int_0^{\sigma_{11}} \frac{d\sigma_{11}}{E_t{}^P(\sigma_{11})}, \tag{2.115}$$

with the functional dependence of $E_t{}^P$ on σ_{11} noted. The equivalent plastic strain is defined in an analogous way as

$$\epsilon_{\text{eq}}{}^P = \int_0^{\sigma_{\text{eq}}} \frac{d\sigma_{\text{eq}}}{E_t{}^P(\sigma_{\text{eq}})}, \tag{2.116}$$

where the plastic tangent modulus now depends on the equivalent stress σ_{eq}. To obtain a specific equation for $\epsilon_{\text{eq}}{}^P$ in terms of the plastic strain components we first take the differential of (2.116) and then substitute into (2.113). Thus

$$d\epsilon_{ij}{}^P = \frac{3}{2}\frac{d\epsilon_{\text{eq}}{}^P}{\sigma_{\text{eq}}} s_{ij}. \tag{2.117}$$

Taking the scalar product of each side of this equation with itself,

$$\begin{aligned} d\epsilon_{ij}{}^P\, d\epsilon_{ij}{}^P &= \frac{9}{4}\frac{(d\epsilon_{\text{eq}}{}^P)^2}{(\sigma_{\text{eq}})^2} s_{ij}s_{ij} \\ &= \tfrac{3}{2}(d\epsilon_{\text{eq}}{}^P)^2 \end{aligned}$$

from (2.105). Therefore

$$d\epsilon_{\text{eq}}{}^P = (\tfrac{2}{3}\, d\epsilon_{ij}{}^P\, d\epsilon_{ij}{}^P)^{1/2} \tag{2.118}$$

and

$$\epsilon_{\text{eq}}{}^P = \int_0^{\epsilon_{kl}{}^P} (\tfrac{2}{3}\, d\epsilon_{ij}{}^P\, d\epsilon_{ij}{}^P)^{1/2}. \tag{2.119}$$

The plot of σ_{eq} versus $\epsilon_{\text{eq}}{}^P$ as determined from (2.116) is precisely the stress versus plastic strain curve from the tensile test. Alternately, the plot of σ_{eq}

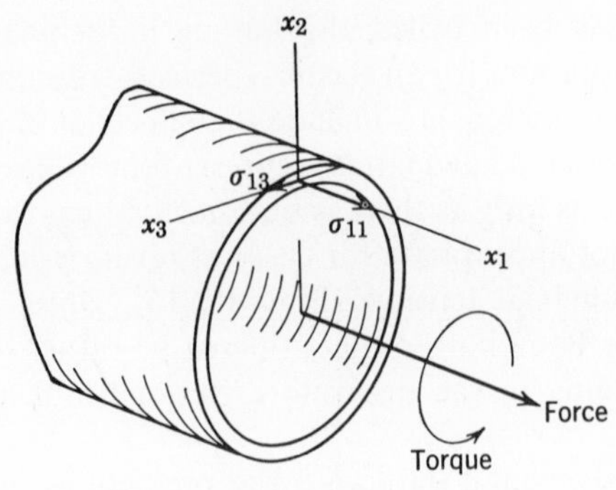

Figure 2.23

versus $\epsilon_{eq}{}^P$, as determined from (2.119), using measured rather than calculated strains, may differ from this curve, and the results will be a measure of the adequacy of the Laning law to predict plastic deformation.

As a simple example of the application of this law, consider the problem of a thin-walled tube subjected to combined axial load and torsion (Figure 2.23). The stress state is statically determined and consists of only two nonzero stress components: the axial stress σ_{11} and the circumferential shearing stress σ_{13}. The components of the deviatoric stress tensor are

$$
\begin{aligned}
s_{11} &= \tfrac{2}{3}\sigma_{11}, & s_{12} &= s_{21} = 0,\\
s_{22} &= -\tfrac{1}{3}\sigma_{11}, & s_{13} &= s_{31} = \sigma_{13},\\
s_{33} &= -\tfrac{1}{3}\sigma_{11}, & s_{23} &= s_{32} = 0.
\end{aligned}
$$

Thus

$$s_{kl}s_{kl} = \tfrac{2}{3}\sigma_{11}{}^2 + 2\sigma_{13}{}^2 = 2J_2,$$

and the loading surface in the two-dimensional stress space reduces to

$$f = \tfrac{1}{3}\sigma_{11}{}^2 + \sigma_{13}{}^3 - \tfrac{1}{3}\sigma_{eq}{}^2 = 0 \qquad (2.120)$$

or

$$\sigma_{11}{}^2 + 3\sigma_{13}{}^2 = \sigma_{eq}{}^2. \qquad (2.121)$$

Equation 2.121 is represented graphically in Figure 2.24. Note that the initial loading surface (the von Mises yield criterion in this reduced stress space) predicts that the material will yield in pure torsion at a stress 0.577 times the yield stress σ_0 in pure tension.

Using the above values of s_{ij} in (2.107),

$$
\begin{aligned}
dJ_2 &= \frac{\partial f}{\partial \sigma_{kl}}\, d\sigma_{kl}\\
&= \tfrac{2}{3}\sigma_{11}\, d\sigma_{11} + 2\sigma_{13}\, d\sigma_{13}.
\end{aligned}
$$

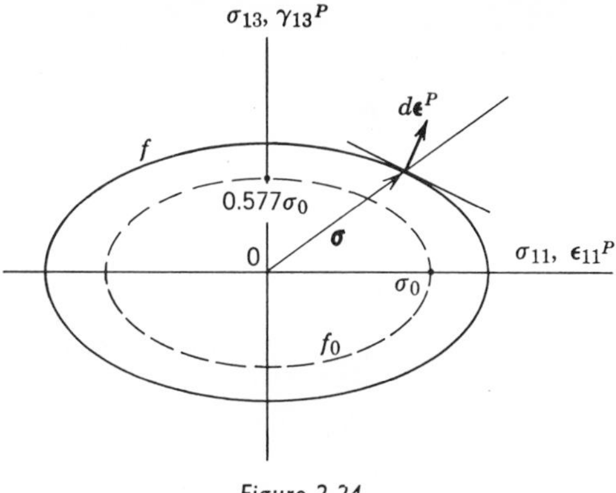

Figure 2.24

Substituting into (2.114), we determine the plastic components of strain as

$$\begin{aligned} d\epsilon_{11}{}^P &= \left(\frac{1}{E_t} - \frac{1}{E}\right) \frac{\sigma_{11}\, d\sigma_{11} + 3\sigma_{13}\, d\sigma_{13}}{\sigma_{11}{}^2 + 3\sigma_{13}{}^2}\, \sigma_{11}, \\ d\gamma_{13}{}^P &= 2\, d\epsilon_{13}{}^P = 3\left(\frac{1}{E_t} - \frac{1}{E}\right) \frac{\sigma_{11}\, d\sigma_{11} + 3\sigma_{13}\, d\sigma_{13}}{\sigma_{11}{}^2 + 3\sigma_{13}{}^2}\, \sigma_{13}. \end{aligned} \tag{2.122}$$

By adding the incremental elastic components the final equations for the total increments in strain become

$$\begin{aligned} d\epsilon_{11} &= \frac{d\sigma_{11}}{E} + \left(\frac{1}{E_t} - \frac{1}{E}\right) \frac{\sigma_{11}\, d\sigma_{11} + 3\sigma_{13}\, d\sigma_{13}}{\sigma_{11}{}^2 + 3\sigma_{13}{}^2}\, \sigma_{11}, \\ d\gamma_{13} &= \frac{d\sigma_{13}}{G} + 3\left(\frac{1}{E_t} - \frac{1}{E}\right) \frac{\sigma_{11}\, d\sigma_{11} + 3\sigma_{13}\, d\sigma_{13}}{\sigma_{11}{}^2 + 3\sigma_{13}{}^2}\, \sigma_{13}. \end{aligned} \tag{2.123}$$

The value of $E_t = E_t(\sigma_{eq})$ entering the computations at a particular increment

$$d\sigma_{eq} = \frac{\sigma_{11}\, d\sigma_{11} + 3\sigma_{13}\, d\sigma_{13}}{\sqrt{\sigma_{11}{}^2 + 3\sigma_{13}{}^2}}\,.$$

in equivalent stress is calculated as the slope of the stress-strain curve (from the tensile test) at that stress level. The total plastic strain corresponding to arbitrary axial and torsional loads can be determined by summing the incremental values, beginning the calculations at inception of yielding in the tube. The conditions under which the incremental stress-strain law can be integrated directly are discussed later in this section.

The second stress-strain law to be considered is that of kinematic hardening, wherein subsequent loading surfaces are determined by rigidly translating the initial yield surface in stress space. Kinematic hardening has the advantage of being able to represent a Bauschinger effect and thus conceivably can be applied to more complex loading paths than isotropic hardening. Restricting our attention to the von Mises criterion (2.102) for the initial yield surface, we find that the equation of the loading function corresponding to a translation of this surface is

$$f = \tfrac{1}{2}(s_{kl} - \alpha_{kl})(s_{kl} - \alpha_{kl}) - \tfrac{1}{3}\sigma_0^2 = 0, \tag{2.124}$$

where the tensor α_{kl} represents the displacement vector to the center of the yield surface in stress space as plastic straining takes place. The simplest representation of this displacement is that proposed by Reuss [10] and used by Prager [11], for which α_{kl} is taken to be a linear function of plastic strain

$$\alpha_{kl} = c\epsilon_{kl}{}^P, \tag{2.125}$$

where c is a material constant. Thus, substituting into (2.124),

$$f = \tfrac{1}{2}(s_{kl} - c\epsilon_{kl}{}^P)(s_{kl} - c\epsilon_{kl}{}^P) - \tfrac{1}{3}\sigma_0^2 = 0, \tag{2.126}$$

which, of course, reduces to the von Mises yield criterion when the plastic strain is zero. Equation 2.126 is referred to as the Reuss loading function. To determine the corresponding stress-strain law we first derive a special form of the general linear incremental law (2.92). Since f as given in (2.126) is dependent only on the material's present state of stress and plastic strain, the total differential of the equation $f = 0$ can be written

$$\frac{\partial f}{\partial \sigma_{kl}}\, d\sigma_{kl} + \frac{\partial f}{\partial \epsilon_{kl}{}^P}\, d\epsilon_{kl}{}^P = 0. \tag{2.127}$$

Taking the scalar product of each side of (2.92) with respect to $\partial f/\partial \epsilon_{ij}{}^P$,

$$\frac{\partial f}{\partial \epsilon_{ij}{}^P}\, d\epsilon_{ij}{}^P = G\left(\frac{\partial f}{\partial \sigma_{ij}} \frac{\partial f}{\partial \epsilon_{ij}{}^P}\right)\left(\frac{\partial f}{\partial \sigma_{kl}}\, d\sigma_{kl}\right).$$

Substituting for the left-hand side the equivalent scalar product from (2.127) (changing the former's summation index ij to kl),

$$-\frac{\partial f}{\partial \sigma_{kl}}\, d\sigma_{kl} = G\left(\frac{\partial f}{\partial \sigma_{ij}} \frac{\partial f}{\partial \epsilon_{ij}{}^P}\right)\left(\frac{\partial f}{\partial \sigma_{kl}}\, d\sigma_{kl}\right).$$

Dividing through by the scalar product on the left, we obtain an equation for G:

$$G = -\left(\frac{\partial f}{\partial \sigma_{ij}} \frac{\partial f}{\partial \epsilon_{ij}{}^P}\right)^{-1}. \tag{2.128}$$

Substitution of this relation into (2.92) yields the desired law, first given by Prager [11]:

$$d\epsilon_{ij}{}^P = -\frac{(\partial f/\partial \sigma_{kl})\, d\sigma_{kl}}{(\partial f/\partial \sigma_{mn})(\partial f/\partial \epsilon_{mn}{}^P)} \frac{\partial f}{\partial \sigma_{ij}}. \tag{2.129}$$

The specific stress-strain law for the Reuss loading function is derived as follows: from (2.126)

$$\frac{\partial f}{\partial \epsilon_{ij}{}^P} = -c\frac{\partial f}{\partial s_{ij}} = -c\frac{\partial f}{\partial \sigma_{ij}} \tag{2.130}$$

and (2.129) becomes

$$d\epsilon_{ij}{}^P = \frac{1}{c}\frac{(\partial f/\partial \sigma_{kl})\, d\sigma_{kl}}{(\partial f/\partial \sigma_{mn})(\partial f/\partial \sigma_{mn})} \frac{\partial f}{\partial \sigma_{ij}}. \tag{2.131}$$

Again substituting (2.126),

$$\frac{\partial f}{\partial \sigma_{ij}} = \frac{\partial f}{\partial s_{ij}} = s_{ij} - c\epsilon_{ij}{}^P,$$

$$\frac{\partial f}{\partial \sigma_{kl}}\, d\sigma_{kl} = \frac{\partial f}{\partial s_{kl}}\, ds_{kl} = (s_{kl} - c\epsilon_{kl}{}^P)\, ds_{kl},$$

and

$$\frac{\partial f}{\partial \sigma_{mn}}\frac{\partial f}{\partial \sigma_{mn}} = (s_{mn} - c\epsilon_{mn}{}^P)(s_{mn} - c\epsilon_{mn}{}^P) = \tfrac{2}{3}\sigma_0{}^2.$$

Thus the stress-strain law is

$$d\epsilon_{ij}{}^P = \frac{3}{2}\frac{(s_{kl} - c\epsilon_{kl}{}^P)\, ds_{kl}}{\sigma_0{}^2 c}(s_{ij} - c\epsilon_{ij}{}^P). \tag{2.132}$$

The material parameter c is determined from a tensile test. Proceeding as in the case of isotropic hardening (2.109),

$$d\epsilon_{11}{}^P = \frac{3}{2}\frac{(2/3)\sigma_{11}\, d\sigma_{11} - c\epsilon_{11}{}^P\, d\sigma_{11}}{\sigma_0{}^2 c}(\tfrac{2}{3}\sigma_{11} - c\epsilon_{11}{}^P) = \frac{d\sigma_{11}}{E_t{}^P},$$

which reduces to

$$\frac{3}{2}\frac{1}{\sigma_0{}^2}(\tfrac{2}{3}\sigma_{11} - c\epsilon_{11}{}^P)^2 = \frac{c}{E_t{}^P}.$$

Combining with the equation

$$\frac{\partial f}{\partial \sigma_{mn}}\frac{\partial f}{\partial \sigma_{mn}} = \tfrac{3}{2}(\tfrac{2}{3}\sigma_{11} - c\epsilon_{11}{}^P)^2 = \tfrac{2}{3}\sigma_0{}^2,$$

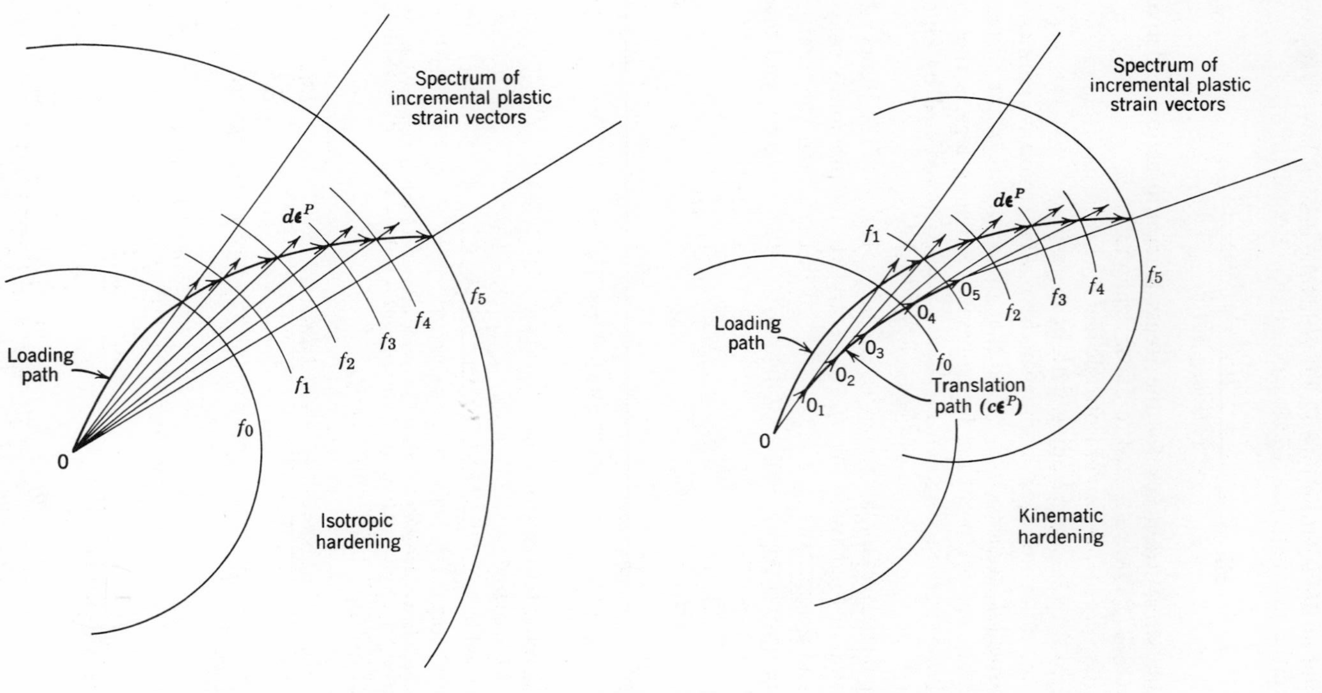

Spectrum of incremental plastic strain vectors
$d\epsilon^P$
Loading path
0
f_0
f_1
f_2
f_3
f_4
f_5
Isotropic hardening
Spectrum of incremental plastic strain vectors
$d\epsilon^P$
f_1
Loading path
0
0_1
0_2
0_3
0_4
0_5
f_0
f_2
f_3
f_4
f_5
Translation path ($c\epsilon^P$)
Kinematic hardening

Figure 2.25

we see that c equals two thirds of the slope $E_t{}^P$ of the stress-plastic strain curve and the stress-strain law can be written in final form as

$$d\epsilon_{ij}{}^P = \frac{9}{4}\frac{1}{E_t{}^P}\frac{(s_{kl} - (2/3)E_t{}^P\epsilon_{kl}{}^P)\, ds_{kl}}{\sigma_0{}^2}(s_{ij} - \tfrac{2}{3}E_t{}^P\epsilon_{ij}{}^P). \tag{2.133}$$

This is the Reuss-Prager stress-strain law of kinematic hardening. Since c is a constant, (2.133) is restricted to linear strain-hardening as in Figure 2.15.

A qualitative comparison of isotropic and kinematic hardening is shown in Figure 2.25 for the same loading path in the π-plane. For the purpose of illustration, the continuous loading in the plastic region is represented by five incremental stress vectors. Note that the resultant of the incremental plastic strain vectors predicted by isotropic hardening lags farther behind the resultant stress vector than the strain resultant predicted by kinematic hardening.

The final stress strain law to be presented here is that of the Hencky-Nadai deformation or total strain theory. In contrast to the previously considered general and special laws, the Hencky-Nadai theory relates the total (rather than incremental) plastic strain to the current state of stress, and thus is a path-independent theory of plasticity. It is given as [12]

$$\epsilon_{ij}{}^P = \frac{3}{2}\frac{1}{E_{\text{sec}}^P}\, s_{ij} \tag{2.134}$$

or

$$\epsilon_{ij}{}^P = \frac{3}{2}\left(\frac{1}{E_{\text{sec}}} - \frac{1}{E}\right)s_{ij}, \tag{2.135}$$

where E_{sec}^P is the secant modulus of the stress-plastic strain curve (Figure 2.22) and E_{sec} is the secant modulus of the stress-total strain curve. Equation 2.135 is not derivable from the general linear stress-strain law of (2.92), for in its incremental form it is actually a nonlinear law. To see this clearly and to compare this law with the Laning hardening law, which it resembles, we take the total differential of (2.135):

$$d\epsilon_{ij}{}^P = \frac{3}{2}\left(\frac{1}{E_{\text{sec}}} - \frac{1}{E}\right) ds_{ij} + \frac{3}{2}\, d\left(\frac{1}{E_{\text{sec}}^P}\right)s_{ij}. \tag{2.136}$$

From Figure 2.22, in terms of equivalent stress and a newly defined equivalent plastic strain,

$$\epsilon_{\text{eq}}{}^P = \sqrt{\tfrac{2}{3}\epsilon_{kl}{}^P\epsilon_{kl}{}^P}, \tag{2.137}$$

we have

$$d\left(\frac{1}{E_{\text{sec}}^P}\right) = d\left(\frac{\epsilon_{\text{eq}}{}^P}{\sigma_{\text{eq}}{}^P}\right) = \frac{d\epsilon_{\text{eq}}{}^P}{\sigma_{\text{eq}}} - \frac{\epsilon_{\text{eq}}{}^P}{\sigma_{\text{eq}}{}^2}\, d\sigma_{\text{eq}}$$

$$= \frac{d\sigma_{\text{eq}}}{\sigma_{\text{eq}}}\left(\frac{1}{E_t{}^P} - \frac{1}{E_{\text{sec}}^P}\right) = \left(\frac{1}{E_t} - \frac{1}{E_{\text{sec}}}\right)\frac{d\sigma_{\text{eq}}}{\sigma_{\text{eq}}}. \tag{2.138}$$

Therefore the final equation for the incremental plastic strain is

$$d\epsilon_{ij}{}^P = \frac{3}{2}\left(\frac{1}{E_{\text{sec}}} - \frac{1}{E}\right) ds_{ij} + \frac{3}{2}\left(\frac{1}{E_t} - \frac{1}{E_{\text{sec}}}\right)\frac{d\sigma_{\text{eq}}}{\sigma_{\text{eq}}} s_{ij}. \tag{2.139}$$

For proportional loading (i.e., when the components of the stress tensor increase in constant ratio to one another), $d\sigma_{\text{eq}}/\sigma_{\text{eq}} = ds_{ij}/s_{ij}$ (no summation) for each deviatoric stress component and (2.139) simplifies to

$$d\epsilon_{ij}{}^P = \frac{3}{2}\left(\frac{1}{E_t} - \frac{1}{E}\right)\frac{d\sigma_{\text{eq}}}{\sigma_{\text{eq}}} s_{ij}, \tag{2.140}$$

which is identical to (2.113) or (2.114) for isotropic hardening. Consequently (2.135) satisfies Drucker's requirements for proportional loading, for it is then simply the integrated form of the Laning incremental law derived from those requirements. Conversely, it is seen that the conditons under which the incremental stress-strain law of (2.114) can be integrated directly are those of proportional loading.

For other loading paths (2.139) expresses a dependence of the incremental plastic strain vector on the direction of the incremental stress vector in stress space. Hence it is nonlinear, and from Drucker's postulate the yield surface must develop a corner at the point of loading as plastic straining takes place. This is shown in Figure 2.26 for a typical loading path in the π-plane. At each incremental stress point the two components of the incremental plastic strain vector are represented: one in the direction of the deviatoric stress s_{ij} and the other in the direction of the increment in deviatoric stress ds_{ij}. Since

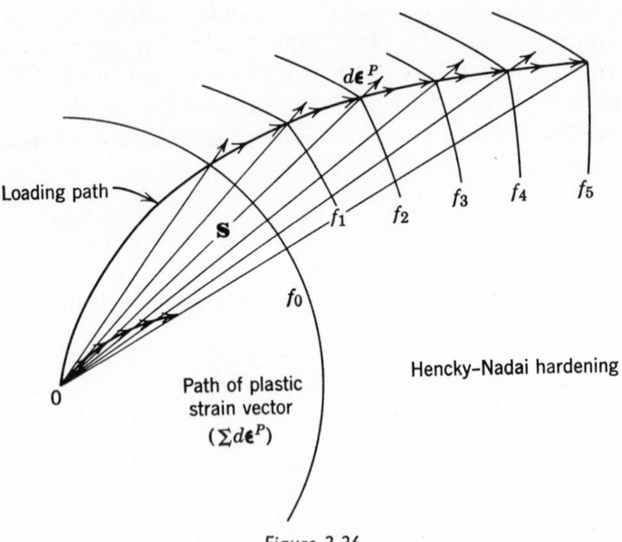

Figure 2.26

(2.139) is an exact differential (i.e., integrable), the resultant of these two vector components is such that the total plastic strain vector at any point is always collinear with the total stress vector. Budiansky [13], who developed equations for determining the range of loading paths for which the Hencky-Nadai law is admissible, used Drucker's requirements as criteria. The law unquestionably is limited and clearly cannot be used in a problem involving unloading from a plastic state followed by reloading. Nevertheless, it is a useful theory that has found wide application in the stress analysis of structural elements subjected to noncyclic loading [14].

2.5 THEORIES OF STATIC FRACTURE

The phenomenon of fracture in metals under static load has been studied at all levels from the atomic to the purely empirical (i.e., the correlation of stress and critical crack length in sheet specimens). For the prediction of macroscopic fracture in a structural element, however, a single point of view and one criterion related to the macroscopic concepts of stress and strain is desired. One such criterion that arises from the classic work of Griffith [15] on fracture in glass and has been widely used is that of strain energy balance. In these closing sections we will restrict attention to a conceptual study of energy balance criteria, considering fracture problems that involve both elastic and plastic strains.

2.5.1 Necking Instability in the Tensile Test

Consider a tensile specimen of a structural metal which exhibits pronounced necking prior to fracture in a constant strain rate test (Figure 2.27). We

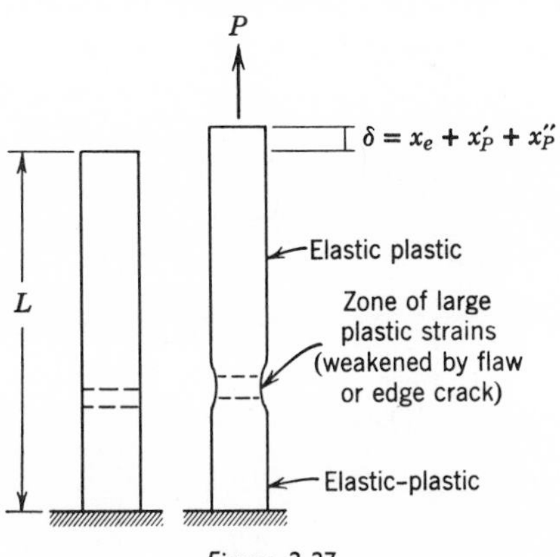

Figure 2.27

hypothesize that the necking occurs in a narrow zone (relative to the specimen length) weakened by a flaw or edge crack. Denoting the displacement of the moving head by δ, the total elastic extension of the bar by x_e, the plastic extension of the bar outside the flawed zone by x'_p, and the permanent extension of the flawed zone by x''_p, an incremental work equation can be written which is valid during the elastic loading phase of the test. (The elasticity of the testing machine can be taken into account by defining the length L as an effective length of specimen determined from the compliance of the heads.) This work equation is

$$dWk_{\text{ext}} = dWk_{\text{int}} \equiv dU_e + dU'_p + dU''_p, \tag{2.141}$$

where U_e is the elastic (recoverable) strain energy stored in the bar and U'_p and U''_p are the plastic energies dissipated without and within the flawed zone, respectively. From the relations $dWk_{\text{ext}} = Pd\delta$, $dU_e = Pdx_e$, $dU'_p = Pdx'_p$, and $dU''_p = Pdx''_p$, (2.141) can be expressed in terms of the extensions as

$$\frac{dx_e}{dx''_p} = \left(\frac{d\delta}{dx''_p} - \frac{dx'_p}{dx''_p}\right) - 1, \tag{2.142}$$

which is greater than or equal to zero during the elastic loading phase.

At the inception of necking unloading begins outside the flawed zone. Thus $dx'_p \equiv 0$ and the incremental work equation becomes

$$P\,d\delta = dU_e + dU''_p \tag{2.143}$$

or

$$\frac{dx_e}{dx''_p} = \frac{d\delta}{dx''_p} - 1, \tag{2.144}$$

which is valid during the elastic unloading or necking phase of the tensile test. Consider Figure 2.28, which represents a plot of load P versus permanent

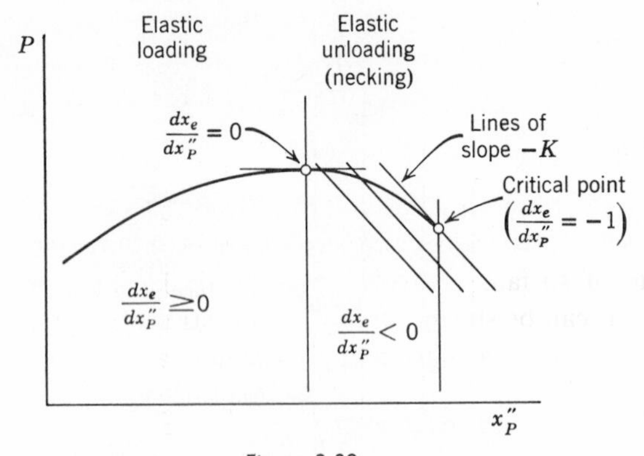

Figure 2.28

extension x_p'' of the flawed region. A criterion for necking instability and instantaneous fracture has been deduced from such a diagram by Orowan [16]. Since the response of the bar outside the flawed region is elastic during necking, an elastic spring constant K can be introduced:

$$K = \frac{dP}{dx_e}. \tag{2.145}$$

Orowan proposed that a catastrophic, self-propagating fracture occurs when the slope of the load-plastic elongation diagram becomes equal to $-K$, that is, when

$$\frac{dP}{dx_p''} = -\frac{dP}{dx_e}. \tag{2.146}$$

Thus at instability

$$\frac{dx_e}{dx_p''} = -1, \tag{2.147}$$

and, from (2.143) and (2.144),

$$-\frac{dU_e}{dx_p''} = \frac{dU_p''}{dx_p''}. \tag{2.148}$$

Equation 2.148 is an appropriate mathematical representation, expressed as an energy balance, of Orowan's criterion. This equation and (2.147) are consistent with the concept that at the instant of necking instability it is no longer necessary for external work to be done on the specimen in order for necking to continue and head movement may cease. The unloading elastic portion of the bar releases sufficient energy to the necking region to enable the self-propagation of the plastic strains. Thus

$$\frac{dx_e}{dx_p''} \geq -1 \quad \text{and} \quad \frac{d\delta}{dx_p''} \geq 0$$

throughout the tensile test, with the equalities satisfied only at instability (see Figure 2.29).

2.5.2 *Griffith Criterion of Brittle Fracture*

Before investigating the extension of (2.148) to ductile fracture in cracked sheet specimens first consider the elastic case of purely brittle fracture. For a wide plate of surface area A with a relatively short crack of length $2l$ (Figure 2.30) it can be shown that the elastic strain energy per unit thickness under an edge loading σ is given approximately as

$$U_E = (A + 2\pi l^2)\frac{\sigma^2}{2E} = \frac{\sigma^2 A}{2\bar{E}}, \tag{2.149}$$

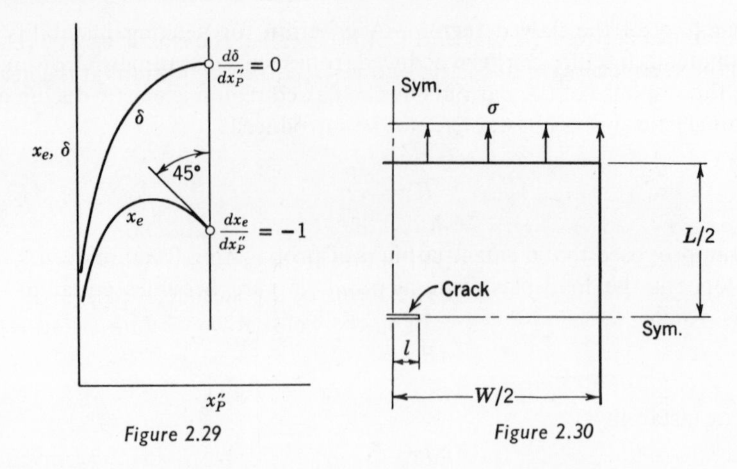

Figure 2.29

Figure 2.30

with the equivalent tensile modulus $\bar{E}$ of the cracked plate defined as

$$\bar{E} = \frac{E}{1 + 2\pi l^2/A}. \tag{2.150}$$

Assuming that the crack will open by an amount dl under a stress increment $d\sigma$, the increase in strain energy is

$$dU_E = (A + 2\pi l^2)\frac{\sigma}{E}\,d\sigma + \frac{2\pi\sigma^2 l}{E}\,dl. \tag{2.151}$$

The differential external work done is

$$dWk^{(\text{ext})} = \sigma A\,d\bar{\epsilon} = \sigma A\left(\frac{\partial\bar{\epsilon}}{\partial\sigma}\,d\sigma + \frac{\partial\bar{\epsilon}}{\partial l}\,dl\right), \tag{2.152}$$

where

$$\bar{\epsilon} = \frac{\sigma}{\bar{E}} = \frac{\delta}{L}, \tag{2.153}$$

and δ is the total elongation of the plate. From (2.150) and (2.153),

$$\frac{\partial\bar{\epsilon}}{\partial\sigma} = \frac{1}{\bar{E}}, \qquad \frac{\partial\bar{\epsilon}}{\partial l} = \frac{4\pi\sigma l}{AE}. \tag{2.154}$$

Thus the work rate at inception of crack propagation is

$$\frac{dWk^{(\text{ext})}}{dl} = \frac{\sigma A}{\bar{E}}\frac{d\sigma}{dl} + \frac{4\pi\sigma^2 l}{E}. \tag{2.155}$$

From (2.150) and (2.151) the rate of increase in elastic strain energy is

$$\frac{dU_E}{dl} = \frac{\sigma A}{\bar{E}}\frac{d\sigma}{dl} + \frac{2\pi\sigma^2 l}{E}. \tag{2.156}$$

The difference between (2.155) and (2.156) is called the elastic energy release rate in the special cases of fixed grip or constant stress conditions at the instant of fracture. Thus

$$\frac{dWk^{(\text{ext})}}{dl} - \frac{dU_E}{dl} = \frac{2\pi\sigma^2 l}{E} = \frac{dW}{dl}, \tag{2.157}$$

independent of the load rate $d\sigma/dl$ (taking l to be the primary variable), with dW/dl defined as the energy dissipation rate at the crack tip required to extend the crack. Expression of this rate in terms of the surface energy T per unit area of the new crack surfaces leads to the well-known Griffith criterion

$$\sigma^2 l = \frac{2ET}{\pi} \equiv \text{constant} \tag{2.158}$$

for fracture in perfectly brittle materials. The quantity T was experimentally determined by Griffith for glass. For high-strength metals exhibiting relatively brittle fracture, this equation has been modified by Irwin [17] and Orowan [18] to read

$$\sigma^2 l = \frac{EG_c}{\pi} \equiv \text{constant.} \tag{2.159}$$

The quantity G_c, called the crack extension force, is an order of magnitude greater than the surface energy T, reflecting the fact that considerable permanent deformation adjacent to the crack surfaces precedes fracture even in the most brittle of metals. In the case of more ductile behavior, G_c is not a constant and neither (2.159) nor (2.157) is appropriate as a starting point for the formulation of a fracture criterion. A more general statement of energy balance which distinguishes slow crack growth from crack instability and is consistent with experimental results is presented in the following section.

2.5.3 *Energy Balance Criteria in Ductile Fracture*

To consider the energy balance of (2.148) as a criterion for ductile fracture in cracked sheets, wherein catastrophic rupture is preceded by slow crack growth under increasing load, we write the counterpart to this equation as

$$-\frac{dU_E}{dl} = \frac{dU_p}{dl} \tag{2.160}$$

in which $-dU_E/dl$ and dU_p/dl are the elastic energy release rate and plastic energy dissipation rate, respectively, and the permanent extension x_p'' of the necking region in the tensile test has been replaced by the half crack length l. Equation 2.160 has been suggested in the literature as the basic criterion for catastrophic rupture. However, a conceptual argument pointing

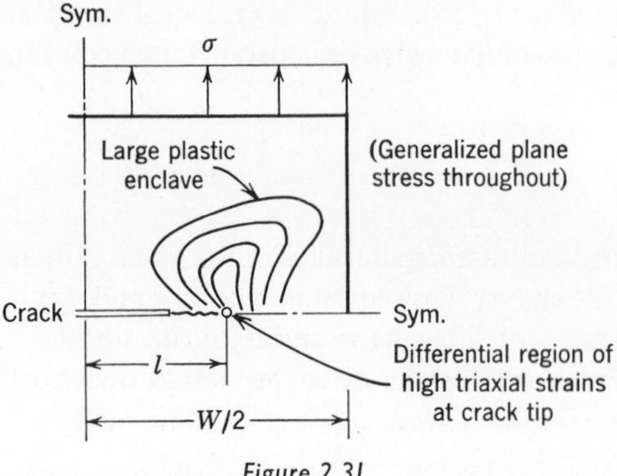

Figure 2.31

out the inconsistency between this equation and the observed phenomena of ductile fracture can be made, as follows. Turning again to necking instability in the tensile test, we see from Figure 2.28 and (2.148) that, at the instant of instability, the entire bar outside the necking region is releasing (dumping) energy into this region at a rate sufficient for self-propagation of the large plastic strains, and the rate of external work is zero. In a cracked sheet with a slowly growing crack, the physical situation is altogether different. The entire specimen outside the small region at the tip of the opening crack (Figure 2.31) is being loaded up to the very moment of fracture; for example, from results reported by Gerberich and Swedlow [19], the plastic energy dissipated in each of several ductile metals can be represented as a continuously increasing function of gross stress and crack length and is an order of magnitude greater than the classical elastic energy release rate $2\pi\sigma^2 l/E$. The increase in the gross stress is shown in Figure 2.32, wherein data of load versus crack length is

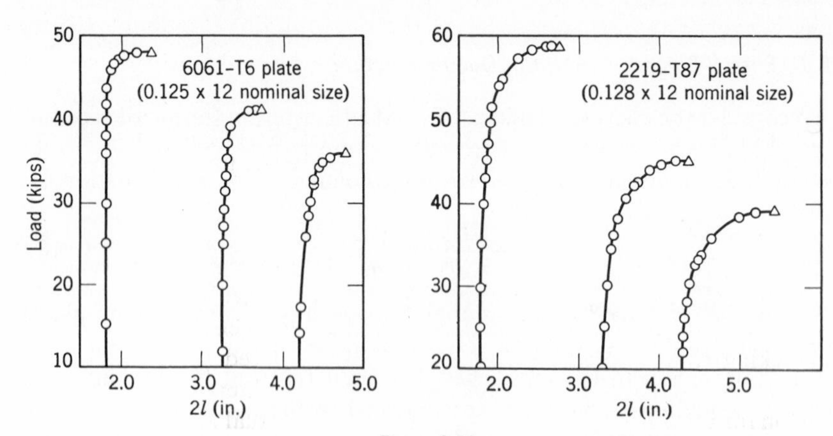

Figure 2.32

shown for slowly growing cracks in several aluminum alloy specimens. Thus the external work done on the cracked-sheet specimen effects increases in both elastic and plastic strains throughout. The plastic energy dissipation rate is determined by this work and not by the elastic energy released, since the total plate is not dumping energy into the plastic zones as in the tensile specimen. The energy release rate is that of the essentially local elastic strain energy released as the crack extends under increasing load. The energy exchange at instability must take place between this quantity and the high rate of plastic straining in the small volume at the crack tip. In light of this discussion the following equation is an appropriate statement of the energy balance relation that must hold throughout slow crack growth:

$$\frac{dWk^{(\text{ext})}}{dl} - \left(\frac{dU_E}{dl} + \frac{dU_p}{dl}\right) \equiv \frac{dU_R}{dl} = \frac{dW}{dl}. \tag{2.161}$$

The change in residual energy dU_R is defined as the difference between the increment in external work and the sum of the increments in plastic energy (outside the small region at the crack tip) and elastic energy, integrated over the plate. As before, the quantity dW/dl is defined as the rate at which work is done in the small volume at the crack tip as the crack extends. This rate is taken to be dependent only upon material properties, thickness, and the amount of crack growth that has already occurred and is assumed to be independent of the specimen configuration and the distribution of external loading. (The dimensions of the volume in which this high rate of straining takes place are considered to be several orders of magnitude less than the crack length and the other plate dimensions.) As a general criterion for inception of crack growth, (2.161) was given independently in 1965 by Rice [21] and by the writer [20].

To establish a criterion for crack instability (catastrophic fracture), consider the conceptual representation of the variation of dU_R/dl and $dW/dl \equiv W'(l)$ with σ and l during slow crack growth, as given in Figure 2.33. From

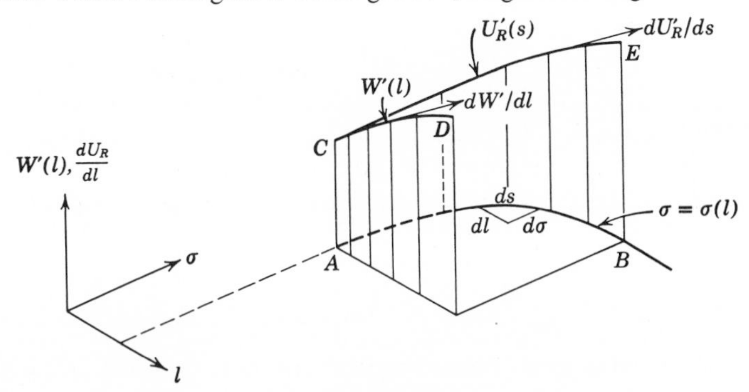

Figure 2.33

(2.161), dU_R/dl equals $W'(l)$ throughout this interval. Thus the residual energy release rate curve CE (denoted as U'_R for simplicity) is a space curve whose projection on the $U'_R - l$ plane coincides with the $W'(l)$ curve CD [20,22]. Figure 2.33 clearly indicates the distinction between the left- and right-hand sides of (2.161). The left side is a function of crack length and applied stress, whereas the local energy dissipation rate dW/dl on the right side is assumed to be dependent only on crack growth, as previously emphasized, and is therefore a planar curve in the diagram.

A second derivative criterion can now be determined. The slope of the U'_R curve at any point is

$$\frac{dU'_R}{ds} = \frac{\partial U'_R}{\partial \sigma}\frac{d\sigma}{ds} + \frac{\partial U'_R}{\partial l}\frac{dl}{ds}, \tag{2.162}$$

where s is the arc-length parameter (the projection of the curve in the σ-l plane). From the equation

$$ds = \sqrt{1 + (\sigma')^2}\, dl, \qquad \left(\sigma' = \frac{d\sigma}{dl}\right) \tag{2.163}$$

(2.162) can be expressed

$$\frac{\partial U'_R}{\partial l} = \sqrt{1 + (\sigma')^2}\,\frac{dU'_R}{ds} - \sigma'\,\frac{\partial U'_R}{\partial \sigma}. \tag{2.164}$$

Since $U'_R(s) = W'(l)$, $dU'_R(s) = dW'(l)$, from which, combining with (2.163),

$$\frac{dU'_R(s)}{ds} = \frac{1}{\sqrt{1 + (\sigma')^2}}\,\frac{dW'(l)}{dl}. \tag{2.165}$$

Thus during slow crack growth

$$\frac{\partial}{\partial l}\left(\frac{dU_R}{dl}\right) = \frac{d}{dl}\,W'(l) - \frac{\partial}{\partial \sigma}\left(\frac{dU_R}{dl}\right)\frac{d\sigma}{dl}. \tag{2.166}$$

At instability $d\sigma/dl = 0$ from the concept that no further increase in load is required for continued extension of the crack (as observed in the test results of Figure 2.32). Hence in the moment just before catastrophic fracture

$$\frac{\partial}{\partial l}\left(\frac{dU_R}{dl}\right) = \frac{d^2W}{dl^2}. \tag{2.167}$$

Equation 2.167 is a general mathematical criterion for crack instability and rupture in ductile metals.

REFERENCES

[1] William Prager, *Introduction to Mechanics of Continua*, Ginn, Boston, 1961.

[2] R. Hill, *The Mathematical Theory of Plasticity*, Oxford at the Clarendon Press, 1950.

[3] P. M. Naghdi, "Stress-Strain Relations in Plasticity and Thermoplasticity," *Plasticity* (Proceedings of the Second Symposium on Naval Structural Mechanics), E. H. Lee and P. S. Symonds, eds., Pergamon, New York, 1960, pp. 121–169.

[4] V. D. Kliushnikov, "On Plasticity Laws for Work-Hardening Materials," *PMM J. Appl. Math. Mech.*, **22,** 129 (1958).

[5] D. C. Drucker, "A More Fundamental Approach to Plastic Stress-Strain Relations," *Proc. First U.S. Nat. Congr. Appl. Mech.*, 1951, pp. 487–491.

[6] D. C. Drucker, "A Definition of Stable Inelastic Material," *J. Appl. Mech.*, **26,** 101 (1959).

[7] P. W. Bridgman, "The Compressibility of Thirty Metals as a Function of Pressure and Temperature," *Proc. Nat. Acad. Arts Sci. U.S.*, **58,** 165 (1923).

[8] R. von Mises, "Mechanik der festen Koerper im plastisch-deformablen Zustand," *Göttinger Nachrichten, mathematisch physikalische Klasse*, 1913, pp. 582–592.

[9] W. Prager, "The Stress-Strain Laws of the Mathematical Theory of Plasticity—a Survey of Recent Progress," *J. Appl. Mech.*, **15,** 226 (1948).

[10] E. Reuss, "Anisotropy Caused by Strain," *Proc. of the Fourth Intern. Congr. Appl. Mech.*, Cambridge, England, 1934, p. 241.

[11] W. Prager, "Recent Developments in the Mathematical Theory of Plasticity," *J. Appl. Phys.*, **20,** 235 (1949).

[12] A. Nadai, "Plastic Behavior of Metals in the Strain-Hardening Range. Part I," *J. Appl. Phys.*, **8,** 205 (1937).

[13] B. Budiansky, "A Reassessment of Deformation Theories of Plasticity," *J. Appl. Mech.* **26,** 259 (1959).

[14] K. S. Havner, "On the Formulation and Iterative Solution of Small Strain Plasticity Problems," *Quart. Appl. Math.*, **23,** 323 (1966).

[15] A. Griffith, "The Phenomenon of Rupture and Flow in Solids," *Philos. Trans. Royal Soc. (London)*, Ser. A, **221,** 163 (1921).

[16] E. Orowan, "Condition of High-Velocity Ductile Fracture," *J. Appl. Phys.*, **26.** 900 (1955).

[17] G. R. Irwin, "Fracture Dynamics," *Fracturing of Metals*, American Society of Metals, Cleveland, 1948, pp. 147–166.

[18] E. Orowan, "Fundamentals of Brittle Behavior in Metals," *Fatigue and Fracture of Metals*, Wiley, New York, 1952, pp. 139–167.

[19] W. W. Gerberich and J. L. Swedlow, "Plastic Strains and Energy Density in Cracked Plates. Parts I and II," *Experim. Mech.*, **4,** 335 (1964).

[20] K. S. Havner and J. B. Glassco, "On Energy Balance Criteria in Ductile Fracture," Douglas Aircraft Co., Inc., Paper No. 3575, July 1965. [*Intern. J. Fracture Mech.*, **2,** 506 (1966)].

[21] J. R. Rice, "An Examination of the Fracture Mechanics Energy Balance from the Point of View of Continuum Mechanics," *Intern. Conf. Fracture*, Sendia, Japan, September 1965, Paper No. A-18.

[22] D. Broek, "The Residual Strength of Cracked Sheet and Structures," Nationaal Lucht—En Ruimlevaart—laboratorium, Amsterdam, Rept. M. 2135, August 1964.

3

The Influence of Processing and Metallurgical Factors on Fatigue

JAMES A. SCOTT

A wide variety of factors affect the behavior of a member or assembly under conditions of fatigue loading. The most obvious parameters are those that deal with the sign, magnitude and frequency of loading, the geometry and material strength level of the structure and the ambient service temperature. Often ill-considered are those processing and metallurgical factors that determine the cleanliness and homogeneity of materials, the sign and distribution of residual stresses, and the surface finish. These processing and metallurgical factors, however, may have an overriding influence on the fatigue performance of the structure, to its benefit or detriment. These parameters and their effect on fatigue properties are discussed in this chapter.

3.1 PROCESSING FACTORS

Fatigue usually initiates at a surface because stresses are normally higher there, particularly since most parts experience bending loads resulting in substantially higher stresses in the outermost fibers. In addition, stress raisers are more likely to be present as a result of surface irregularities inherent in the design, produced in service, or resulting from processing. The detrimental or beneficial effect of processing on fatigue properties is usually manifested in its effect on the strength level or residual stress condition, or both, of the surface material. Recognizing that residual and service

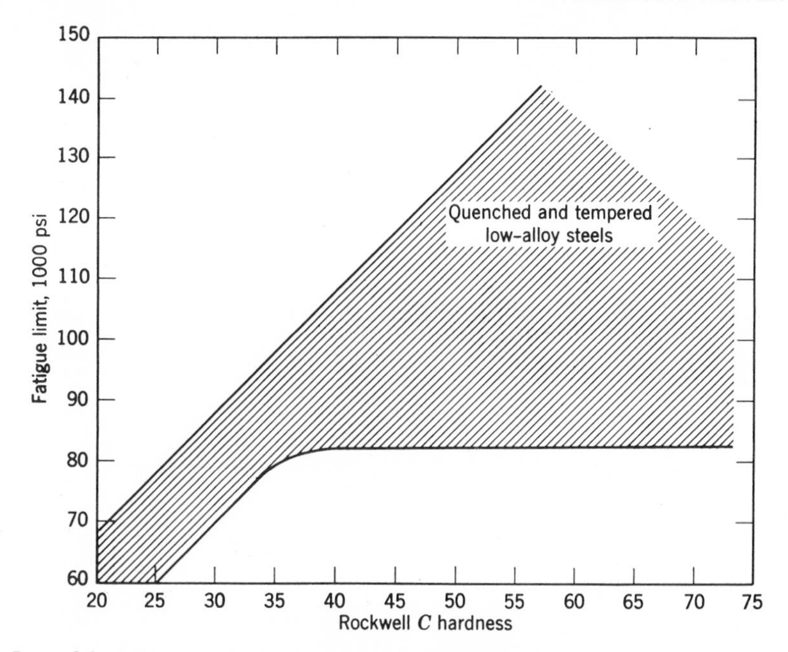

Figure 3.1 Effect of hardness of rotating beam fatigue limit of through-hardened and tempered low-alloy steel: fatigue limit versus Rockwell C hardness [26].

stresses are essentially additive, algebraically, the benefit of residual compressive stresses in the critical surface layers is easily seen; conversely, the detriment of residual surface tensile stresses follows. Since fatigue strength is ordinarily proportional to tensile strength (other factors being equal), the effect of processing on the mechanical properties of material—particularly surface material—directly affects fatigue properties. The relation of hardness to fatigue for steels is shown in Figure 3.1.

3.1.1 Forming

By definition, the forming process produces plastic deformation (and residual stresses) in a part to achieve a permanent change in configuration. The effect of residual forming stresses on fatigue properties is dependent upon their intensity and sign compared with the intensity and sign of the applied load. Occasionally these residual stresses may prove beneficial; however, usually there is some loss in fatigue. Consequently, the residual stresses produced in forming (and their effect on fatigue) often dictate the forming limits for materials.

Residual forming stresses in the completed part are dependent on at least

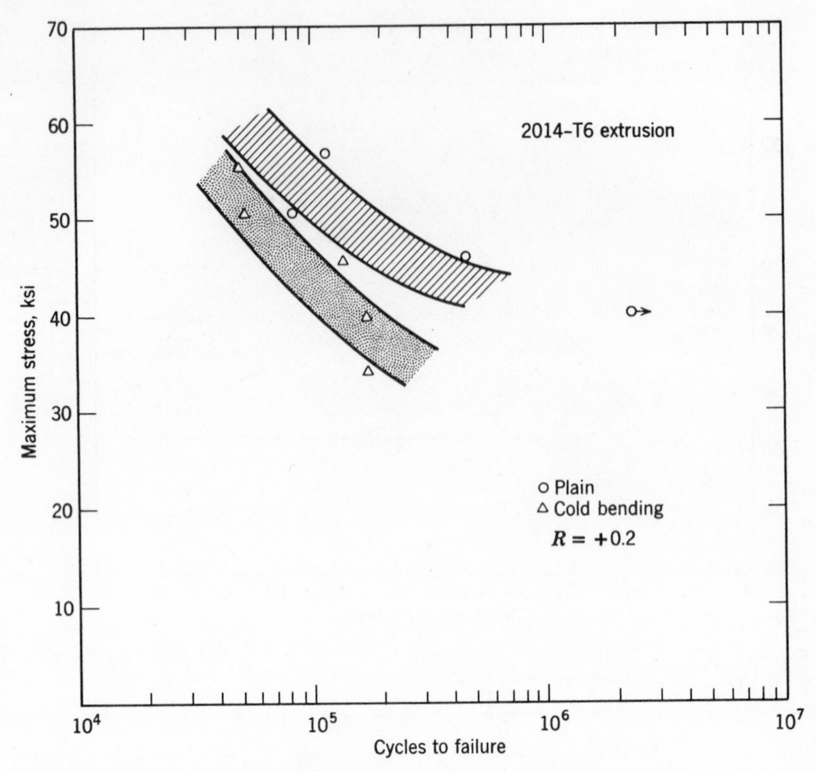

Figure 3.2 Fatigue coupons (open triangles) were cold bent 90° in the annealed condition over an 8T radius and then straightened in the T6 condition [1].

three additional factors: the heat treatment-forming sequence in processing, the temper of the material, and the forming temperature. Parts formed and subsequently completely heat-treated are free of prior forming stresses. Parts formed and stress relieved contain reduced forming stresses, depending upon the stress relieving temperature. The forming temperature and the material temper, e.g., "*AQ*," "*T*-4," or "*T*-6" for aluminum alloys, also influence the magnitude of forming stresses to the extent that they affect the yield strength of the material at the forming temperature. In general the lower the yield strength when forming occurs, the weaker the residual stress field generated. Typical data showing the effect of room temperature forming on fatigue [1] are shown in Figure 3.2.

To effectively design structures for high fatigue life, some knowledge of residual forming stresses in the components must be secured. A sample print-out sheet from a computer program [2] which utilizes a mathematical method for determining the residual stresses induced in the bend area of

TABLE 3.1

SAMPLE COMPUTER PRINT-OUT SHEET

Initial Radius	Axial Load	Linear Neutral Axis	Area	Moment of Inertia
50.00	0.	1.3287	1.2523	0.36019
Final Radius	**Axial Strain**	**Modulus**	**Depth**	**Initial Moment**
98.74	0.	10400000.	2.0100	36983.

Distance from Neutral Axis	Initial Strain	Initial Stress	Final Stress
0.6813	0.0108365	75137.	5181.
0.5755	0.0087207	71041.	11948.
0.4697	0.0066049	64472.	16241.
0.3639	0.0044891	46687.	9318.
0.2582	0.0023733	24683.	−1824.
0.1524	0.0002576	2679.	−12966.
0.0466	−0.0018582	−19398.	−24180.
−0.0592	−0.0039740	−41660.	−35580.
−0.1650	−0.0060898	−62365.	−45423.
−0.2708	−0.0082056	−71170.	−43365.
−0.3766	−0.0103214	−75791.	−37125.
−0.4824	−0.0124372	−78670.	−29141.
−0.5882	−0.0145530	−80605.	−20214.
−0.6940	−0.0166688	−82473.	−11220.
−0.7997	−0.0187845	−83619.	−1504.
−0.9055	−0.0209003	−84635.	8342.
−1.0113	−0.0230161	−85553.	18287.
−1.1171	−0.0251319	−86361.	28341.
−1.2229	−0.0272477	−87152.	38412.
−1.3287	−0.0293635	−87781.	48645.

structural members is shown in Table 3.1. Programs of this type are invaluable in setting-up processing forming limitations. Check and straightening operations and stresses induced in assembly, however, further make difficult residual stress control.

3.1.2 Heat Treatment

Residual stresses are both produced and relieved in many of the common heat treat cycles for both ferrous and nonferrous alloys. The principal source of residual stress occurs in quenching from high-temperature solutioning or austenitizing treatments. Residual stresses are built up by nonuniform cooling rates between surface and core. For aluminum alloys and other single-matrix phase systems, differential cooling produces residual surface compression and core tensile stresses. It has been reported [3] that the surface compressive stresses are of sufficient magnitude to produce slightly higher fatigue strengths. For martensitic ferrous alloys the simultaneous differential cooling and expansion on transformation develops surface tensile and core compressive stresses, as quenched.

Aging temperatures for aluminum alloys are too low to produce any appreciable stress-relieving; however, most steels are tempered at temperatures sufficiently high to affect residual quench stresses. Consequently, for steels, after tempering, quenching stresses are not recognized as a detrimental factor. Quenching stresses in aluminum alloys, however, persist after completion of heat treatment, as indicated by distortion in machining, increased susceptibility to stress-corrosion and possible detrimental effects on fatigue. To minimize these effects in aluminum alloys, special processing techniques have been developed, such as: reducing section sizes by rough machining before heat treatment, use of less severe quenches where possible, and stress relief/equalization by cold working of quenched materials; for example, stretch-stress relief tempers.

3.1.3 Case Hardening

Case hardening of ferrous alloys is usually accompanied by an increase in fatigue resistance resulting from an increase in hardness and addition of

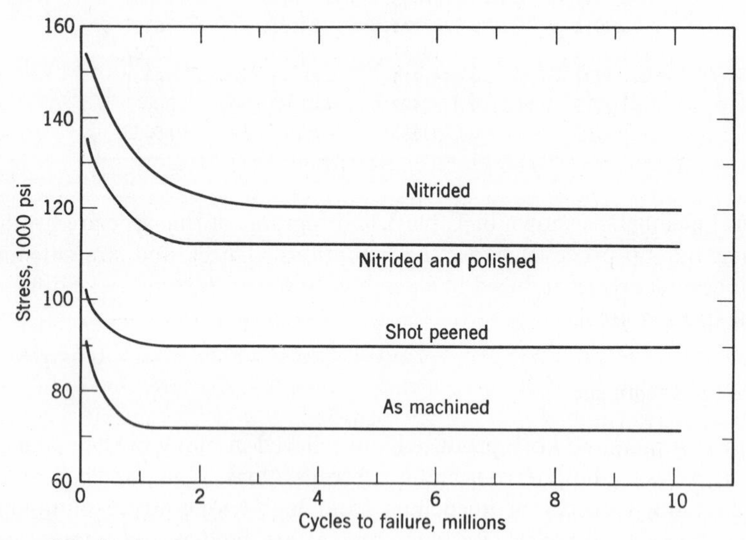

Figure 3.3 Bending fatigue test results on sections from crankshafts: endurance limit versus surface treatment.

compressively stressed surface layers. This effect has been observed in the carburizing, nitriding, and induction hardening processes. Typical examples of improved fatigue resistance resulting from case hardening are shown in Figure 3.3.

Case hardening, however, also increases the notch sensitivity of the surface material and therefore cannot be used indiscriminately. If the case-to-core juncture is located too near the surface or at an area of stress concentration, subsurface cracks may develop. In addition, if the case hardening process is not closely controlled, a defective condition may arise, such as carbide networks in a carburized case, which will impair fatigue properties. Unintentional carburization through improper control of heat treatment atmosphere can have catastrophic effects on the serviceability of ferrous parts. Similarly, failure to remove the oxygen enriched layer from titanium parts which have been heat treated in air is detrimental to fatigue properties as shown in Figure 3.4.

Perhaps the most common heat treatment problem is the generation of a decarburized layer on steel parts through inadequate atmosphere control. The resultant reduced tensile properties directly reduce the fatigue strength as shown in Figure 3.1. A typical example of this effect is shown in Figure 3.5. Carbon restoration processes often further increase the detrimental effect and, in any event, are generally not acceptable for aerospace hardware.

3.1.4 Surface Finish

A given surface finishing process influences the fatigue properties of a part by affecting at least one of the following surface characteristics: smoothness, residual stress level, and metallurgical structure. Many investigators have considered the effect of súrface finish on fatigue (eg., [4] and [5]). The results of one such investigation [4] are shown in Figure 3.6. Here it can be seen that, in general, fatigue life increases as the magnitude of surface roughness decreases. Decreasing surface roughness is seen as a method of minimizing local stress raisers.

Most mechanically finished metallic parts have a shallow surface layer in residual compression which extends down for several tenths of a thousandth to several thousandths of an inch. Aside from effect on surface roughness, the final surface finishing process will be beneficial to fatigue when it increases the depth and intensity of the compressively stressed layer and detrimental when it decreases or removes this desirable layer. Thus sandblasting, glass bead peening, burnishing, and other similar operations generally improve fatigue properties [6]. Conversely, such processes as electropolishing, chem-milling, and electrical chemical machining (ECM), which remove

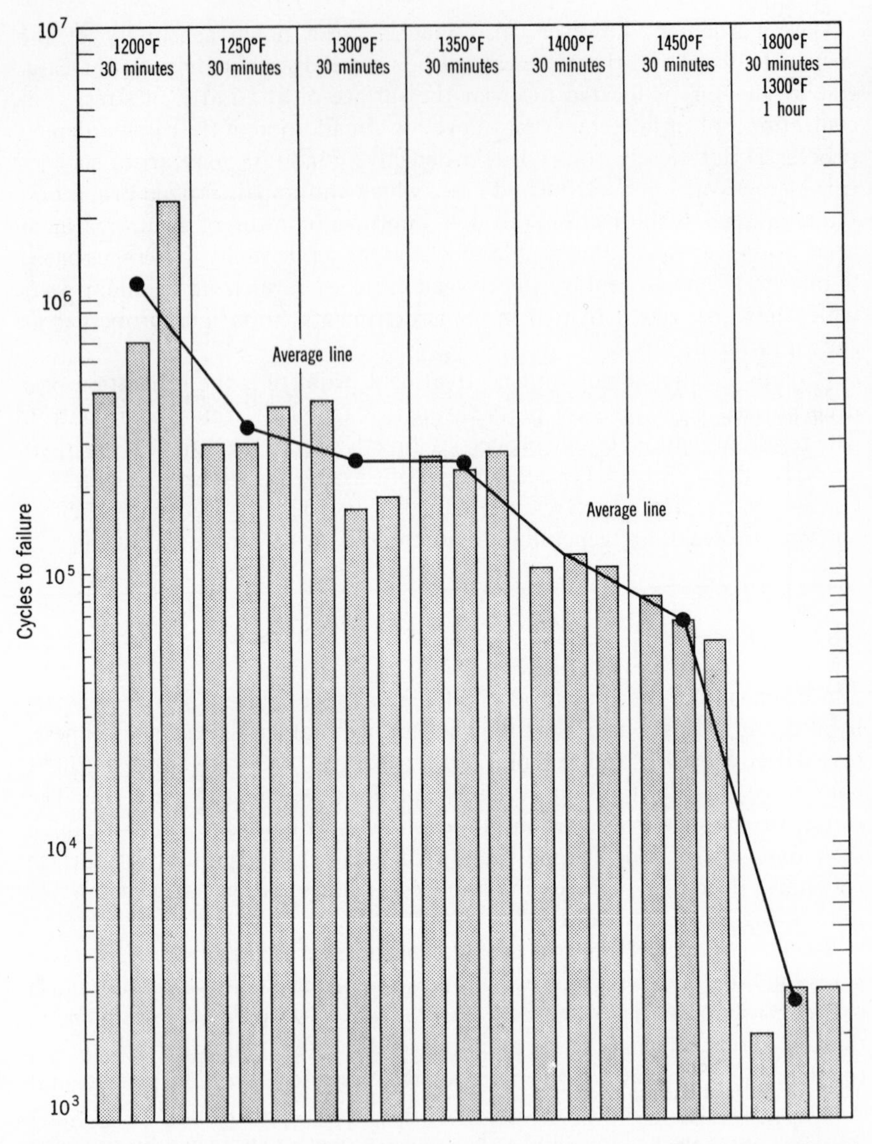

Figure 3.4 Exposure temperature is plotted against cycles to failure. All tests were conducted at room temperature: titanium flexural fatigue (Ti-GAl-4V ann.); 0.025-in. Coil strip; maximum stress = 70,000 psi; R = −1 (flexure).

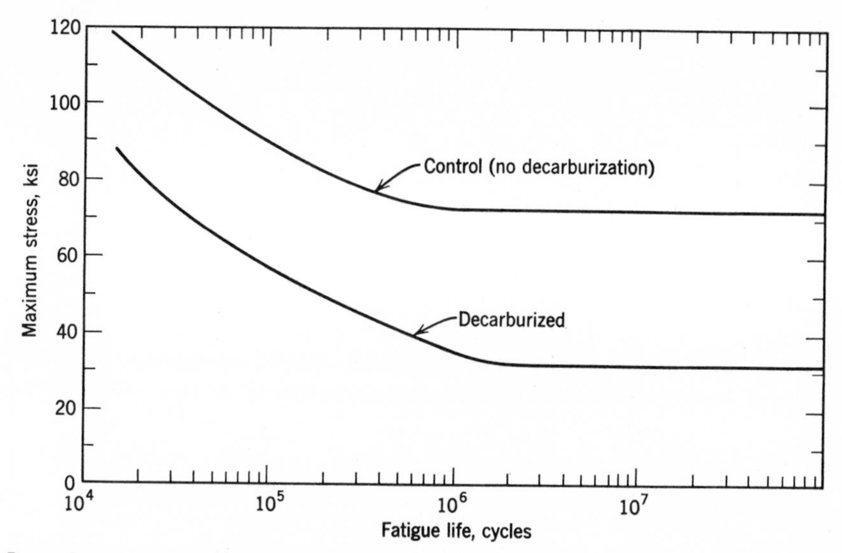

Figure 3.5 Effect of decarbonization on fatigue strength of rotating beam specimens of SAE 4140 steel, tempered for normal hardness of R_C-48 [8].

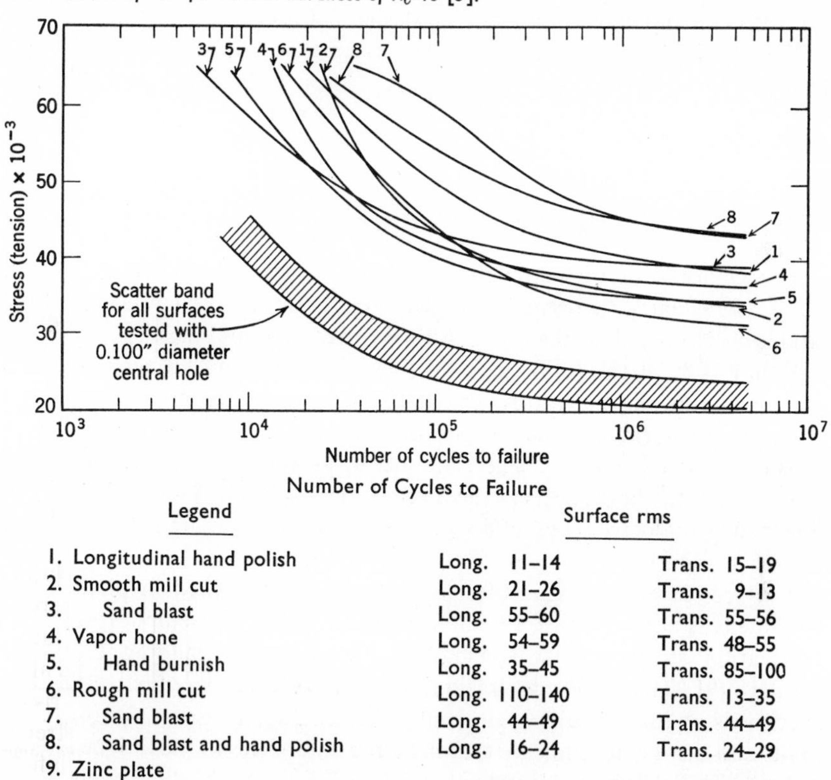

Figure 3.6 Surface finish versus fatigue life for 7075-76 extrusions. The longitudinal and transverse surface finish (rms) are given in the legend: lower limit of scatter bonds—surface RMS—maximum surface stress for each finish tested.

metal without plastic deformation at the tool point, may reduce fatigue properties [7].

Because of the high temperatures generated locally, such unconventional metal removal techniques as electrical discharge machining (EDM) can be extremely detrimental to fatigue properties without proper control and subsequent processing because of surface and subsurface microstructural changes. Improperly controlled grinding can have similar effects (see Section 3.2).

Many local surface defects and irregularities occur in fabrication and service that are difficult to anticipate, inspect for, or control. Stress-concentrations resulting from small indentations on an otherwise smooth surface in the form of accidental tool marks, grinding scratches, corrosion pits, or other service related minor damage are occasionally as effective in reducing the fatigue life of structural parts as are large scale stress concentrations resulting from design deficiencies—depending on the material, heat-treat range, design margins, etc. Such unintentional stress-raisers are damaging in a structure only if their notch effect is more severe than the most severe stress concentration arising from design, unless they are located so as to intensify the stress raising effect of the critical design feature.

Note in Figure 3.6 that the fatigue specimens with a 0.100-in. diameter central hole have the same fatigue life regardless of the surface finish. Here the effect of design (the central hole) far outweighs the effect produced by surface finish.

3.1.5 *Cladding, Plating, Chemical Conversion Coatings, and Anodizing*

The cladding of high-strength aluminum alloys reduces the fatigue life of the core material, because the cladding is usually of significantly lower strength material than the core. Although the cladding usually amounts to less that 10% of the total sheet thickness, its effect on fatigue life can be quite pronounced. The ratios of axial loading fatigue strength to static tensile strength are normally comparable for bare and clad sheets under loading causing short-life failure. The clad sheets, however, become progressively weaker than the bare sheets as the lifetime increases [8]. Typical fatigue data illustrating the detrimental effect of cladding on fatigue life for 2024-$T3$ are shown in Figure 3.7.

The reduction in fatigue life caused by cladding noted in plain fatigue specimens is not present to the same degree in built up assemblies, where the fabrication stress raisers overshadow the effect of cladding on fatigue strength [9]. The -1.0 stress ratio results for several types of bare and alclad specimens of 2024-$T3$ alloy are given in Figure 3.8. These results for riveted joints and box beams indicate that the large differences in the fatigue strength of bare and alclad material sighted in Figure 3.7 are not likely to be significant in riveted assemblies.

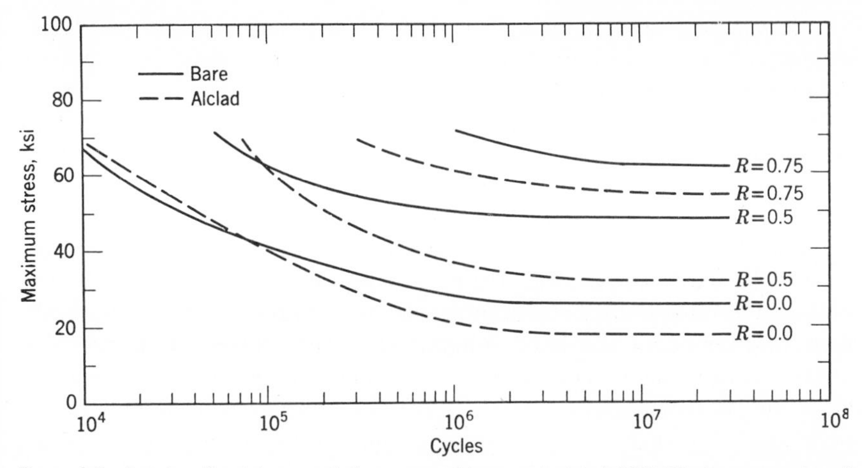

Figure 3.7 Results of axial stress fatigue tests of bare and alclad 2024-T3 sheet at various R values [9].

Since cladding is available only for the sheet, strip, plate, wire, and tubing forms of aluminum alloys, aluminum metal spraying is sometimes employed for other forms such as extrusions and forgings for added corrosion protection. For plain fatigue specimens, the detrimental effect on fatigue properties resulting from a surface layer of sprayed aluminum is similar to that observed for cladding. The likelihood that fabrication stress-raisers would overshadow the effect of sprayed aluminum coatings on forms such as forgings

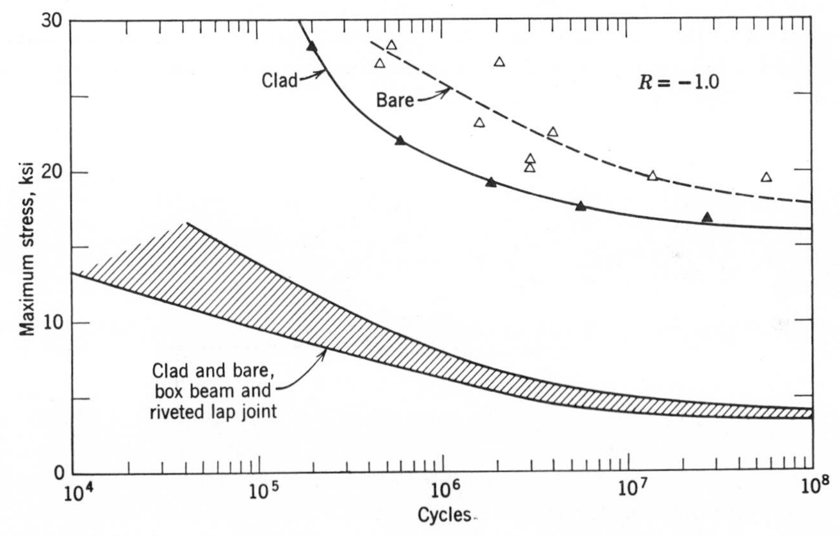

Figure 3.8 Results of flexural fatigue tests on bare and alclad plain specimens, box beams and riveted joints.

and extrusions, however, is not as great as for sheet metal assemblies. Sprayed zinc finishes on aluminum alloys have been reported as not producing any measurable reduction in fatigue life [6].

Three potentially harmful effects of plating on the fatigue properties of the substrate are: notch effect of fissures in the plating, induced residual tensile stresses in the substrate, and hydrogen embrittlement. These factors are often so interrelated that even qualitative observations of the effect of individual factors is difficult. For a discussion of the effect of specific platings on specific alloys, processing parameters such as prior surface condition, plating solutions, bath temperatures, current densities, post plating treatment, and the like would have to be known and evaluated. It is generally agreed, however, that neither zinc nor cadmium plating has any appreciable effect on fatigue properties. Nickel plating is believed to impose some moderate reduction on fatigue life, with indications that electroless nickel plating may reduce fatigue life by as much as 25%. Figure 3.9 shows the effect of electro-deposited nickel on the fatigue strength of steel [28].

The effect of chromium plating has been more widely studied. The fatigue strength of chromium-plated steel parts decreases appreciably as the plating thickness increases. It has been suggested [10] that the plating effect is a notching effect and the effect of plating thickness is essentially an effect of notch depth, or at least that the notch effect is a dominating factor. Figure 3.10 shows the effect of chrome plate thickness on fatigue properties.

Chemical conversion and anodic coatings which are applied to aluminum alloys for corrosion protection or wear resistance usually produce a reduction in fatigue life ranging from a negligible amount up to 10 or 15% of the endurance limit. This reduction is attributed to one or a combination of the following factors: increased surface roughness, cracking in the coating, and induced residual tensile stresses in the material immediately beneath the

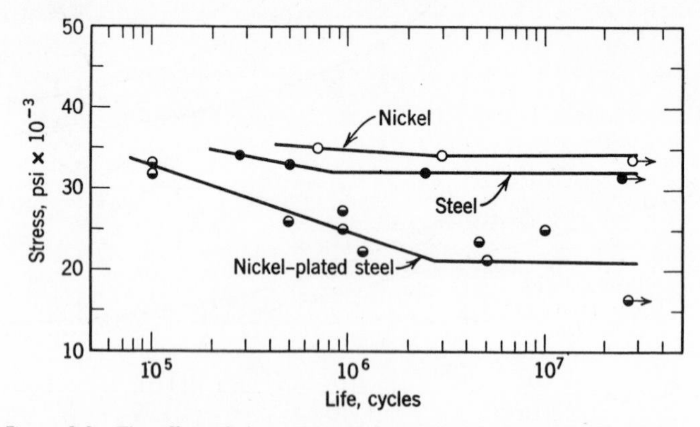

Figure 3.9 The effect of electrodeposited nickel on fatigue strength of steel.

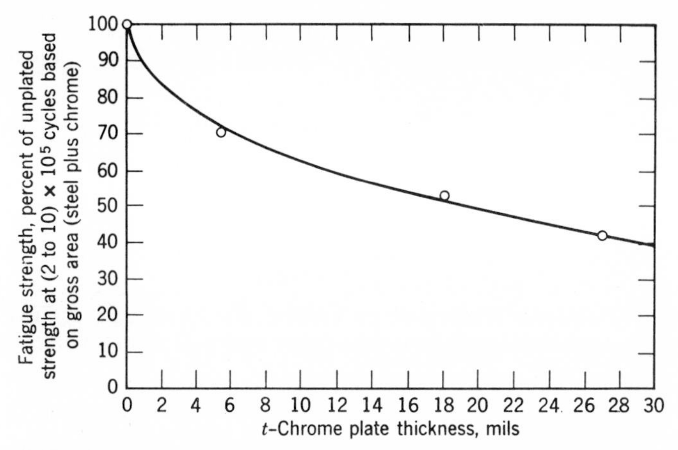

Figure 3.10 Fatigue strength versus chrome plate thickness for $\frac{1}{4}$-in. diameter specimens.

coating. The method of producing the coating further affects fatigue properties. As an example, fatigue tests run on chromic acid anodized 7075-*T*6 indicated that anodizing with a 5% dichromate seal effected a "slight but definite lowering in fatigue life as compared to that of unanodized metal" [11], whereas a sulfuric type anodizing treatment resulted in a substantial reduction in fatigue life as shown in Figure 3.11. The thickness of the anodic coating

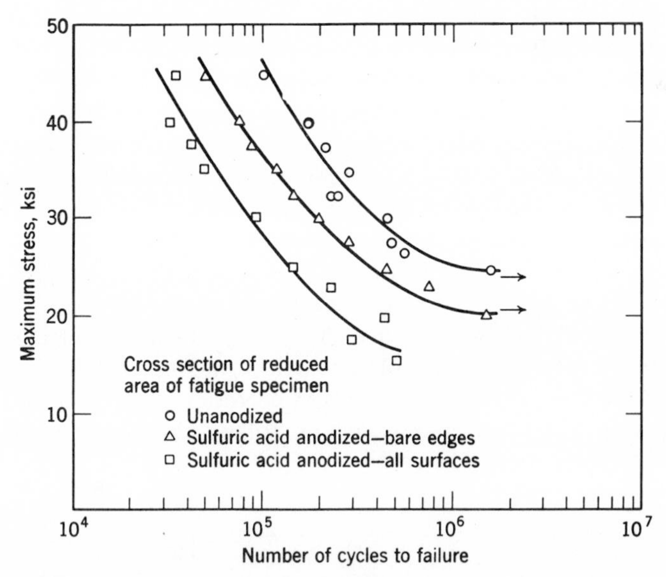

Figure 3.11 Effect of sulfuric acid anodize on fatigue properties: 7075-T6 alclad sheet 0.090-in. thickness; longitudinal stress ratio +0.2.

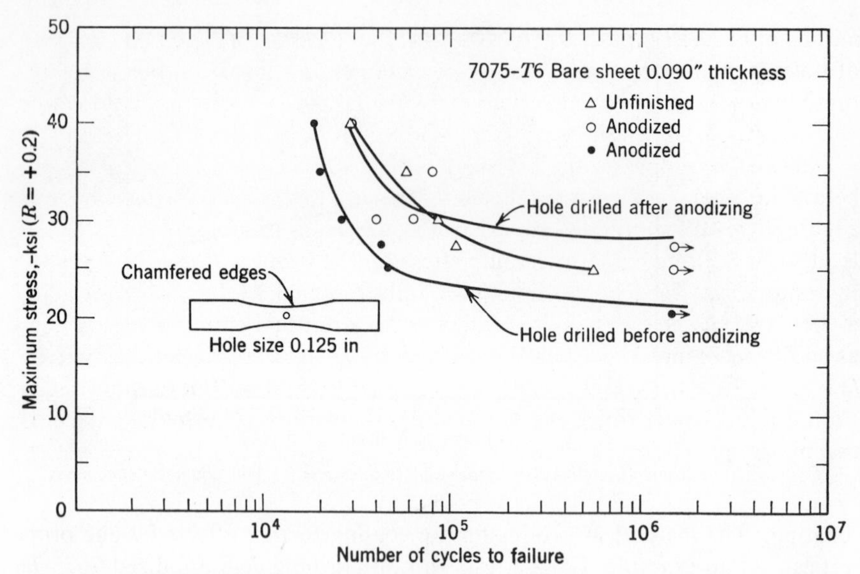

Figure 3.12 S-N curves $\frac{1}{8}$ in. hole specimens anodized surface versus unfinished surface.

also is an important factor, with fatigue properties varying inversely with coating thickness [12]. The thickest films used in "hard" anodizing for wear surfaces are recognized as being detrimental to fatigue properties.

Fatigue tests of 7075-*T*6 bare sheet with an Alodine #1200 coating showed no loss in fatigue properties compared to untreated material [13].

When a stress-raiser of considerable severity is introduced into the fatigue test specimen (such as a central hole), the geometric stress raiser becomes dominant, eliminating the effect on fatigue of the chemical conversion coating, *provided* the hole is drilled after the coating process. If, for instance, a control hole fatigue specimen is anodized after the hole is drilled, the effect of hole and anodic coating is additive, as shown in Figure 3.12. A similar phenomenon can be observed in Figure 3.10 in which in this case the geometric stress raiser is the edge of the specimen.

Here the designer is faced with a dilemma: which is controlling, the corrosion protection within the hole or the fatigue properties in the area of the hole? Of course, there are methods of corrosion protection other than anodizing, and in some instances, such as drilling holes on assembly, it is impractical to apply an anodic coating in the holes.

3.1.6 Cold Working

Cold working of parts to induce residual surface compressive stresses has been found to be an effective tool for improving the fatigue life of both simple and complicated shapes. Methods of imparting cold work include: coining of holes, thread rolling, fillet rolling, peening, hole expansion, pre-stressing, tumbling, and grit blasting. Before discussing further specific methods of cold working let us examine why these induced residual stresses are beneficial.

Consider a simple beam externally loaded in bending to a stress well below the yield stress. Such a beam will have a stress gradient as shown in Figure 3.13. If no other stress exists and the beam is subjected to a flexing load, then the calculated maximum stress on the beam will oscillate from f_b in compression to f_b in tension under fully reversed loading. If, however, a thin layer of the outer shell is subjected to a residual compressive stress f_r as shown in Figure 3.13, then the net maximum stress on the surface will be $f_b - f_r$ on the tension side and $f_b + f_r$ on the compression side as indicated in Figure 3.13. Under fully reversed loading, the surface will now be subjected to a maximum tension stress of $f_b - f_r$ which, as far as the surface is concerned

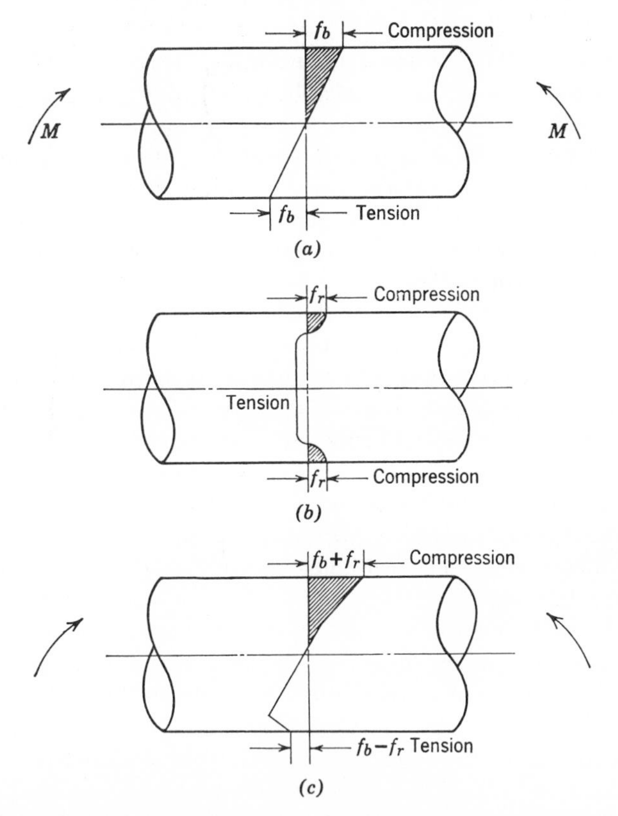

Figure 3.13 Stress gradient of a simple beam with: (a) bending, no residual stress, (b) residual stress, no bending, (c) residual stress plus bending [14].

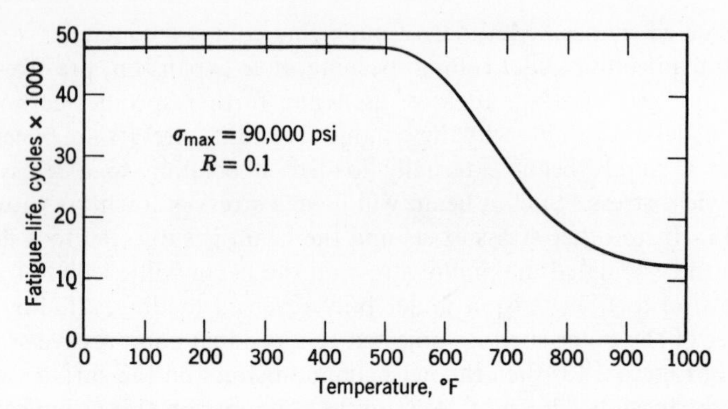

Figure 3.14 The effect of elevated temperature exposure for one hour on the fatigue properties of AISI 8740 bolts [14]

means longer life because a lower mean stress is being applied. It follows that residual surface compressive stresses are most beneficial under loading conditions which produce high surface tensile stresses, as in bending [8]. Residual compressive stresses, however, are also beneficial to the fatigue life of axial loaded specimens.

The most important features of a residual stress distribution curve (as a result of prestressing) are (a) the value of stress at the surface, (b) the maximum compressive stress, (c) the depth of the compressive layer, (d) the compressive residual stress area, and (e) the location and magnitude of the maximum residual tensile stress.

In order for the desirable effect of surface cold working to be maintained, the cold-working process must be accomplished in the final heat-treated condition and subsequent thermal treatment eliminated when feasible and closely controlled when they are essential. Exposure of cold-worked surfaces to elevated temperatures initially results in stress relief of the plastically deformed zone and ultimately in recovery or perhaps recrystallization of the work-hardened area, with complete loss of the desirable residual stress gradient. The effect of elevated temperature exposure on the room-temperature fatigue life of AISI 8740 steel bolts with threads rolled after heat treatment is shown in Figure 3.14 [14].

Fillet radii are the sight of various degrees of stress concentration due to the change in section occurring there and the notch effect of the radii. In achieving the final contour of these radii by rolling in the heat treated condition, residual surface compressive stresses are established which significantly improve the fatigue life. Figure 3.15 illustrates this improvement for steel bolts heat treated to 125,000 psi minimum ultimate tensile strength [14].

The rolling of threads in material in the final heat-treated condition, in addition to producing a favorable compressive surface stress distribution,

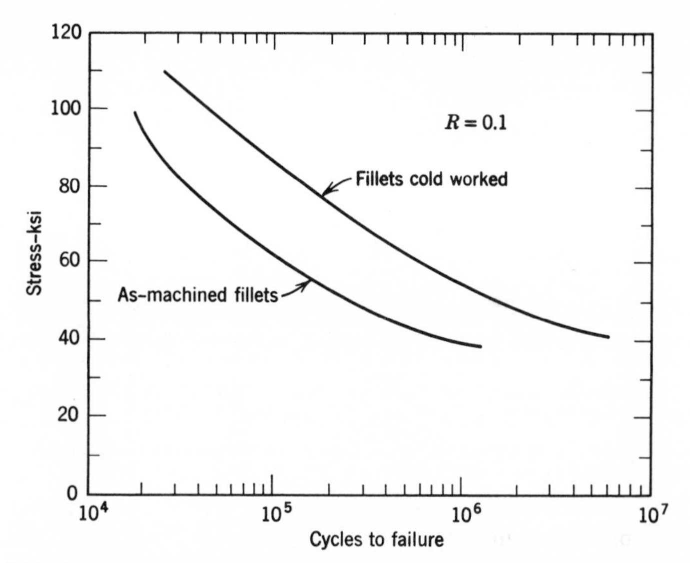

Figure 3.15 Effect of cold rolling fillets on steel bolts heat-treated to 125,000 psi minimum [14].

also generates a smoother root radius and beneficial microstructural flow from root to crest with no cut "fibers."

In Figure 3.16 the thread form of all specimens is identical, the only difference being that one group was rolled before heat treatment while the other was rolled after. Rolling of threads after heat treatment becomes more beneficial as the strength of the part increases (Figure 3.17). Conversely, with low-strength material the desirability of thread rolling after heat treatment diminishes.

Perhaps the most widely used process for inducing residual compressive stresses in surface material is shot peening. Its widespread use can be

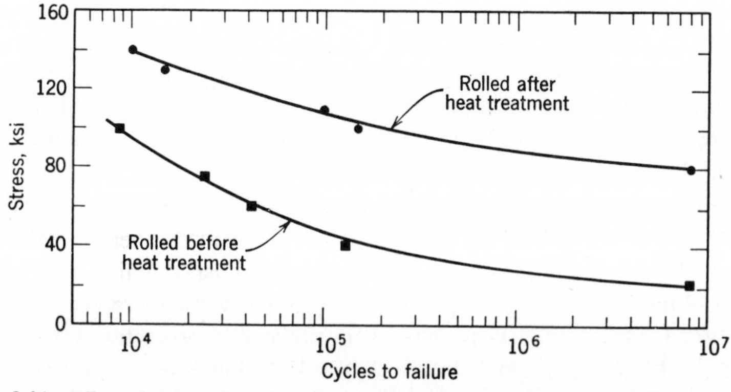

Figure 3.16 Effect of rolling threads before and after heat treatment on 220,000 psi bolts [14].

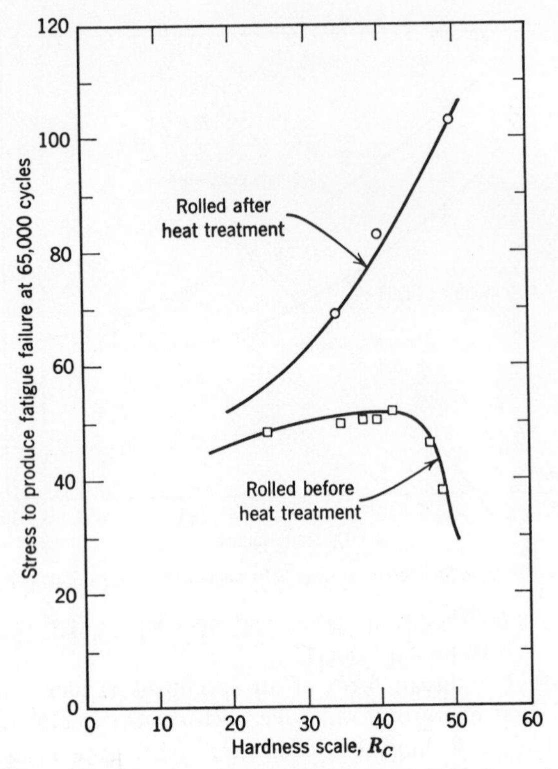

Figure 3.17 Effect of hardness on the fatigue life of threads rolled before and after heat treatment.

accounted for by its low cost and relatively easy application to a wide variety of materials and parts of varied size or configuration. The intensity of peening is governed by the size and hardness of the shot, the velocity of the shot, the angle of impact, and the peening time. The intensity is normally measured in terms of the arc height of an Almen strip or similar device in which the increased length (curvature) of a thin strip peened on one side is used as a measure of peening severity. For thick sections in which there is adequate material to absorb the subsurface tension stress (see Figure 3.13) a relatively high peening intensity is desirable.

In thin sections the arc height should be limited to avoid excessive residual subsurface tensile stresses. It is known that the Almen intensity alone does not measure the peening effect on aluminum alloys and it is doubtful whether by itself alone it measures the effect on very hard steel and on soft materials. Therefore it has become customary to specify shot size and shot material besides the intensity. Specification of results rather than of process details would be more desirable [29]. For sheet metal parts and other thin sections

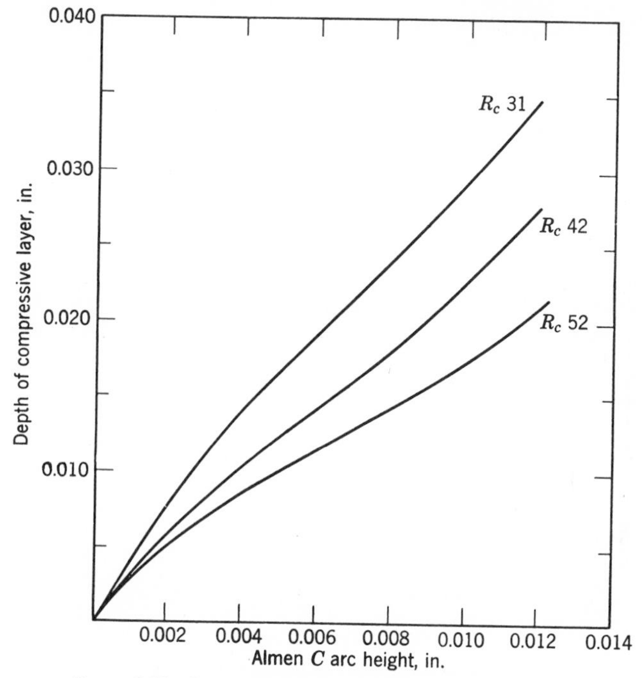

Figure 3.18 Depth of compression versus Almen arc height [16].

best results are often obtained by the use of very fine beads suspended in a liquid blast.

As a general rule the depth of the compressively stressed layer should be two to five times the depth of any anticipated surface defects such as roughness, decarburization, abrasion, and galling [15]. Figure 3.18 gives the relationship between Almen "*C*" height and depth of residual compressive layer for AISI 4340 steel of various strength levels [16].

By definition the optimum intensity is achieved at the point at which the inferior properties of the skin have been raised by residual compressive stresses to the same fatigue life to which the superior properties of the core are reduced by residual tension stresses.

Shot peening sees any number of applications on parts that are critical in fatigue. Peening is also employed on surfaces on which an unfavorable residual stress pattern is anticipated or on which inferior mechanical properties are believed to exist. As an example, peening might be used before chrome plating or hard anodizing, or in special situations, after a heat treatment which produced decarburization of a steel part. Figure 3.19 illustrates the beneficial effect of peening on bare surfaces and surfaces that are subsequently chrome plated [17]. Table 3.2 shows some typical effects of peening

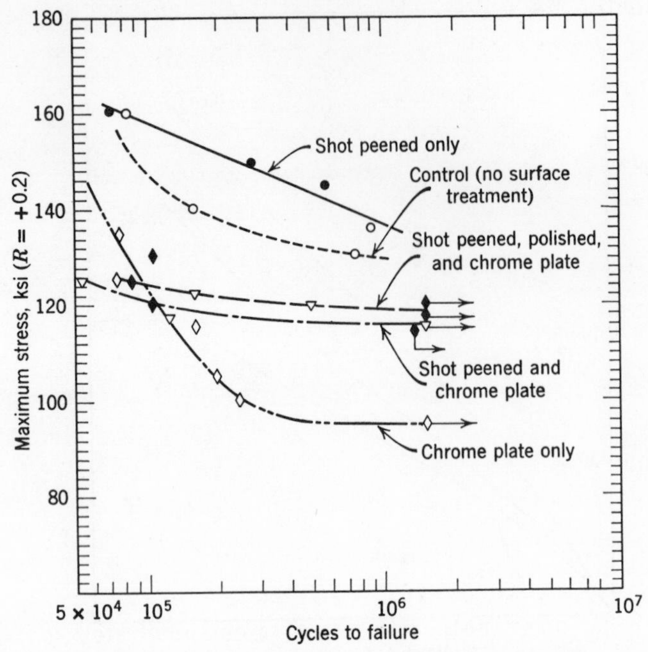

Figure 3.19 Fatigue life of axially loaded, round 4130 steel specimens heat-treated to 160 to 180 ksi.

of decarburized surfaces. As with chrome plating, the detrimental effect is not completely eradicated by the peening.

The consideration of the fatigue properties of material adjacent to a hole is often controlling, where this is the region of highest stress under external loading, by virtue of the geometric stress concentration associated with the hole itself. Consequently, the introduction of residual compressive stresses on the surface of the hole is highly desirable. Hole expanding, that is, plastically increasing the diameter of the hole by the passage of a tapered pin

TABLE 3.2

EFFECT OF PEENING ON A DECARBURIZED SURFACE
(4340, 240–200 ksi)

Depth of Decarburization	Peening	Maximum Stress	Cycles
None	No	120,000 psi	1,000,000
0.003 to 0.030 in.	No	120,000 psi	40,000
0.003 to 0.030 in.	Yes	120,000 psi	95,000

Shot size P-28, intensity A-0.012A2

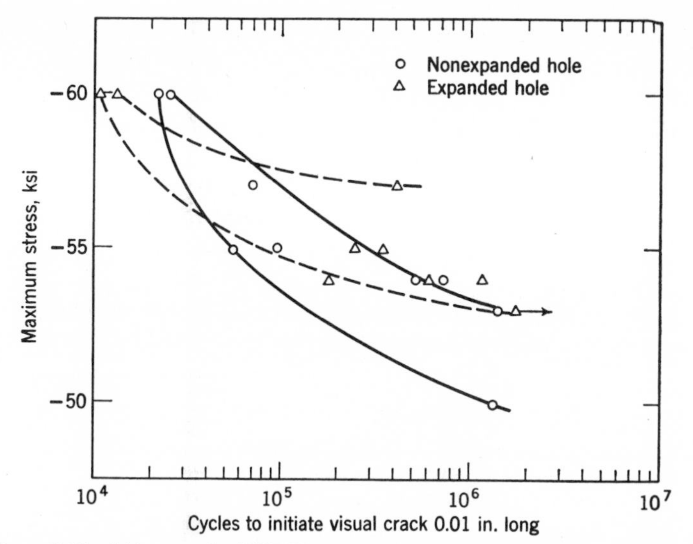

Figure 3.20 S-N curves for 7075-T6 nonclad sheet showing the affect of hole expansion.

whose maximum diameter exceeds the hole diameter, has been shown to increase the fatigue life of specimens containing holes when tested by either alternating tensile or compressive stresses. Figure 3.20 shows the results of compressive fatigue tests on bare 7075-*T*6 sheet with and without hole expansion [18]. The expense and inconvenience associated with the implementation of this process, however, severely limits its application in production.

When a part experiences loads that are much higher in one direction than in the opposite direction, prestressing the part by applying an excessive load in the direction of major service loading will often produce beneficial residual surface compressive stresses and working hardening that will significantly improve fatigue properties. In reverse loading this effect is lost and a detriment to fatigue life may result. In addition, when the part configuration includes grooves or notches or other areas of high local stress concentration, the material in these areas may be subject to tension along all three principal axes, precluding ductile behavior. Prestressing under these circumstances would, of course, produce rupture.

Prestressing has long been used outside the aircraft industry on such diverse parts as automotive leaf springs, railroad wheels and gun tubes (where the process is known as autofrettage). In the aircraft industry prestressing is often used for torsion bars and bomb hooks, to name two examples. Laboratory test data on the effect of prestressing on the notched fatigue properties of 7075-*T*6 rod are given in Figure 3.21 [3]. Tensile prestressing to 90% of the tensile strength increased the endurance limit about

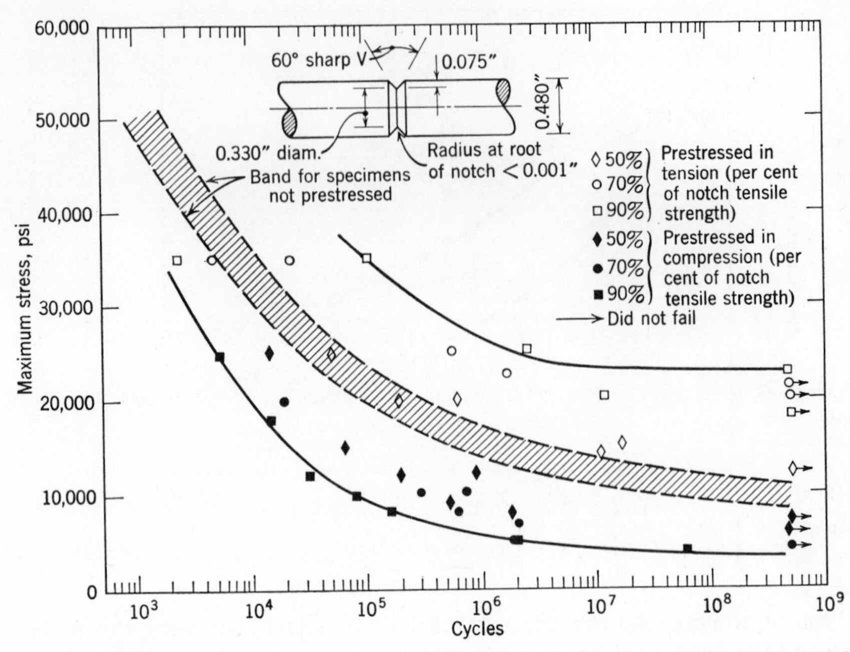

Figure 3.21 Rotating-beam fatigue curve for 7075-T6 rod, $\frac{3}{4}$-in diameter rolled and drawn with various amounts of prestress.

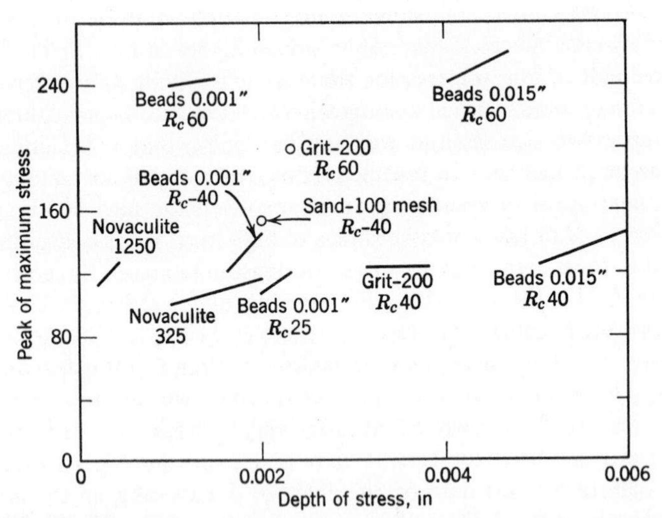

Figure 3.22 Peak stress plotted against depth of stress. Lines identified with hardness key show the effect of blasting pressure 60 to 90 psi. Individual points represent use of 90 psi.

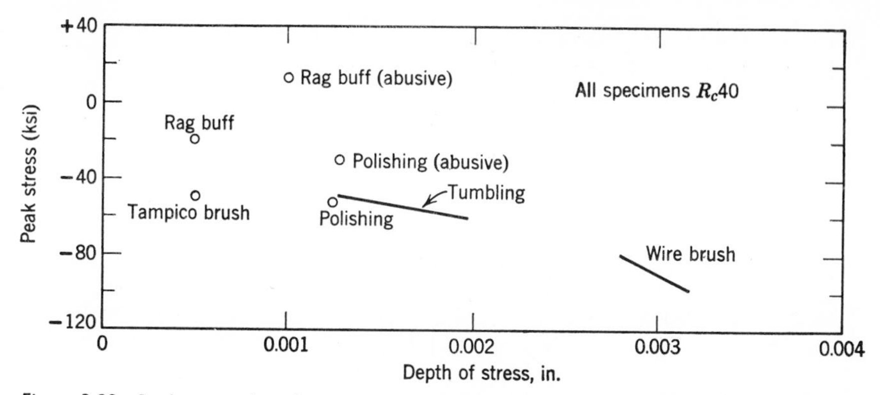

Figure 3.23 Peak stress plotted against depth of stress obtained by typical finishing procedures [20].

100%. It is interesting to note that compressive preloading reduced the endurance limit of notched specimens about 70%. Preloading to 50% or less of the tensile strength produced little or no effect on the fatigue properties.

Barrel tumbling has been considered for use on small parts to produce the beneficial surface residual stress effect of shot peening with smoother surfaces at lower cost. At least one investigator [19] reported that this process exhibited poor reproducibility from batch to batch, resulted in a wide range of stresses in parts tumbled in the same batch and did not produce any benefit to fatigue life. Another source, however, concluded that tumbling provides a "consistent method of producing compressive stress" [20].

Other techniques used to improve surface finishes, to remove heat treat oxides and foreign material, or to prepare a surface for a subsequent coating or plating also produce changes in the sign and/or magnitude of residual surface stresses. The depth and magnitude of compressive stresses generated in various abrasive blasting processes are shown in Figure 3.22. The individual points represent a blasting pressure of 90 psi, whereas the lines represent 60 to 90 psi pressure. It was found that, as the particle size for a given material increases, the depth of stress increases but not the peak stress; for constant particle size, with increase of density or specific gravity of the blasting material, the depth of stress increases with no significant effect on peak stress [20].

Processes for developing very fine surface finishes also have their effect on surface stresses, as shown in Figure 3.23. It is suggested by the investigator that many surface finishing processes, blasting in particular, can and should be controlled where engineering properties are of concern. Specific methods for process control are offered [20].

3.2 METALLURGICAL FACTORS

The distinction between processing factors and metallurgical factors is not always clear. In fact, it is rather arbitrary in some areas. In this section, however, the focus is on regions *within* the material, either at the surface or core, which adversely affect fatigue properties. These regions may arise from melting practices or primary or secondary working of the material or may be characteristic of a particular alloy system. In nearly every instance the detriment to fatigue properties results from a local stress-raising effect.

3.2.1 Surface Defects

Primary and secondary working are often responsible for a variety of surface defects that occur during the hot plastic working of material when lapping or folding or turbulent flow are experienced. The resultant surface defects bear such names as laps, seams, cold shuts, or metal flow through. Similar defects are also noted in cold working, such as fillet and thread rolling, in which the terms lap and crest cracks apply. Other surface defects develop from the embedding of foreign material under high pressures during the working process. Oxides, slivers, or chips of the base material are occasionally rolled or forged into the surface. The surface defects in castings might include entrapped die material, porosity, or shrinkage; in the extrusion or drawing processes such surface defects as tears and seams are not uncommon.

All of the aforementioned surface defects produce a notch of varying intensity which acts as a stress-raiser under load to the detriment of fatigue properties. Because most of these defects are present prior to final processing and are open to the surface, standard nondestructive testing procedures such as penetrant and magnetic particle inspection will readily reveal their presence. If they are not detected, however, the defects may serve as a site for corrosion or crack initiation during processing (in heat treating, cleaning, etc.) further compounding the deleterious effect on fatigue strength.

3.2.2 Sub-Surface and Core Defects, Inhomogeneity, and Anisotropy

Subsurface and core defects considered here are those which originate in the as-cast ingot. Voids in cast materials due to gas entrapment (porosity) and improper metal fill (shrinkage) are not uncommon. In castings (ingots) that are to be subsequently hot and cold reduced, the portion of the ingot containing the preponderance of voids is often removed and discarded. The remaining internal defects normally weld shut under the combination of temperature and pressure experienced in the reduction of the ingot, resulting in a continuous, homogeneous product. Occasionally, when the surfaces of the defects are oxidized or otherwise contaminated, healing (welding) of the opposite surfaces is precluded and the defective area is retained in the wrought product. Terms such as "unhealed porosity" and "laminations" are applied to this condition. Since these defects existed before working, in the final wrought product the major diameter of the now plate or rod-shaped flaw is parallel with the direction of plastic deformation.

Fatigue testing of high-strength aluminum alloy specimens containing defects of the type discussed in this section revealed the following trends [21]:

1. Stressing parallel to the defect plane has a small effect on the fatigue strength, provided the defect does not intersect a free surface.
2. The effect of defect size on the fatigue strength in the short transverse direction of testing, that is, with the plane of the grain flow normal to the direction of loading, is shown in Figure 3.24.
3. An internal defect adversely affects fatigue by introducing a stress concentrator into the material and reducing the load resisting cross-sectional area.
4. With respect to fatigue properties, when the edge of one defect is within approximately two diameters of the center of another defect, these should be

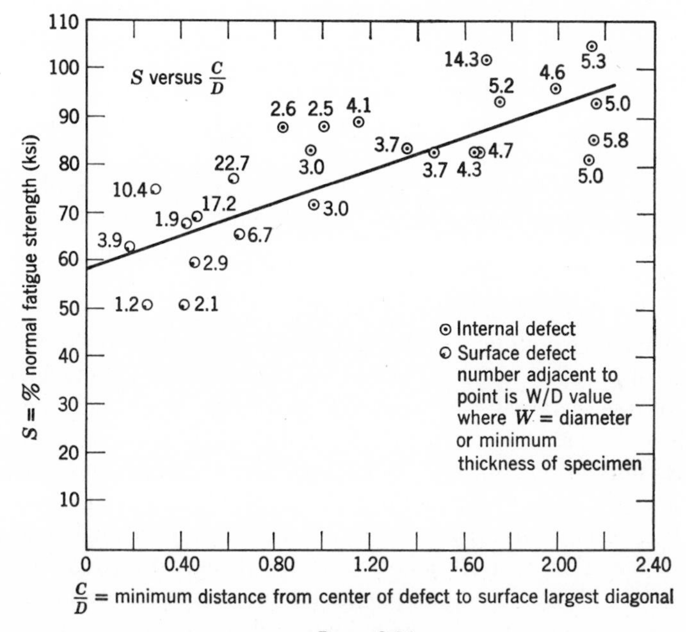

Figure 3.24

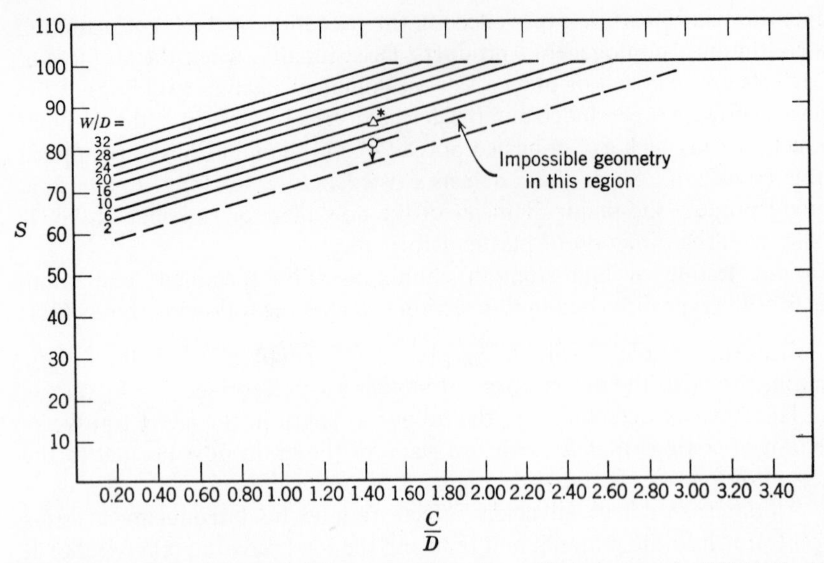

Figure 3.25 Effect of material defect size and location on fatigue strength of 7075-T6. Note. Actual position of any line may be as much as 5.4% S from value shown. S = % normal fatigue strength, C = minimum distance from center of defect to surface, D = largest diagonal of defect, W = diameter or minimum thickness of specimen.

considered as one large defect having a diameter equal to the extreme distance which will include both defects.

A correlative summary of the relation of defect size, location in the test specimen and specimen size on fatigue properties of high-strength aluminum alloys is given in Figure 3.25.

Inasmuch as most subsurface defects do not intersect a surface of a part, inspection is somewhat more difficult. For wrought product ultrasonic or eddy-current testing might be used, whereas for castings fluoroscopic or radiographic inspection is preferred.

There are two types of inclusion in metals, nonmetallic and intermetallic. The amount and distribution of these inclusions is determined by the chemical composition of the alloy, the melting and working practice and the final heat treatment of the material. Nonmetallic inclusions are usually complex compounds of the metallic alloying elements with oxygen, nitrogen, carbon, phosphorus, sulphur and silicon. In wrought forms nonmetallic inclusions occur in discontinuous or semicontinuous stringers parallel to the direction of working. Consequently, their deleterious effect is most strongly felt when the principal direction of stress is normal to the stringer direction. The size of the inclusion is also an important parameter in assessing its effect on fatigue properties, as shown in Figure 3.26 for 4340 steel heat treated to the 260–310

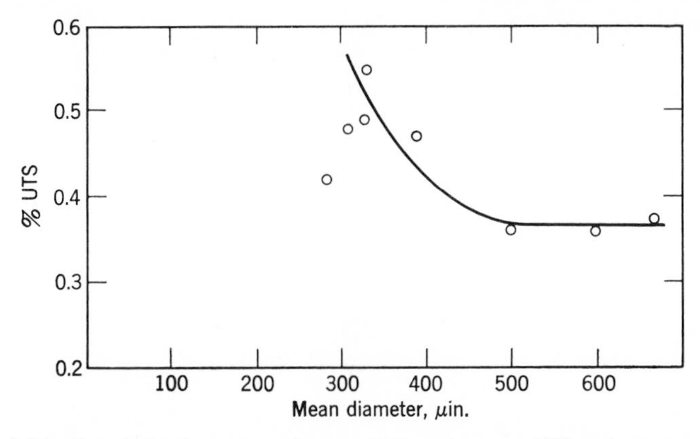

Figure 3.26 Correlation between endurance limit as per cent UTS and average large inclusion arithmetic mean diameter.

ksi tensile range [22]. It was demonstrated in this investigation that in general the endurance limit varied inversely with the size of the nonmetallic inclusion. Although this relation did not apply to all inclusion types, it was suggested that a separate curve exists for each predominant type of nonmetallic inclusion. In many alloy systems the melting practice has been found to be the controlling influence on the "cleanliness" of the materials produced. Vacuum or inert atmosphere melting practices and vacuum pouring practices for ferrous based alloys have been observed to be most effective in reducing the level of inclusions. Some severe effects of inclusions on fatigue properties are shown in Figure 3.27.

Intermetallic inclusions may be either complex metallic compounds or second phases with variable compositions. The type of intermetallic constituent is believed to be an important consideration in determining their effect

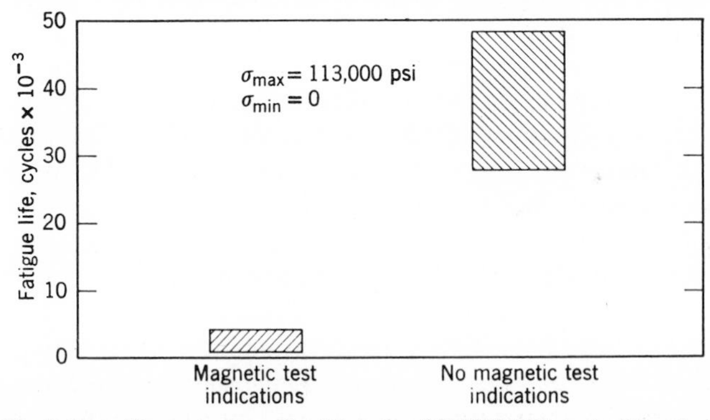

Figure 3.27 Fatigue-life comparison of axial shafts of SAE86B45H steel with and without magnetic test indications. The indications were caused by nonmetallic stringers [25].

on fatigue life, although the mechanism is not clearly understood. The site of such an inclusion, however, is a discontinuous region with physical and mechanical properties different than the matrix phase. Under load these areas would serve as stress-raisers. Once a crack has initiated, the effect of intermetallics on crack propagation is thought to be as follows (based on 7178-*T*6 aluminum alloy sheet) [23]:

1. When the width of the plastic zone preceding the crack tip is less than the size of the second-phase particle, the crack growth rate is controlled primarily by the properties of the matrix.
2. When the width of the plastic zone is of the order of the size of the constituent particles, but less than the interparticle spacing, such particles are usually much harder than the surrounding matrix, and because of their brittle nature they lack the ability to accommodate large plastic strain and so must either separate from the matrix or fracture during plastic deformation. If the particle fractures, the additional fracture surface so produced would slow down the crack tip acquiring the profile of the particle; such a reduction in sharpness would cause a lowering of the local stress intensity factor.
3. When the width of the plastic zone is greater than the interparticle spacing, the crack growth rate is primarily controlled by fracturing in the vicinity of second-phase particles. This fracturing, which occurs at several locations within the plastic zone, produces internal voids that coalesce by a process of internal and localized necking [23].

Inclusion near the surface of a part may be revealed by magnetic particle inspection (for ferromagnetic alloys) if they are of sufficient size and difference in magnetic permeability from the matrix material. Penetrant inspection will not indicate their presence, for there is no void for penetrant entrapment. Microscopic examination of a metallographically prepared sample is the usual method of inspection for inclusions.

Some alloys are subject to microstructural banding which often has an adverse effect on fatigue properties. The banding is usually produced by local chemical segregation which stabilizes a phase not normally present in the alloy at room temperature. The severity of the loss in fatigue properties is dependent on the direction of the banding relative to the maximum stress direction (the banding is always in the direction of prior working) and the degree of compatibility between the banded and matrix phase. Some examples of banded microstructures are shown in Figure 3.28 and 3.29. Banded retained austenite and delta ferrite are occasionally seen in a large number of low-alloy and stainless steels. In some of the stainless steels the presence of ferrite is intentional; in others it is not. The loss in fatigue properties produced by ferrite stringers in 431 stainless steel is shown in Figure 3.30.

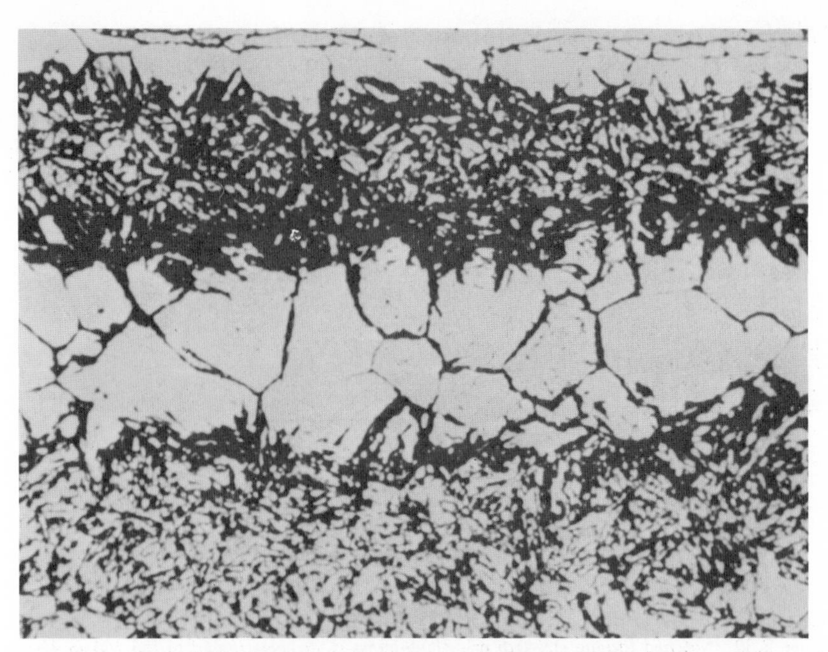

Figure 3.28 Banded retained austenite in 431 stainless-steel bar.

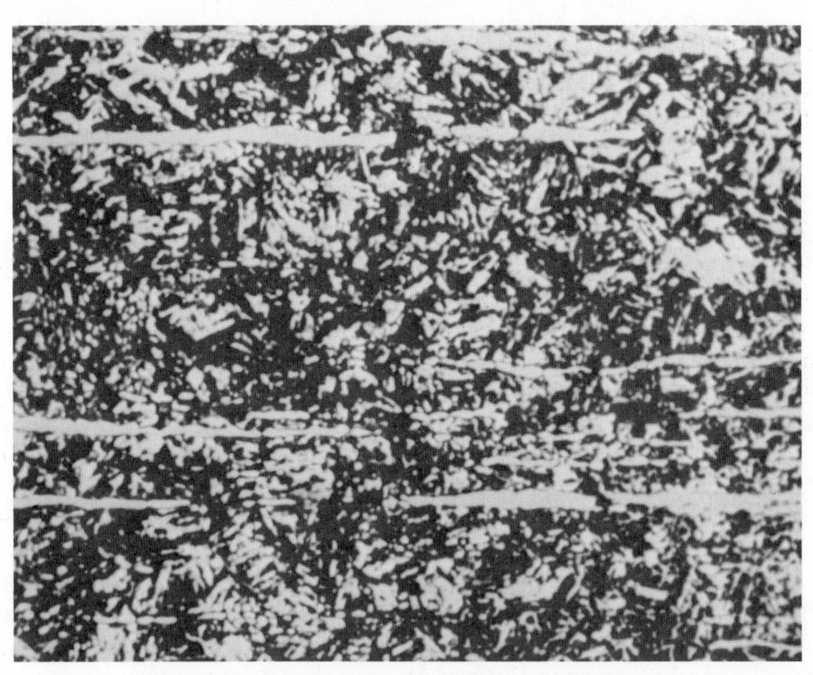

Figure 3.29 Delta ferrite in 431 stainless-steel forging.

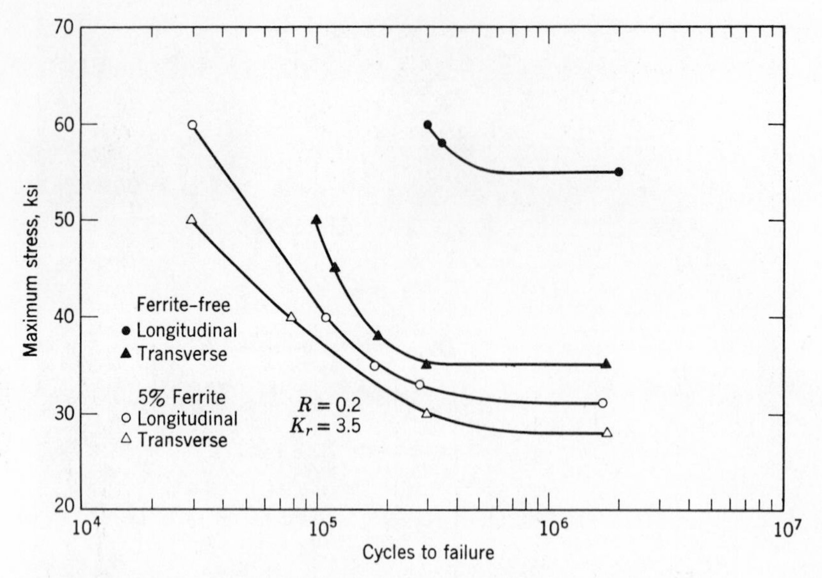

Figure 3.30 Notched fatigue strength of 431 stainless-steel heat-treated to the 180 to 200 ksi range with no ferrite and 5% ferrite.

In this instance the relatively low-strength ferrite phase in the harder martensitic matrix produced a stress concentration at the ferrite-martensite interface. The improper balance of austenite and ferrite stabilizing elements in the alloy composition caused the room temperature retention of the ferrite phase.

The condition seen in Figure 3.29 was caused by segregation that stabilized the ferrite phase. This sharply reduced the fatigue properties of this alloy. Most wrought alloys exhibit a significant variation in mechanical properties with grain direction, that is, longitudinal, long transverse, and short transverse. There are several reasons for this phenomenon, some of which have already been discussed. Material defects, inclusions and segregation all flow with the matrix material during plastic deformation. This plastic deformation may also trigger or effect a subsequent thermal treatment where preferential strain aging or transformation may occur along specific crystallographic planes. Finally, the grain and subgrain structure may also reflect a preferential alignment. As previously indicated, anisotrophy is most pronounced in the short transverse grain direction. It has been shown in tests on 7075-*T*6 aluminum alloy forgings that the endurance limit is reduced by approximately 20% when testing in the short transverse direction as opposed to the longitudinal direction [24]. The effect of directionality on fatigue properties of low-alloy steel is shown in Figure 3.31.

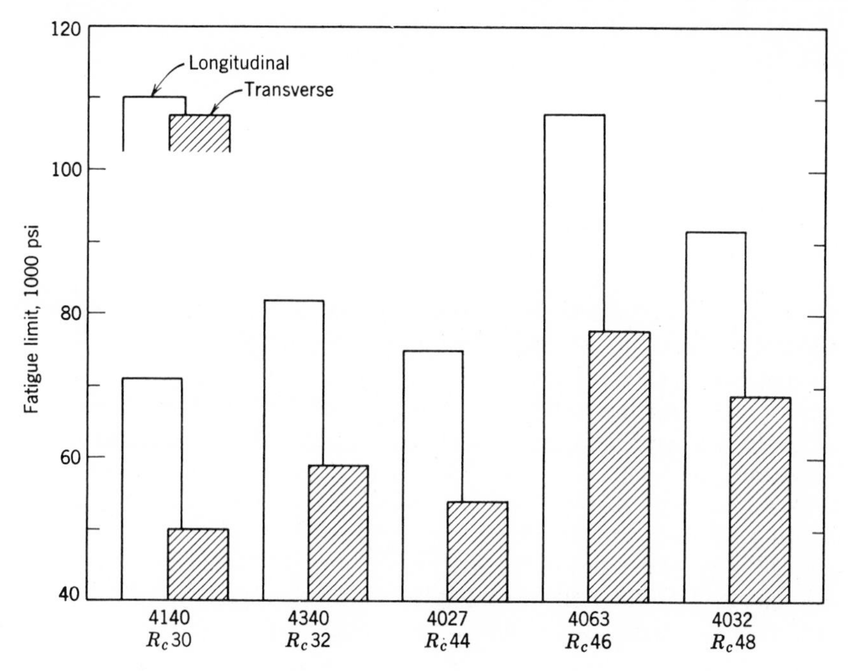

Figure 3.31 Anisotrophic effect on fatigue limit of low-alloy steels.

For many material forms such as sheet, light plate and extrusion, the loading normal to the short transverse direction is low such that fatigue properties in this direction are not critical. For heavy plate, bar, and forgings, however, directionality or anisotrophy can be a crucial design consideration. The flashline in a closed die forging is a typical example. The location of the flashline is usually specified on the engineering drawing along with grain flow in critical areas. The effect of directionality is also seen in the comparison of bolts with upset heads versus machined heads. In the former case tensile loads are taken by material in the longitudinal direction, as opposed to the transverse direction for the machined head. The fatigue characteristics of bolts with upset heads are far superior to those of bolts with machined heads.

3.2.3 *Improper Heat Treatment*

The heat-treatment processes are potentially a source of hazard to a material because at the elevated temperatures encountered many diffusion controlled mechanisms are operative that could harm the integrity of the alloy if not properly controlled. If the furnace atmosphere is not controlled, the chemical composition of the surface layer might be altered and thus produce a low strength or brittle surface skin. Decarburization of low-alloy steel and carburization of low-alloy steels or stainless steels fall in this category, as do interstitial enriched cases in titanium alloys. The diffusion of hydrogen into alloys during heat treatment has long been recognized as a serious problem area. Hydrogen embrittlement of low-alloy steels and titanium alloys can produce disastrous results in subsequent processing or in service. Hydrogen is also suspect in the blistering mechanism in aluminum alloys. With respect specifically to fatigue properties, a brittle case will render an alloy susceptible to surface cracking. The introduction of a shallow crack produces a notch effect, so that the detriment to fatigue is essentially one of a high-surface stress raiser in a layer of material with low fracture toughness.

If the heat-treating temperature is not properly controlled, grain coarsening may occur which lowers fatigue properties for some but by no means all alloys [27]. Overheating of high-strength aluminum alloys is particularly disastrous, since most of these alloys are subject to eutectic melting at temperatures only marginally higher than the solution heat treatment temperature. Eutectic melting results in a gross embrittlement of the alloy coupled with reduction in strength, see Figure 3.32. The difficulties with austenitizing or solution heat treating at too low a temperature are associated with a lack of hardening potential for the subsequent quench and age or temper treatments.

In order to develop full strength, most martensitic and age hardening alloys must be rapidly cooled from high temperatures by quenching into a liquid medium. There are at least two considerations in the quenching process that could affect fatigue properties. High residual quench stresses are built up in most materials and, if the geometry of the part being quenched is highly irregular, the tensile strength of the material may be exceeded at points of high stresses resulting in the not too uncommon quench cracks. On the other hand, if the quenching rate is for some reason retarded, preferential precipitation may occur which adversely affects fatigue properties. The photomicrograph shown in Figure 3.33 illustrates the preferential precipitation at grain boundaries of chromium carbides in 321 stainless steel. The nearly continuous envelope of carbides lowers the corrosion resistance, and fatigue and toughness properties.

3.2.4 Localized Overheating

There are some processes that are capable of developing high localized surface temperatures which are often difficult to detect and occasionally are responsible for a failure in service. Grinding is one of these processes.

The effect of severe grinding on the fatigue properties of high-strength steel is shown in Figure 3.34. The rapid quenching of the material immediately

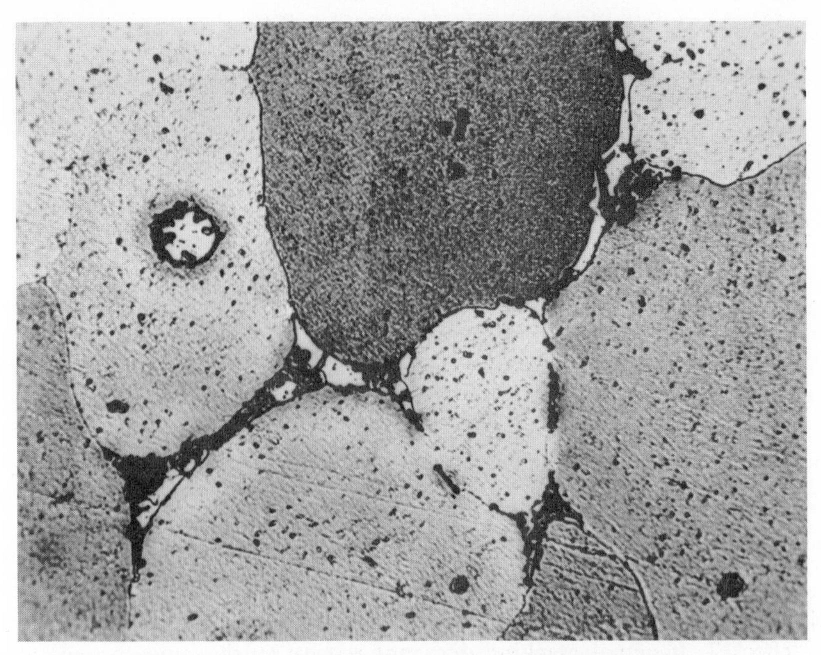

Figure 3.32 Photomicrograph of overheated 7075-T6 material.

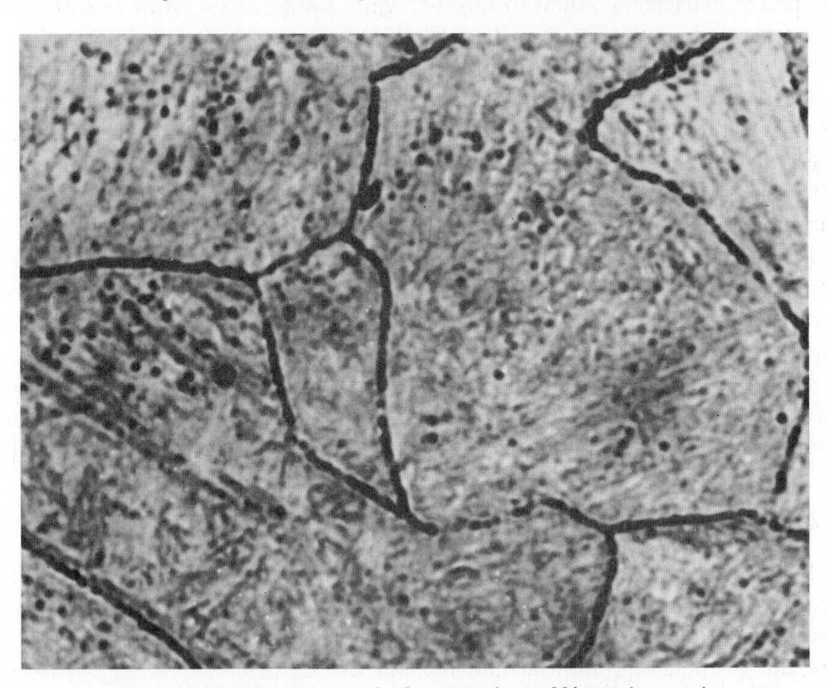

Figure 3.33 Photomicrograph of sensitized type-321 stainless steel.

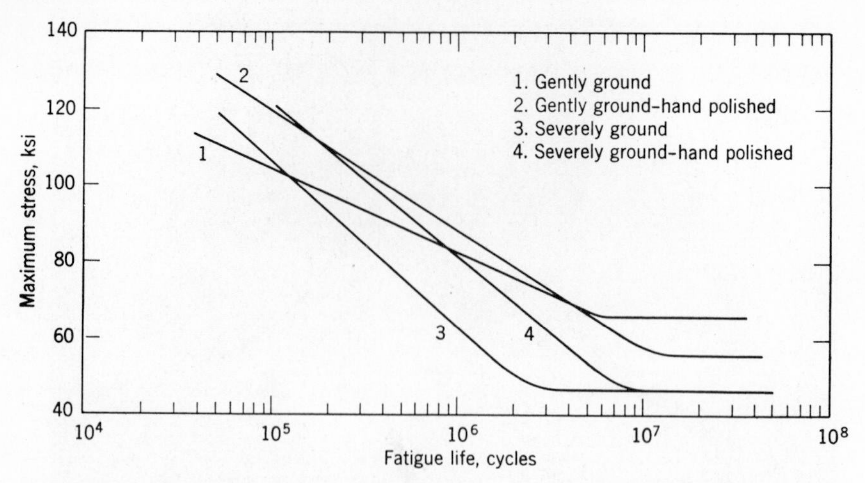

Figure 3.34 S-N curves for flat steel bars (R_c-59 hardness), showing effect of grinding severity [8].

below the grinding wheel by the large mass of cold metal can produce cracks or "check." If actual cracking does not result, brittle, crack-prone, untempered martensite might result or, with lower temperatures, softened overtempered martensite. High-strength steels (for which grinding is most often used) are particularly sensitive to grinding techniques.

In the electroplating processes a plating "burn" sometimes is observed as the result of arcing between the anode and the work piece. Such a burn generally produces a larger heat-affected zone than improper grinding and is often characterized by evidence of surface melting. The potential damage to the substrate is similar to that discussed relative to grinding.

Electrical discharge machine (EDM) is a process of metal removal that employs a spark-erosion principal. The intermittent spark produces highly localized melting on the surface of the workpiece and metal fragments which are swept away by the dielectric coolant. Although the heat-affected zone is shallow, surface cracking and untempered martensite are sometimes observed on martensitic alloys along with eutectic melting and other evidences of overheating in aluminum alloys if the process is not properly controlled [5].

3.2.5 *Corrosion Fatigue*

By strict definition corrosion fatigue is that peculiar interaction of a corrosive environment with an alternating stress field which causes accelerated crack initiation and propagation, possibily where neither the environment nor the stress acting alone would be sufficient to produce a crack. In the practical application of the term the corrosive environment usually serves to

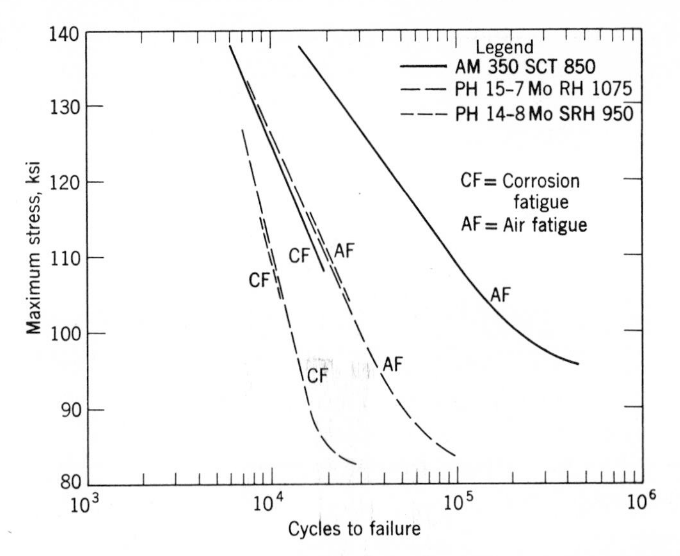

Figure 3.35 Corrosion fatigue and air fatigue S-N curves for precipitation hardening stainless steel tested at room temperature.

introduce stress raisers in the surface in the form of corrosive attack. The irregular surface, in turn, is detrimental to the fatigue properties of the part in a mechanical or geometric sense. For materials susceptible to embrittlement by hydrogen or for parts which are exposed to a fairly continuous corrosive environment with intermittent applications of loading, the cracking mechanism may be some-what more complex. An example of corrosion fatigue testing is presented in Figure 3.35, which illustrates the effect of a corrosive test environment on the fatigue properties of precipitation hardened stainless steels.

3.2.6 *Fretting Corrosion*

The fretting corrosion phenomenon has been defined as that form of damage that arises when two surfaces in contact and nominally at rest experience relative periodic motion. In vacuum or inert atmospheres the process is completely mechanical but in ordinary atmospheres oxidation is also involved. Fretting is potentially dangerous because it can result from extremely small surface movements that often cannot be anticipated or even prevented. Motions with amplitudes as low as 5×10^{-9} in. are sufficient for this mechanism to be operative [31,32].

Soft metals exhibit a higher susceptibility to fretting fatigue than hard metals. Fretting corrosion increases with load-amplitude, number of load cycles, contact pressure and an increase of oxygen in the environment [31,32]. Fretting varies inversely with the relative humidity of the environment. The oxidized particles that accumulate between the fretting surfaces lead to both chemical and mechanical surface disintegration which generate nuclei for fatigue crack initiation. The presence of fretting may reduce fatigue strength by 25 to 50%, depending on the loading conditions [30]. When a part or assembly is known to be critical in fretting, one or a combination of the following factors will be beneficial in reducing or eliminating fretting corrosion:

1. Electroplating critical surfaces.
2. Case-hardening wearing surfaces.
3. Lubricating.
4. Eliminating or dampening vibration.
5. Increasing fastener load or closeness of fit.
6. Bonding elastic material to surface.
7. Excluding atmosphere.

3.2.7 *Analysis of Fatigue Failures*

Fatigue may be defined as the phenomenon leading to cracking under repeated or fluctuating stresses having a maximum value frequently much less than the static tensile yield strength [33]. For components critical in fatigue, designs usually incorporate large safety factors due to the difficulty of estimating the time required to initiate fatigue and the difficulty in crack detection when it does occur. Fatigue failures, in fact, occur in three distinguishable stages. Initially, fatigue damage occurs on a submicroscopic and then microscopic scale consisting of highly localized plastic deformation. This first stage is terminated with fatigue crack nucleation that is believed to involve slip-plane fracture caused by the repetitive (stage one) reversing of the operative slip system in the metal. Stage two includes the period of crack propagation until the point at which sudden final rupture (stage three) occurs.

From a macroscopic point of view fatigue cracking is often a subyield point cracking mechanism. Consequently, fatigue cracks are classified as brittle cracks, characterized by little or no evidence of ductility in the area adjacent to the second stage cracking, that is, devoid of necking or shear lips on the fracture face. In contrast, the final rupture (stage 3) occurs in a ductile manner with the usual evidence of plasticity at fracture. As a result, a useful characteristic in identifying a fatigue failure is evidence of a brittle area of prior cracking leading to a ductile region of final rupture. There are, however, several other fracture mechanisms which propagate in a brittle fashion, such as stress corrosion and hydrogen embrittlement, among others. The true hallmark of a fatigue crack is the striation, the microscopic line of crack arrestment which gives evidence of discontinuous crack propagation. If this is observed in the region of prior cracking, the fatigue cracking mechanism has been identified.

Fatigue striations or families of striations may be observed with the unaided eye (the familar beach marks), macroscopically, with a light microscope or under the electron microscope. For lower strength more ductile materials and for a variety of strength ranges under conditions of low load, high cycle stressing, fatigue striation families are clearly evident with little or no optical magnification, as seen in Figure 3.36. The distinguishable concoidal markings usually represent points of variation in the load environment. For higher strength materials and for high load fatigue the use of an optical or electron microscope is often required. A typical electron microscope fractograph is shown in Figure 3.37.

With the close examination of a fatigue fracture face, much or all of the following information regarding the crack can be determined:

1. Point(s) of crack nucleation.
2. Direction of crack growth.
3. Size of prior crack.
4. Relative magnitude of stress.
5. Direction of loading (axial, bending, reverse bending etc.)

Items 1 and 3 help to define item 4. Generally, when loads are low, only one crack is generated. Conversely, multiple cracking is a sign of high loads. In addition, the ratio of prior crack area to total cross-sectional area gives information about the magnitude of stress at final rupture; and, finally, with the use of an electron microscope, direct measurement of striation spacing offers good insight into the stress environment during crack growth.

The preparation of a metallographic specimen containing a crack is often helpful in identifying the crack mechanism. Fatigue cracks are invariably transgranular or transcrystalline and frequently are branched. Fatigue striations are not evident, however, on profile.

3.2.8 Reworking

The success of any repair or rework procedure is necessarily closely dependent on the analysis of the degrading mechanism. Only with a proper understanding of the cause of failure can a satisfactory permanent rework be accomplished. In the area of service damage caused by fatigue, in-service or engineering test failure of a part usually provides the impetus to rework procedures. In general, these procedures can be separated into two categories for those parts that contain actual cracks and those that are believed to have experienced fatigue damage. In either instance, when a rework procedure is established, it is by the joint efforts of stress, design, product support, and process engineering personnel.

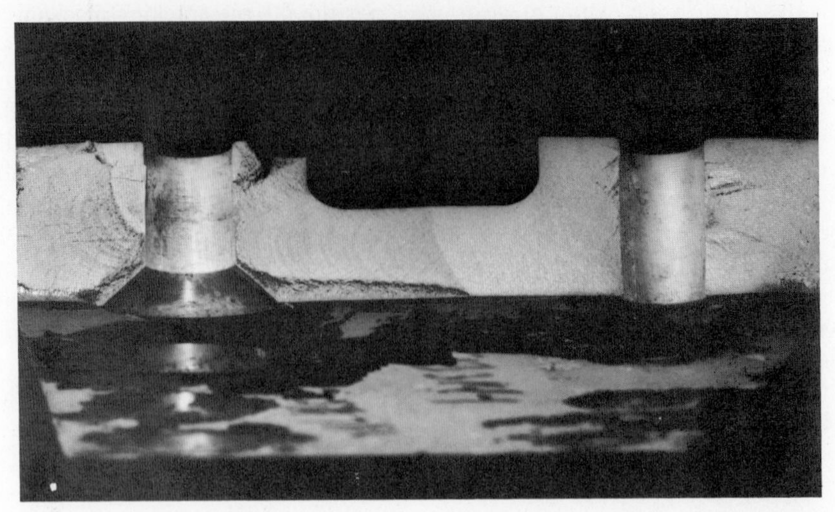

Figure 3.36 Typical fatigue striation under low-load, high-cycle conditions in 7075-T6 plate.

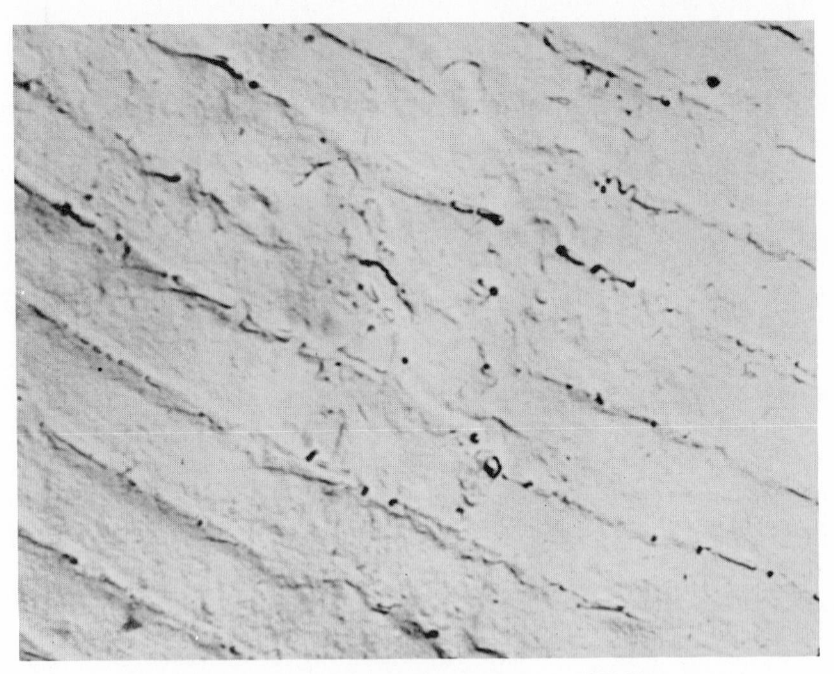

Figure 3.37 Fractograph showing fatigue striation in 7075-T6 extrusion.

Usually, cracked structural parts are scrapped and replaced with a new part. Occasionally, however, because of the location of the crack or other circumstances, such a part is repaired. Repair would consist of removing the crack or blunting its root and supporting or strengthening the damaged area by means of doublers, straps, etc. Care must be taken in dumping the doubler loads so that new sites of fatigue cracking are avoided. Factors such as fretting corrosion, dissimilar metal corrosion, detrimental stress redistribution, access, and practicality are prime considerations in establishing such a rework method.

The reworking of suspected fatigue-damaged, but uncracked, parts takes cognizance of many strength-effecting fatigue parameters described elsewhere in this chapter. Procedures to remove minor stress concentraters, such as increasing a fillet radius, grinding or buffing out coarse tool marks, nicks, and scratches, or radius a sharp edge or corner, are frequently used. If assembly stresses are high, a joint might be shimmed, mismatched surfaces blended, or improved clearance provided. When fretting is contributing to fatigue cracking, a wear strip or lubricant may be inserted between the working surfaces or the fasteners may be tightened to reduce or eliminate motion. Residual compressive stresses are often introduced into the critical areas of fatigue by shot peening or coining operations.

Estimating the depth of fatigue damage on a surface or below the tip of a fatigue crack is difficult and should be experimentally determined for all alloy-forming-heat-treating conditions and the load spectrum. The results of some work done on a 7075-$T6$ plate are given in Figure 3.38. Preliminary

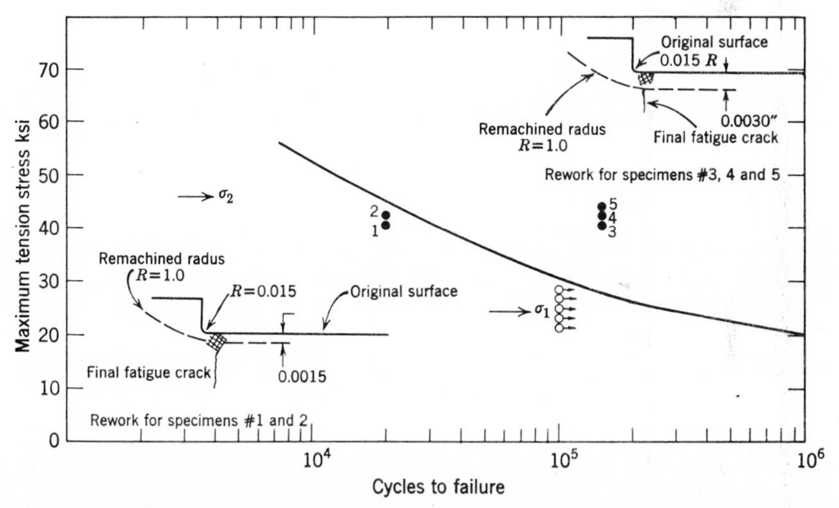

Figure 3.38 Effect of removing surface layer after partial fatigue damage to 7075-T6 plate. Minimum stress in cycle is one-fifth maximum [34].

data, however, indicate that the depth of fatigue damage beneath cracks in high-strength steel may be many times greater than the 0.003 in. for the 7075-*T*6 shown in Figure 3.38.

REFERENCES

[1] R. H. Christensen, "Fatigue Properties of 14S, 24S and 75S Extrusions Cold Bent and Straightened," Tech. Rept. SM-12366, Douglas Aircraft Company, Long Beach, Calif., 1956.

[2] A. E. Ahlman and A. L. Eshleman, 23T Computer Program for Determining Residual Stresses in Bend Areas of Structural Parts, Tech. Rept. LB-31637, Douglas Aircraft Company, Long Beach, Calif.

[3] J. O. Lyst, "The Effect of Residual Strains Upon the Rotating-Beam Fatigue Properties of Some Aluminum Alloys," Tech. Rept. No. 9-60-34, Aluminum Company of America, 1960.

[4] R. N. Hooker, "Surface Finish vs. Fatigue Life for 75 S-T6 Spar Cap Material," Tech. No. Dev-950, Douglas Aircraft Company, Long Beach, Calif.

[5] R. J. Rooney, "The Effect of Various Machining Processes on the Reverse-Bending Fatigue Strength of A-11OAT Titanium Alloy Sheet," WADC Tech. Rept. 57-310, Wright Air Development Center, 1957.

[6] C. S. Yen, "Effects of Zinc-Sprayed Finish and Sandblast Numbering on the Fatigue Life of 75S-T6 Extrusion," Tech. Rept. MP 6216, Douglas Aircraft Company, Long Beach, Calif.

[7] S. Frederick, "Fatigue of Chem-milled Window Frames," Tech. Rept. MP6462, Douglas Aircraft Company, Long Beach, Calif.

[8] H. J. Grover S. A. Gordon, and L. R. Jackson, "Fatigue of Metals and Structures," Bureau of Aeronautics, Department of the Navy, Tech. Rept. NAVAER 00-25-534.

[9] G. E. Nordmark, "Fatigue of Aluminum with Alclad or Sprayed Coatings," *Eng. Notes* **1**, No. 1, 125.

[10] S. Pendleberry and C. S. Yen, "Effect of Chrome Plate Thickness on Fatigue Strength of 4340 Steel," Tech. Rept. MP 20,051, Douglas Aircraft Company, Long Beach, Calif.

[11] F. M. Smith, "The Effect on the Fatigue Life of Aluminum of Anodizing, Sealing and Baking," Tech. Rept. MP 11,364, Douglas Aircraft Company, Long Beach, Calif.

[12] E. C. Supan, "Evaluation of a Chromic Acid Anodizing Process," Tech. Rept. MP 11,938, Douglas Aircraft Company, Long Beach, Calif.

[13] A. Phillips, "Effect of Surface Condition on Fatigue Strength of 7075-T6 Bare Sheet," Tech. Rept. MP 6677, Douglas Aircraft Company, Long Beach, Calif.

[14] T. C. Baumgartner, and R. L. Sproat, "Basic Design and Manufacturing of Aircraft Fasteners for Use Up to 1600°F," Fastener Symposium, 1957–1958, Standard Pressed Steel Company, Jenkintown, Pa.

[15] J. C. Straub, "Choosing the Optimum Method, Intensity and Caverage of Shot Peening, Mechanical Prestressing," 1960 Biennial Meeting, Society of Automotive Engrs., Division 20 on Mechanical Prestressing of Metal, Paper No. SP-181.

[16] R. F. Brodrich, "The Selection of Optimum Conditions of Mechanical Prestressing, Mechanical Prestressing," 1960 Biennial Meeting, Society of Automotive Engrs., Division 20 on Mechanical Prestressing on Metal, Paper No. SP-181.

[17] B. V. Whiteson and C. S. Yen, "Improved Fatigue Life of Chrome Plated Steel by Shot Peening," Tech. Rept. MP 6573, Douglas Aircraft Company, Long Beach, Calif.

[18] T. C. Dvorak and C. S. Yen, "Effect of Hole Expanding on Compression Fatigue Resistance," Tech. Rept. MP 6649, Douglas Aircraft Company, Long Beach, Calif.
[19] J. B. Bowen, "Barrel Tumbling Investigation, Phase IV," Tech. Rept. No. DEV-2581, Douglas Aircraft Company, Long Beach, Calif.
[20] H. J. Noble, "An Evaluation of Fine Particle Abrasive Blasting and Other Methods of Surface Improvement, Mechanical Prestressing," 1960 Biennial Meeting, Society of Automotive Engrs., Division 20 on Mechanical Prestressing of Metals, Paper No. SP-181.
[21] P. W. Kloeris, C. S. Yen, D. V. Del Santo, and J. L. Waisman, "Effect of Internal Flaws on the Fatigue Strength of Aluminum Alloy Rolled Plate and Forgings," Engineering Paper Number 713, Douglas Aircraft Company, Long Beach, Calif.
[22] J. J. Fisher and J. P. Sheehan, "The Effect of Metallurgical Variables on the Fatigue Properties of AISI 4340 Steel Heat Treated in the Tensile Strength Range 260,000–310,000 psi," WADC Tech. Rept. 58-289, Wright Air Development Center, 1958.
[23] D. E. Piper, W. E. Quist, and W. E. Anderson, "The Effect of Composition on the Fracture Properties of 7178-T6 Aluminum Alloy Sheet," AIME Fall Meeting, Philadelphia, October 10, 1964.
[24] D. A. Paul and D. Y. Wang, "Fatigue Behavior of 2014-T6, 7075-T6, and 7079-T6 Aluminum Alloy Regular Hand Forgings," WADC Tech. Rept. 59-591, Wright Air Development Center, 1959.
[25] E. J. Eckert, "Torsional Fatigue Testing of Axle Shafts," ASTM Special Tech. Publication No. 216, "Symposium on Large Fatigue Testing Machines and Their Results."
[26] *Metals Handbook*, Vol. 1, American Society for Metals, Metals Park, Ohio.
[27] C. B. Dittman, G. W. Bauer, and D. Evers, "The Effect of Microstructural Variables and Interstitial Elements on the Fatigue Behavior of Titanium and Commercial Titanium Alloys," WADC Tech. Rept. 56-304, Wright Air Development Center, 1956.
[28] J. O. Almen, "Fatigue Loss and Gain by Electro-plating," *Prod. Engrg.*, **22,** No. 6 (1951).
[29] H. O. Fuchs, "Shot Peening Effects and Specifications," Special Tech. Publications No. 196, ASTM, 1958.
[30] A. M. Freudenthal, "Fatigue Sensitivity and Reliability of Mechanical Systems, Especially Aircraft Structures," WADC Technical Report 61-53, Wright Air Dev. Ctr., 1961.
[31] R. H. Comyn and C. W. Furlani, "Freeting Corrosion, A Literature Survey," AD430908, Defense Documentation Center.
[32] W. L. Starkey, "An Investigation of the Mechanism of the Freeting-Corrosion-Fatigue Phenomenon," Semi-Annual Progress Report to Contract No. AF33(616)-3386, July 20, 1956.
[33] W. F. Payne, "Service Failure Analysis Involving Brittle Cracking Mechanisms, Materials Central," American Society for Metals Metallurgical Education Lectures, May 1962, p-53.
[34] R. H. Christensen, "Fatigue Cracking, Fatigue Damage and Their Detection," *Metal Fatigue*, McGraw-Hill, New York.

ADDITIONAL READING

A. Phillips and M. Russo, "Electron Microfratography," Tech. Rept. PIC-SM-51, Douglas Aircraft Company, Long Beach, Calif.

A. S. Brusunas and E. E. Stansbury, *Proceedings of the Symposium on Corrosion Fundamentals*, The University of Tennessee Press,

G. V. Bennett, "Effect of Elevated-Temperature Exposure on Fatigue Characteristic of 7075-T6 Aluminum Alloy Sheet," Tech. Rept. Met 244, Douglas Aircraft Company, Long Beach, Calif.

H. N. Cummings, F. B. Stulees, and W. C. Schulte, "Research on Ferrous Materials Fatigue," WADC Tech. Rept. 58-43, Wright Air Development Center, 1958.

A. Giuntoli, P. W. Bergstedt, and H. C. Turner, "Materials—Coatings and Finishes—Metal Spray Coatings for Aluminum," Rept. No. 8926-038, General Dynamics/ Convair, San Diego, Calif.

J. G. Weinberg and I. E. Hanna, "An Evaluation of Fatigue Properties of Titanium and Titanium Alloys," TML Report No. 77, July 17, 1957, Columbus, Ohio.

4

Interpretation of Fatigue Data

CHARLES. S. YEN

In this chapter important fatigue data are reviewed and the typical data pattern is interpreted.

Fatigue data provide a record of the basic causes and effects of fatigue phenomena. Two basic causes that contribute to fatigue are repeated stress (which includes stress spectrum and stress distribution) and stressed body which include materials and (structures). The effects include atomic, microscopic, and macroscopic phenomena such as dislocation multiplication, slip formation, crack nucleation, and crack propagation. Engineers are usually interested in macroscopic phenomena, especially crack propagation. A diagram showing the causes and effects of fatigue is shown in Figure 4.1. Further analysis of these factors is provided in Figure 4.2.

Fatigue is basically a property of crystalline solids, and the initiation of fatigue cracking is a problem in dislocation physics. It is the result of the motion and interaction of dislocations activated by cyclic stress. A simple description of the mechanism of fatigue cracking is given in three stages:

Stage 1. During the early cycles of stressing the dislocations originally present in the crystal grains multiply and their density increases sharply. An irregular and disoriented cell wall or subgrain boundary starts to form. The fine slip lines that appear at first in some favorably oriented grains are thin and faint, according to the maximum resolved shear stress law. As the number of stress cycles increases, slip lines become more numerous. Some are localized, some continuously broaden, and the very pronounced become so-called persistent slip bands (i.e., unerasable by electropolishing). Meanwhile, the crystals are distorted and strain-hardened to saturation. Then

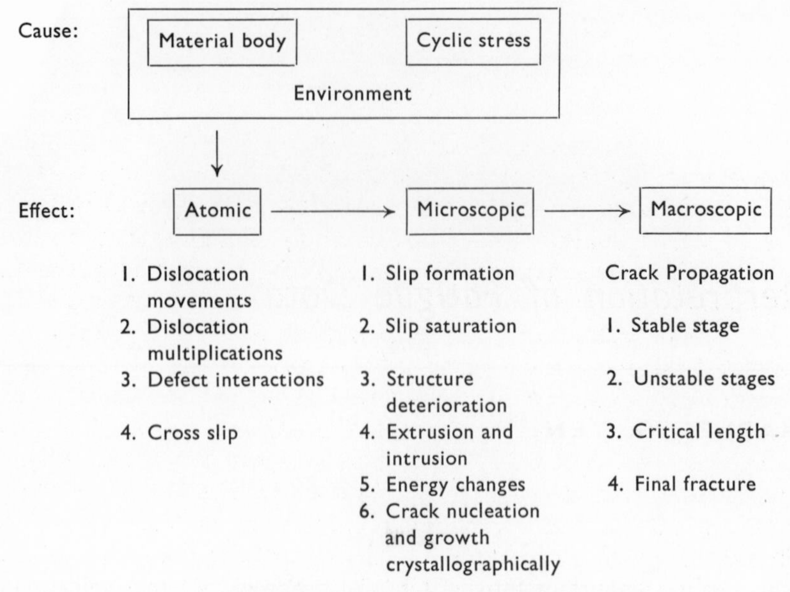

Figure 4.1 Fatigue causes and effects.

dislocation motion in one direction may be fully reversed with the stress. New dislocations and their movements are generated only in some local slip zones in which microstructural features are not the same in both directions of motion. Sometimes annihilation of dislocations or other dislocation mechanisms may lead to relief of lattice strains or strain softening. Strain softening, local recrystallization, overaging, clustering of point defects, and other thermal activation processes are considered to be secondary or side effects.

Stage 2. After the persistent slip bands are fully matured, thin ribbonlike protrusions called extrusions of metal are emitted from the free surface, and fissures called intrusions appear. Both develop along the persistent slip planes. Several dislocation models or mechanisms have been proposed to explain how the extrusions and intrusions are formed. In some of the proposed models dislocation cross slip is considered to be a critical process.

Because the intrusion is the embryo of a crack, so the crack initiates along slip planes according to the maximum resolved shear. Sometimes cracks may initiate at cell walls or grain boundaries.

Stage 3. The crack propagates in a zigzag transgranular path along slip planes and cleavage planes from grain to grain and maintains a general direction perpendicular to the maximum tensile stress. In the fatigue life of a member as much as 99% is spent in the development of fissures into

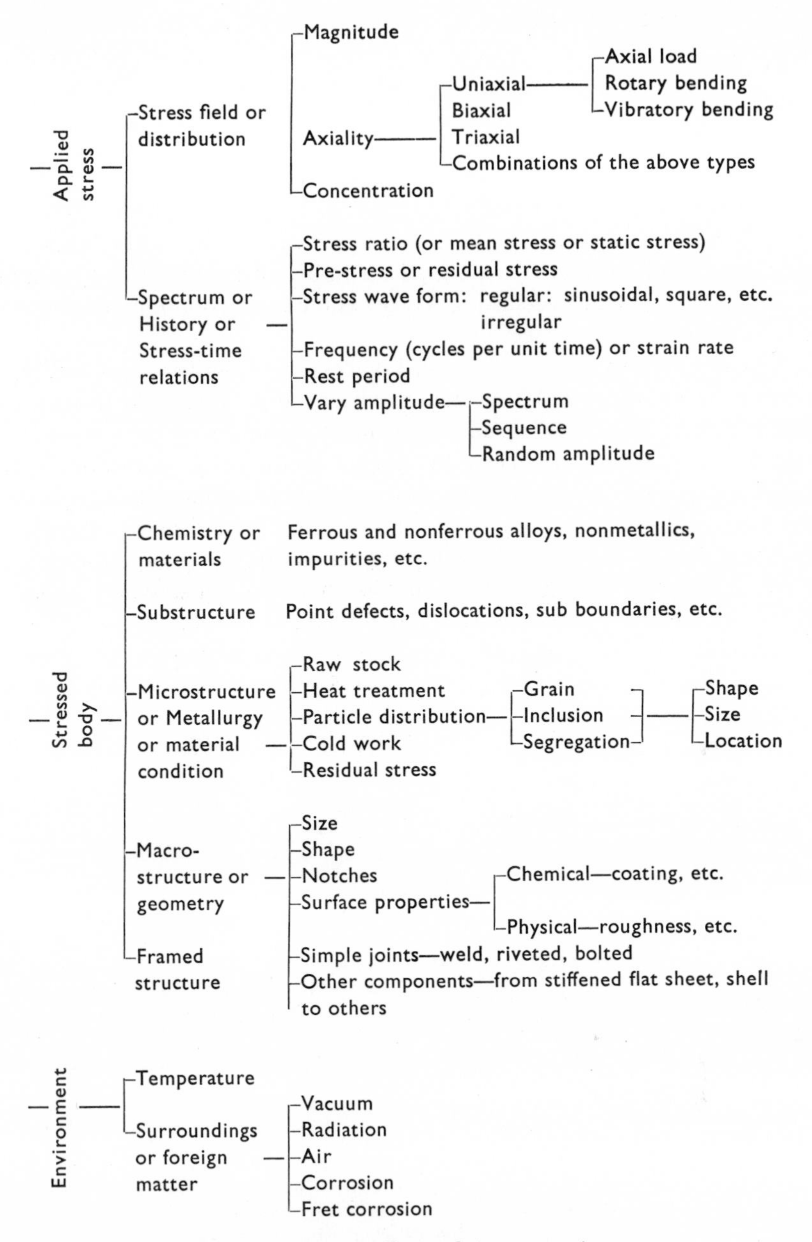

Figure 4.2 Factors affecting fatigue properties.

macroscopic cracks and finally complete fracture. Many factors affecting fatigue properties are those that mainly influence the rate of crack propagation.

4.1 *S-N* DATA

Ever since the beginning of fatigue testing by Wöhler in 1858, *S-N* curves have been the "backbone" of fatigue data: *S* denotes stress amplitude or the maximum cyclic stress; *N* denotes the number of stress cycles to complete fracture. The linear *S* versus log *N* scale is the most common and is used almost exclusively in engineering.

Typical *S-N* curves are shown in Figures 4.3 to 4.5 for axial loading [1] and in Figures 4.6 and 4.7 for rotating beam loading [2]. At the low cycle end the curve is sometimes above tensile strength but is close to its notch strength in notched specimens and to the modulus of rupture in rotating beam specimens. This is because the elementary formulas used to calculate the stresses do not represent a realistic picture of the distribution of stress. In general, *S-N* curves represent progressive structural deterioration and gradual breaking of bonds which may be analyzed as a statistical process in materials under repeated stresses (Section 5.1.2). A similar *S-N* curve for a nonmetallic material—phenolic laminate—shown in Figure 4.8 represents similar structural deterioration [3]. The *S-N* data are discussed further in

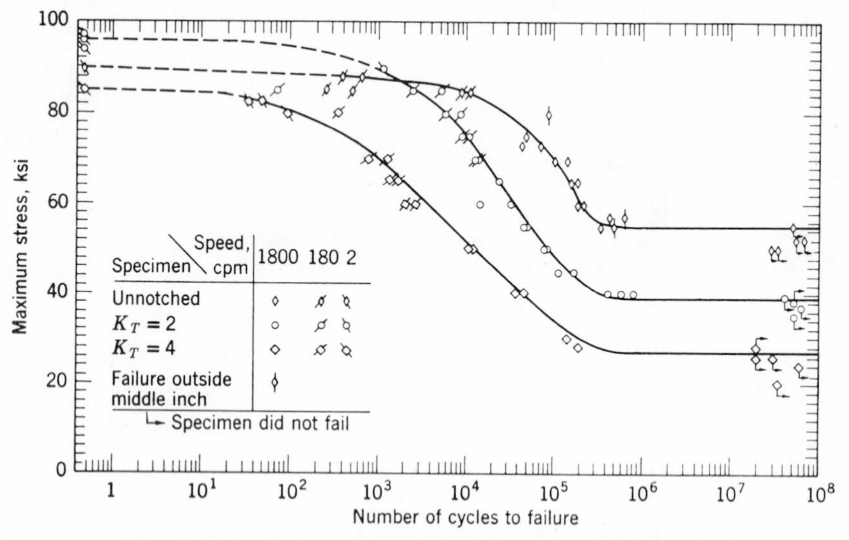

Figure 4.3 Results of fatigue tests on 347 annealed austenitic stainless-steel specimens under axial load at R = 0.

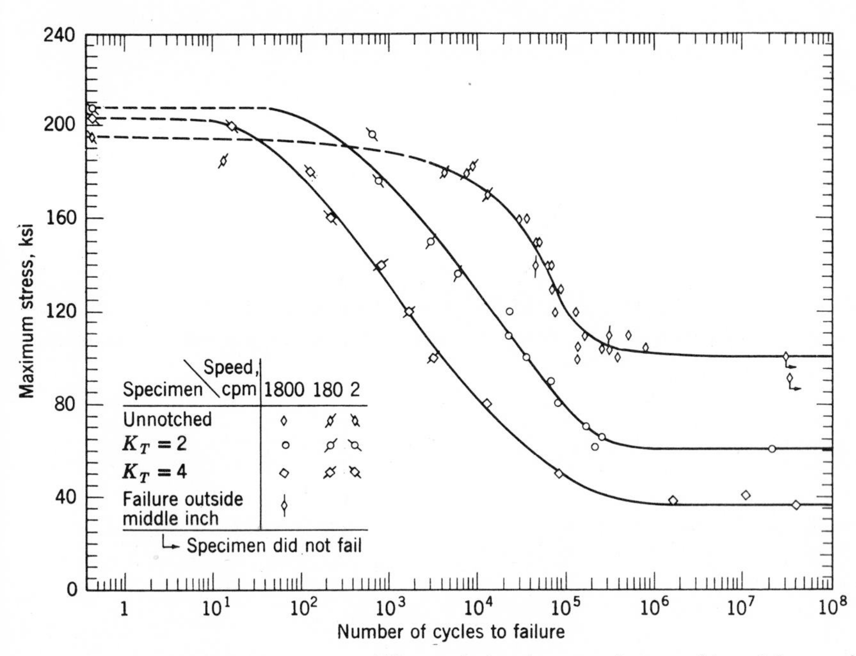

Figure 4.4 Results of fatigue tests on 403 quenched and tempered martensitic stainless-steel specimens under axial load at $R = 0$.

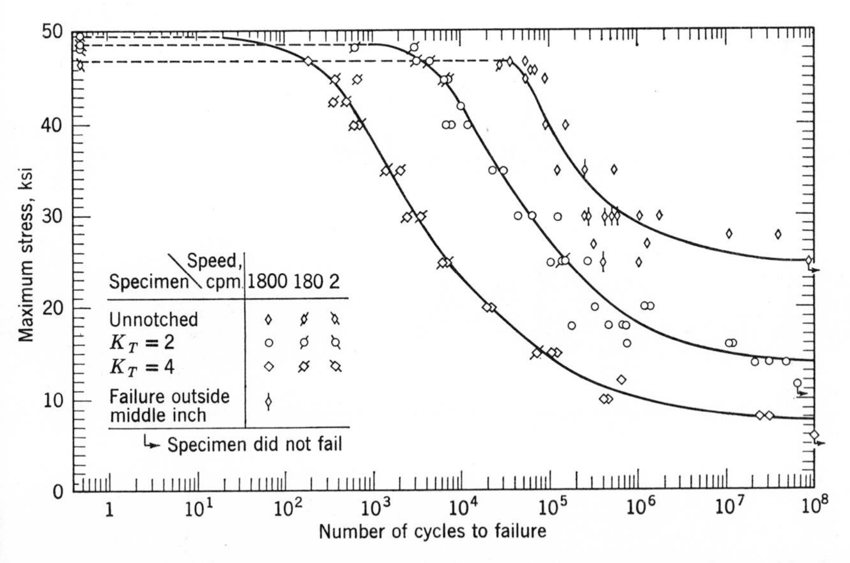

Figure 4.5 Results of fatigue tests on 60 61-T6 Aluminum alloy specimens under axial load at $R = 0$.

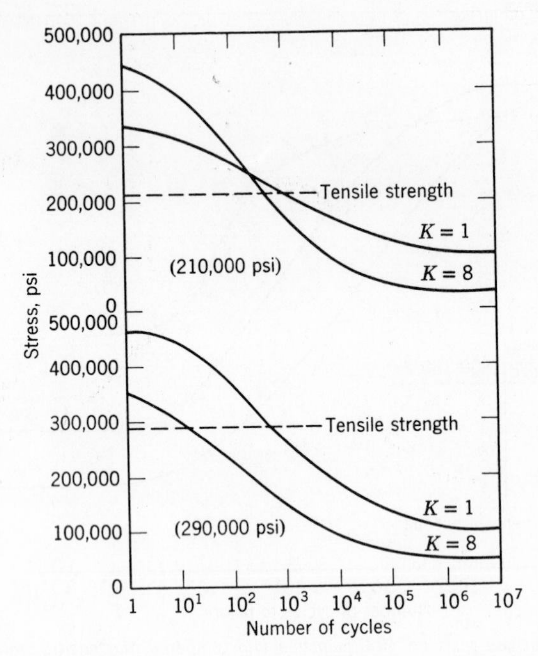

Figure 4.6 S-N curves for smooth and notched rotating beam fatigue specimens of 4340 steel. ASTM [2].

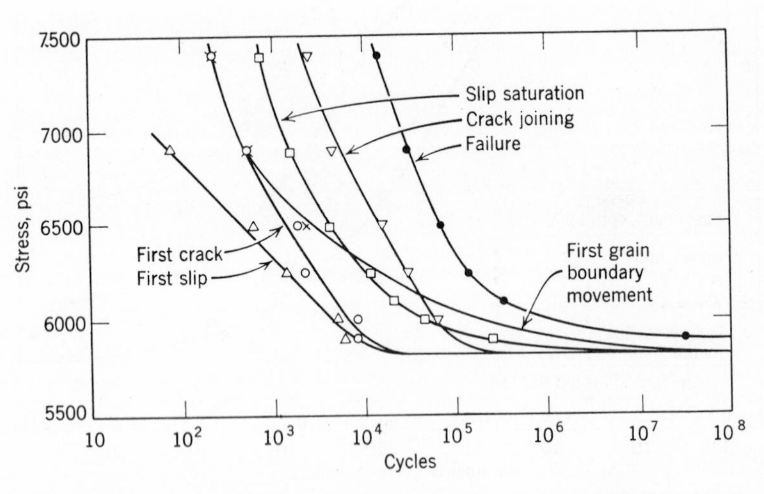

Figure 4.7 Relation between stages of progressive change observed on 0.97% Mg aluminum alloy. ASTM [30].

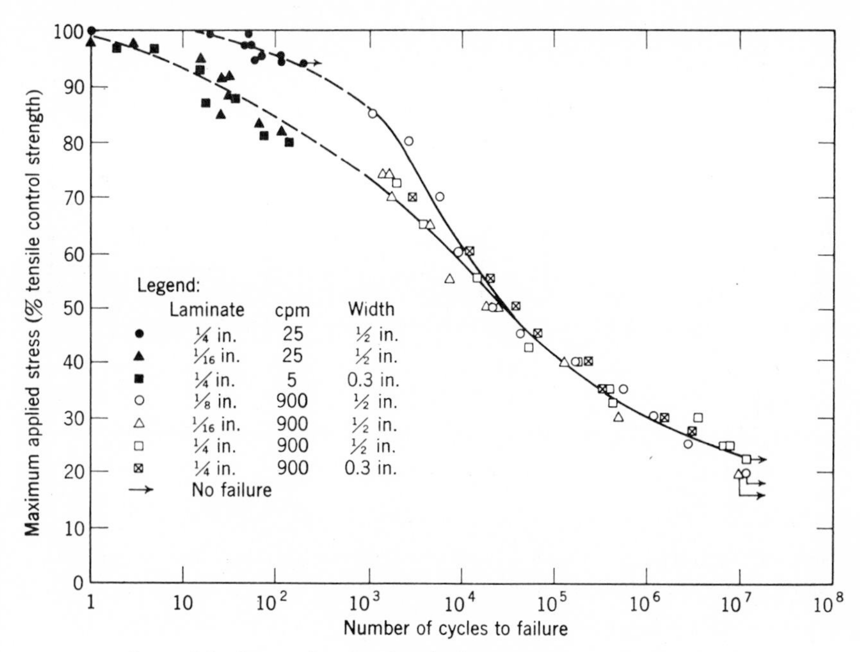

Figure 4.8 Fatigue data for phenolic laminates at room temperature.

later sections; *S-N* data on small rotating-beam polished steel bar specimens is given in Figure 4.9, in which it is shown that the endurance limit is approximately 50% of the tensile strength and the strength at 10^4 cycles is roughly 75% of the tensile strength. This curve may be used sometimes as a rule of thumb for steel when no better fatigue data are available.

Several attempts have been made to find general mathematical laws for the relation between load and life [4,5], and several different equations have been proposed to express the *S-N* relations more or less empirically. Use of these equations will reduce the data to a mathematical form for data reduction, analysis, and standardization of curve-fitting methods. It may also provide some understanding of the *S-N* relations. Two examples are the following:

Weibull proposed an equation in the form [4] of

$$(S - S_e)(N + B)^a = b, \tag{4.1}$$

where S_e is the endurance limit, and a, b, and B are constants. Valluri [5] proposed

$$N = \frac{2}{C} \frac{\ln(\sigma_u/\sigma) \ln[(\sigma - \sigma_i)/K]}{[(\sigma - \sigma_i)/E]^2[(\sigma - \sigma')/\sigma_i]^2}, \tag{4.2}$$

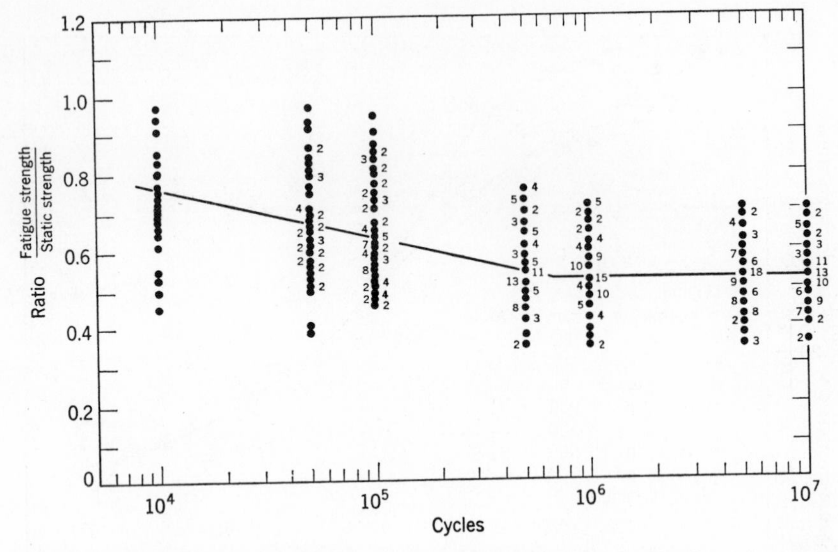

Figure 4.9 Ratio of fatigue strength to tensile strength for various steels. Numbers opposite points indicate number of points at that location.

where σ = maximum cyclic stress,

σ' = minimum cyclic stress,

σ_u = ultimate tensile strength,

σ_i = internal stress = approximate endurance limit,

K, C = material constants,

E = Young's modulus.

4.2 STATISTICAL *P-S-N* DIAGRAMS

When several identical specimens are fatigue-tested at the same stress level, their fatigue lives are generally not the same but may vary or scatter a great deal. When many specimens are tested at several stress levels, the test points will scatter in a band as in Figure 4.10 and 4.11 [6,7]. The centerline of the band is drawn as the mean curve. The meaning of the centerline or mean curve is that 50% of the specimens are expected to fail above this curve and the other 50% below it.

Referring to Figure 4.11, suppose we select a new specimen in the same group to be tested at 17 ksi stress. The life expectancy of this specimen is about 10^4 cycles because it has a 50% chance of failing below 10^4 cycles and a 50% chance of surviving above 10^4 cycles; but if we test this specimen at

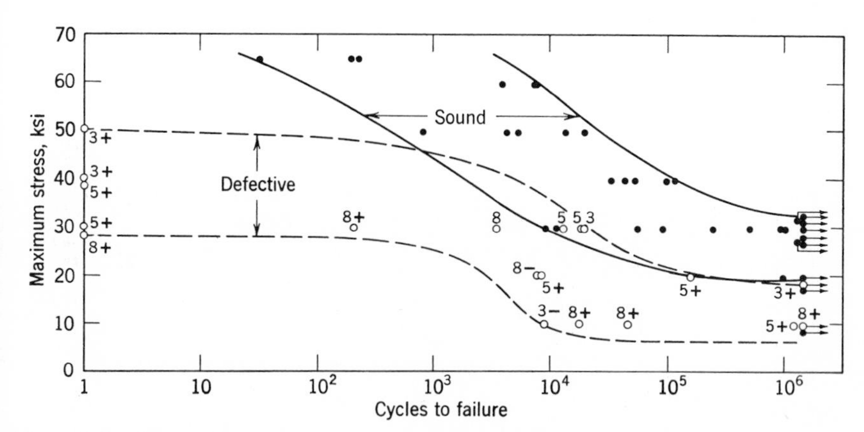

Figure 4.10 Effect of internal defects on fatigue life of 7079-T6 aluminum alloy forged plate stressed in short transverse grain direction. Number adjacent to defect indicates ultrasonic reading in 64ths of an inch.

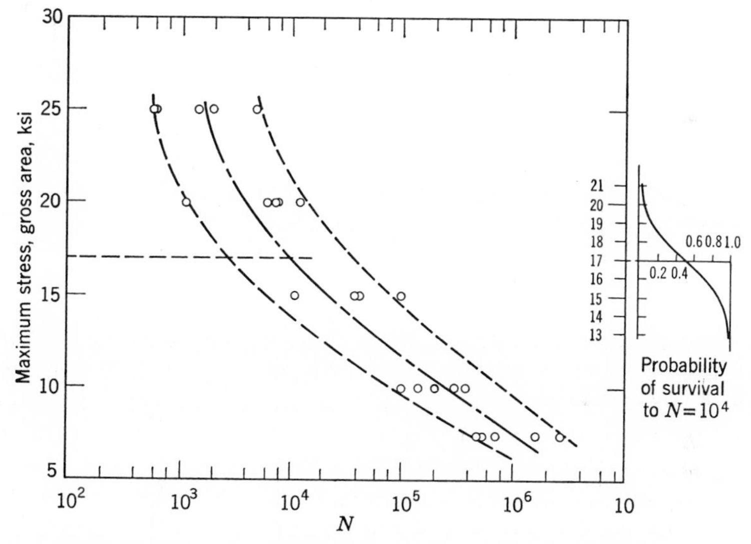

Figure 4.11 S-N scatterband and probability of survival to 10^4 cycles (7075-T6 aluminum alloy with mean stress −10 ksi, $K_t = 3$, specimen 0.040 × 3 × 6, central hole 0.601 D).

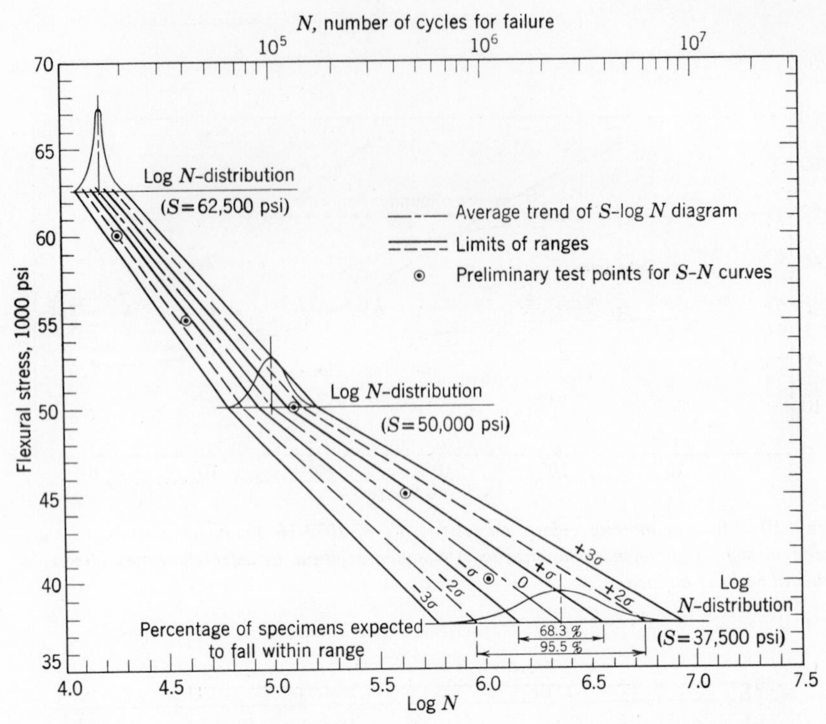

Figure 4.12 Statistical representation of fatigue test results of 7075-T6 aluminum alloy.

18 ksi it probably will fail before it reaches 10^4 cycles.† If we test this specimen at 16 ksi, it will have, say a 70% chance of surviving to 10^4 cycles. The lower the test stress, the higher the probability of survival, as shown in the curve‡ near the right-hand side of Figure 4.11.

In order to draw *P-S-N* diagrams we also need this type of curve or data for other fatigue lives such as 10^5 and 10^6 cycles. When these survival data

† In other words, it will have only a 30% chance of surviving to 10^4 cycles.

‡ There is some evidence to show that this curve is similar to the cumulative normal curve as defined by this equation:

$$P = \int_{-\infty}^{x} \phi(x)\,dx = \int_{-\infty}^{x} \frac{1}{\sqrt{2\pi}} e^{-x^2/2}\,dx \tag{4.3}$$

where P = probability of survival, $y = \phi(x)$ is the normal curve, x = the number of units of standard deviation from the mean. The property of this curve is well-known. If we assume that this survival curve for any given fatigue life is a cumulative normal curve, it is easy to draw a *P-S-N* diagram according to the standard statistical procedure. The procedure used in *A.I.A. Aircraft Fatigue Handbook* (Section 5.5) [8] is an example. It should be noted that there are several other methods, based on different assumptions and different statistical theories, of drawing *P-S-N* diagrams (Section 5.3 and [9, 10, and 11]).

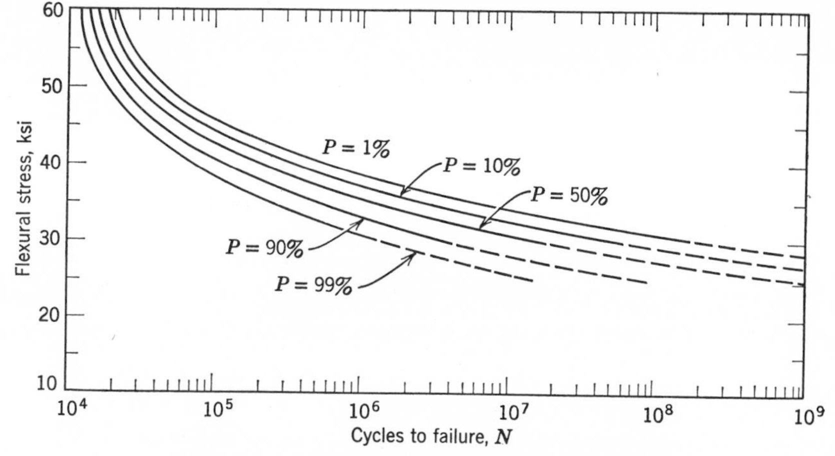

Figure 4.13 P-S-N diagram for aluminum alloy 7075-T6.

for several fatigue lives are plotted, the points with the same probability of survival, say P, are connected as a curve. This curve is the S-N curve for the probability of survival P.

This description is based on the survival data for a given fatigue life (as used in Section 5.5 and [8]). If the survival data for a given stress level are used (as in [9] and [10]), the P-S-N diagrams in Figures 4.12 and 4.13 will be obtained.

4.3 LOW-CYCLE AND HIGH-CYCLE FATIGUE

A complete S-N curve may be divided into two portions: the low-cycle range and the high cycle range. There is no sharp dividing line between the two. We might arbitrarily say that 0 to about 10^3 or 10^4 cycles is low cycle and about 10^3 or 10^4 cycles to 10^7 or higher is high cycle.

Until World War II little attention was paid to the low-cycle range, and most of the existing fatigue results were for high cycles only. Then it was realized that for some pressure vessels, pressurized fuselages, mechanisms for extending landing gears and controlling wing flaps, missiles, spaceship launching equipment, etc., only a short fatigue life was required. Consequently, the low-cycle fatigue phenomena began to gain attention.

The initial portion of an S-N curve is usually horizontal and flat. The flat portion is shorter for notched than for plain specimens, as shown in Figure 4.14*a* [12], 4.18, and 4.19 [13]. The difference in the initial behavior between notched and plain specimens appear to be related to the stress-strain curves

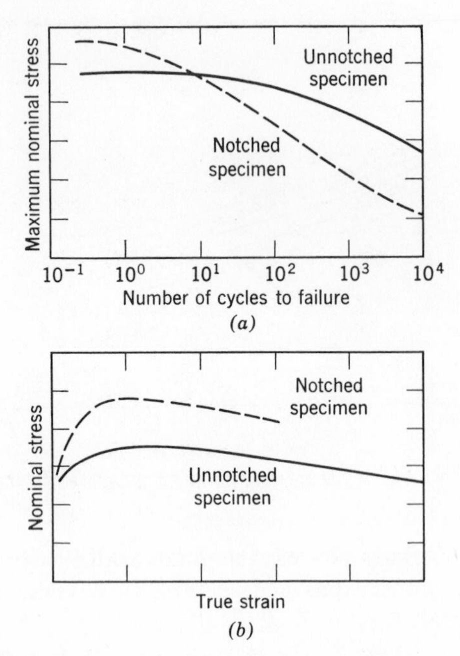

Figure 4.14 (a) Typical S-N curves for constant-load tests; (b) typical stress-strain curves.

of the members as shown in Figure 4.14*a*,*b*. When the cycles are low, the material fatigue strength is closer to the static strength. Consequently, when the notched static strength is higher than the plain tensile strength, the notched fatigue strength at the upper end of an *S-N* curve is also higher than the plain fatigue strength. When the cycles are increased, notched fatigue strength drops rapidly to a lower level than the unnotched fatigue strength.

Another point is worth noting in comparing Figure 4.14*a* with Figure 4.14*b*. In both static and fatigue curves, in the vicinity of the top of the curves, the change in the slope or curvature for the notched curve is often greater than for the unnotched. Since the notched specimens show reduced *plastic* strain in static test and shorter *life* in fatigue test, it is indicated that plastic strain is related to fatigue life, and the plastic strain seems to be a more sensitive measurement of life than the nominal stress. Further, in constant deformation fatigue tests a straight-line relationship exists between the logarithmic values of either the maximum plastic strain or the range of plastic strain and the lives of the members, as shown typically in Figure 4.15 [14]. The line in this figure can be expressed by this equation known as Coffin's law:

$$\Delta\epsilon_p\sqrt{N} = C, \tag{4.4}$$

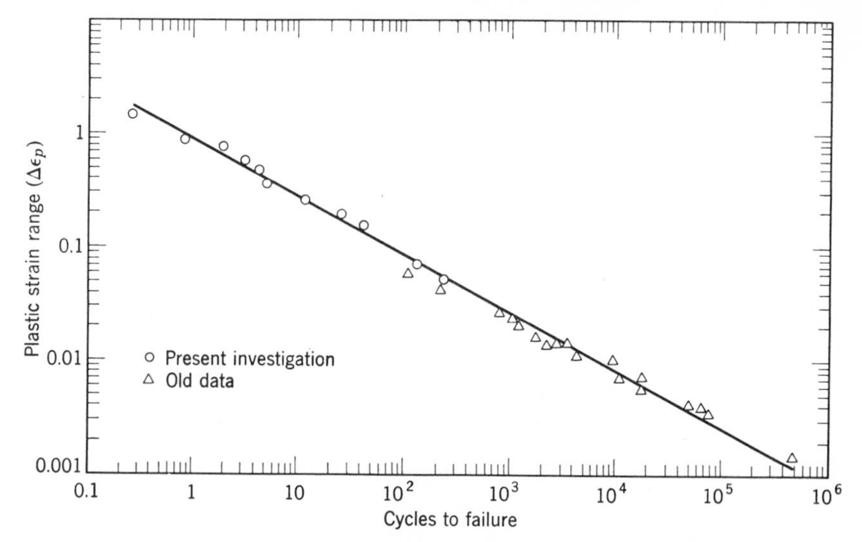

Figure 4.15 Plastic strain range versus cycles to failure, 2S aluminum.

where the constant C may be related to the reduction of area R in static tension test.

$$C = \tfrac{1}{2} \ln \frac{100}{100 - R}. \tag{4.5}$$

Coffin's law has been interpreted by Gilman [15] in terms of a dislocation model and by Grosskeutz [29] in terms of a continuum model.

Many reports have used the static tensile strength as the upper limit of the S-N curve, located arbitrarily as $\frac{1}{4}$, $\frac{1}{2}$, $\frac{3}{4}$, or 1 cycle. It should be noted that the tensile strength is usually obtained from a static tension testing machine running at a small strain rate, whereas fatigue test results are obtained in fatigue machines at higher strain rate.

The failure mechanism in the low-cycle range is close to that in static tension, but the failure mechanism in the high-cycle range is different and may be termed "true fatigue." A comparison between the two mechanisms is made in Figure 4.16 [15].

4.4 EFFECT OF MEAN STRESS AND COMBINED STRESS

The stresses created in the laboratory for fatigue study are usually sinusoidal or vibrational. Vibrational stresses are usually expressed in terms of a pair of variables, such as the mean stress S_m and the amplitude S_a, as shown

	Low-cycle	High-cycle
Internal stresses and strain hardening	High	Low
Net sum of plastic flow $\sum \epsilon$	Macro size	Micro size
Gross sum of plastic flow $\sum \lvert\epsilon\rvert$	Small	Large
X-ray disorientation	Large	Small
Slip	Coarse (10^3–10^4 A)	Fine (10 A)
Slip plane distortion	Normal	Persistent
Crack origin	Interior	Surface
Crack path	Along maximum shear	Cross maximum tensile stress
Fracture	Delayed static	Structure deterioration

Figure 4.16. Comparison of low-cycle and high-cycle fatigue.

in Figure 4.17. The fatigue *S-N* data are defined by a pair of stress variables, as in Figures 4.18 and 4.19, in which the maximum stress $S_{\max}$ and the ratio R are used, R being the ratio of minimum to maximum stress. Figure 4.20 shows clearly the relation between different pairs of stress variables [16]. The same data (Figures 4.18 and 4.19) can also be plotted in terms of S_m and S_a, as in Figure 4.21*a*, *b* [13].

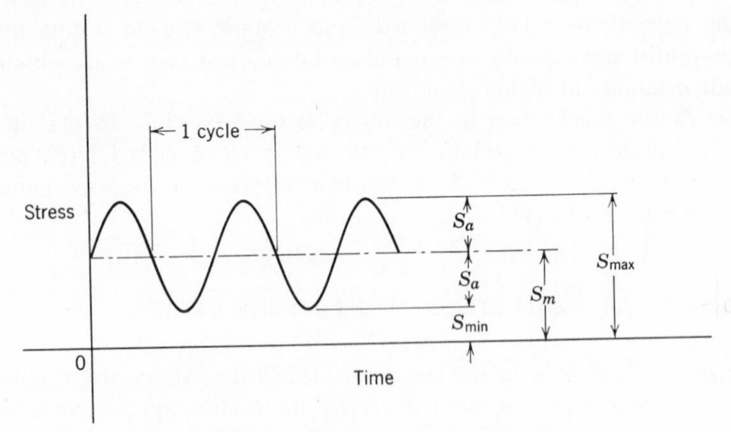

Figure 4.17 Fatigue stress components.

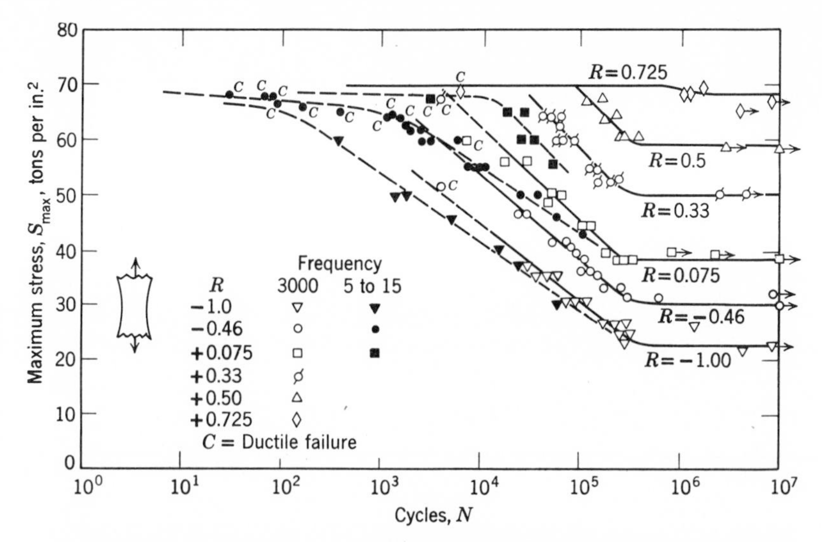

Figure 4.18 S-N curves for various stress ratios, unnotched specimens. ASTM [13].

Figures 4.20 and 4.21 show the effect of mean stress, which may sometimes be termed static or superimposed stress. When two or more nonparallel forces act on a body, two- or three-dimensional stresses are created. So we have combined or complex stresses. The amount of fatigue data of different combinations or states of stresses are overwhelming. Fortunately there are some general rules on the fatigue effect of combined stress to be used as a

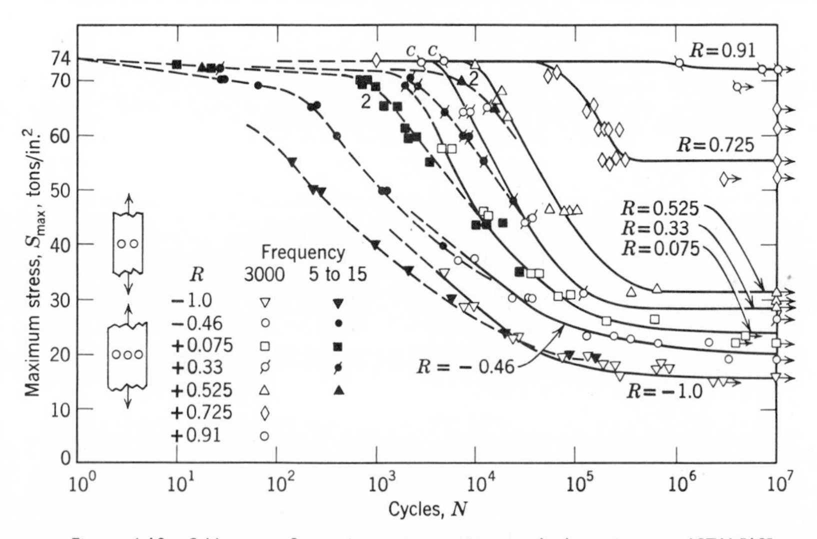

Figure 4.19 S-N curves for various stress ratios, notched specimens. ASTM [13].

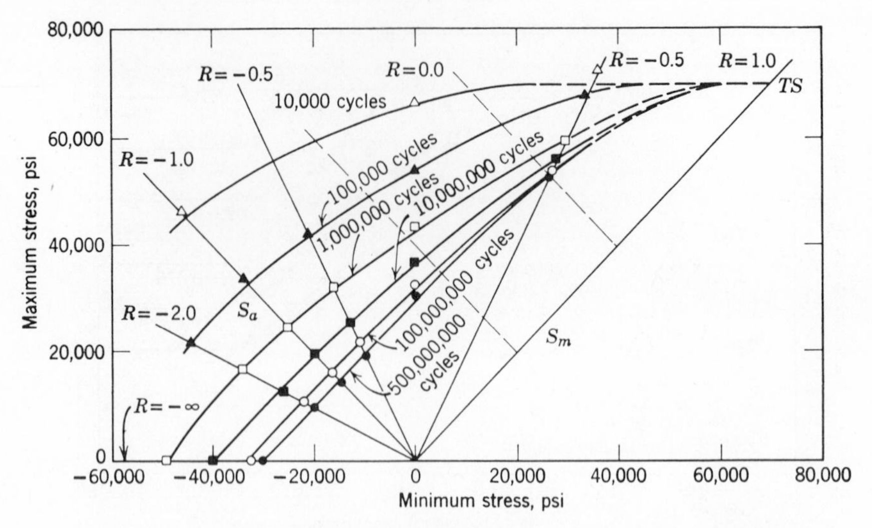

Figure 4.20 Modified Goodman diagram for 2014-T6 aluminum alloy. ASTM [16].

guide. An outstanding rule that explains not only the effect of combined stresses but also the effect of mean and residual stresses is proposed by Sines [17].

Sines proposes that the permissible alternation of the octahedral shear stress is a linear function of the sum of the orthogonal normal static stresses as long as the maximum stress is below the yield strength.

The mathematical equation for this rule is

$$\tfrac{1}{3}\{(P_1 - P_2)^2 + (P_2 - P_3)^2 + (P_3 - P_1)^2\}^{1/2} = A - \alpha(S_x + S_y + S_z + R_{x'} + R_{y'} + R_{z'}), \quad (4.6)$$

where P_1, P_2, and P_3 = the amplitudes of the alternating principal stresses,
S_x, S_y, and S_z = the orthogonal static stresses,
$R_{x'}$, $R_{y'}$, and $R_{z'}$ = the orthogonal residual stresses,
A = a material constant = the octahedral shearing fatigue strength (alternating amplitude) with no superimposed static and residual stresses,
α = a material constant = the proportional factor between static and residual stresses and the permissible alternating octahedral shearing stress.

On the effect of mean stress, in addition to Sine's linear equation, there are Goodman's linear law and Gerber's parabolic law proposed for the uniaxial

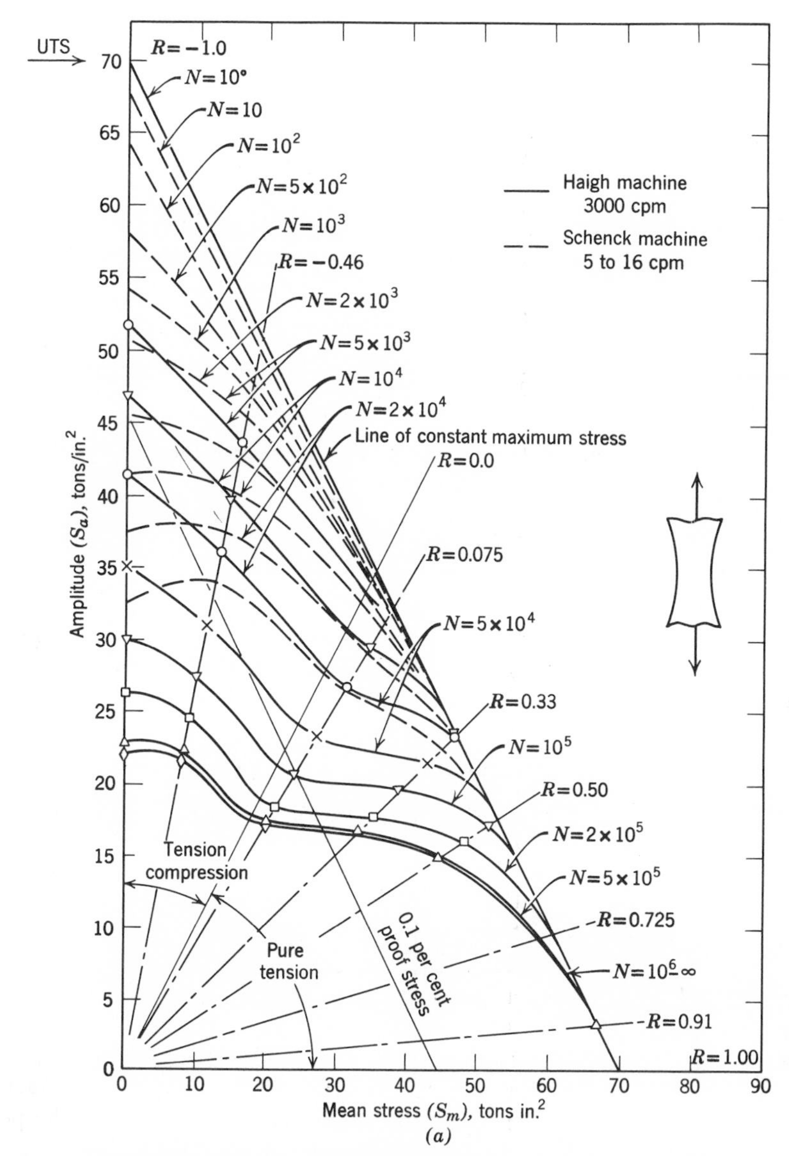

Figure 4.21a S_a-S_m *curves for fatigue specimens, unnotched. ASTM [13].*

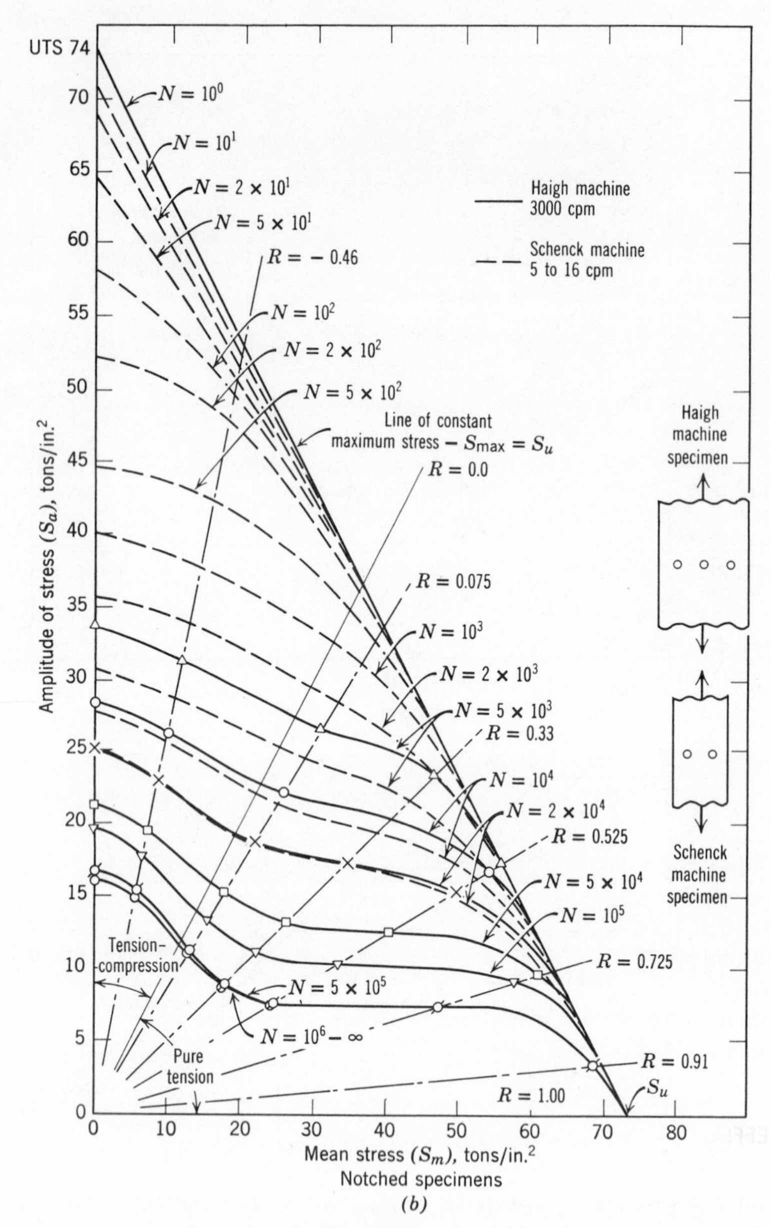

Figure 4.21b S_a-S_m *curves for fatigue specimens, notched. ASTM [13].*

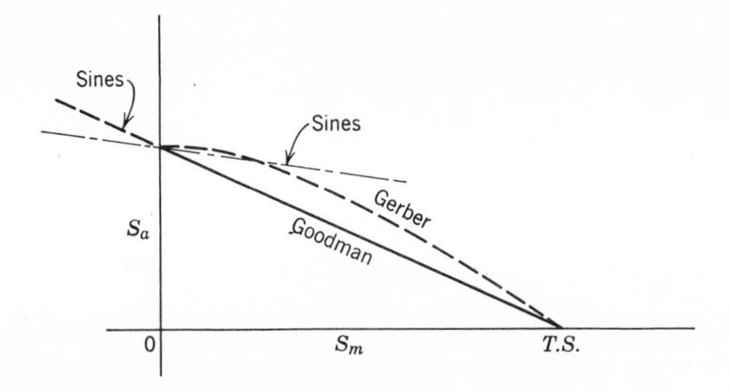

Figure 4.22 Comparison of Sines's, Gerber's, Goodman's relations.

stress system. Their equations and curves are compared here and in Figure 4.22.

Goodman's equation: $$S_a = S_e - S_e\left(\frac{S_m}{T}\right), \tag{4.7}$$

Gerber's equation: $$S_a = S_e - S_e\left[\frac{S_m}{T}\right]^2 \tag{4.8}$$

Sines' equation: $$S_a = S_e - CS_m, \tag{4.9}$$

where S_a = fatigue strength in terms of the stress amplitude,
S_m = superimposed mean stress,
S_e = endurance limit when $S_m = 0$,
T = ultimate tensile strength, and
C = material constant.

In Sines' equation (4.9), S_e corresponds to A in (4.6) and C corresponds to α.

Goodman's and Gerber's relations cover the maximum stress levels up to the ultimate tensile strength, but Sines' relation covers the maximum stress level only up to the yield strength. Sines' relation can be made to coincide with Goodman's and can also be called a linear approximation of Gerber's parabola under the yield strength.

4.5 EFFECT OF STRESS CONCENTRATION

In all types of service failure of aircraft parts and machine parts fatigue failure is a major consideration; nearly all fatigue failures are stress concentration failures. The fatigue crack usually originates at a point at which

Figure 4.23 Fatigue crack starts at ½ × 0.028 S.S. tube O.D.

the stress is highest. The stress peaks often occur at the surface; therefore we may claim that the fatigue failures are surface failures and that the fatigue damage before cracking is only skin deep. Examples of fatigue cracks starting at the surface are shown in Figures 4.23 and 4.24. Depth of fatigue damage beyond crack tip (the reduction in fatigue life of the material in front of the crack tip) has been found to be less than 0.020 or 0.030 in. [21,22].

There are several reasons why stress peaks and fatigue damage often occur on the surface:

1. The applied load is rarely truly axial; some bending or twisting moment nearly always exists. Further, the part often has stress raisers on the surface. As a result, the stress distribution is never uniform, and the maximum stress is usually located at the surface. Even if the applied stress is uniform, the residual stress often reaches a maximum at the surface (Figure 4.25) [18] and [19].
2. The corrosion and erosion damages always roughen the smooth surface and introduce pits and notches that, in turn introduce stress concentrations when under load.
3. Generally speaking, almost all types of fracture—fatigue, creep, ductile, or brittle—start from a microscopic local region with shear stress concentration. In fatigue fractures microscopic plastic deformation occurs more easily on the surface because of less restraint than in the interior. It roughens the

Figure 4.24 Fatigue crack starts at ½ × 0.028 S.S. tube I.D.

surface and introduces extrusions and intrusions. The intrusion is a sharp notch and a stress raiser that starts the crack.

Because of the vulnerable free surface, a fatigue weakening or strengthening process is usually a surface effect, especially near the stress concentration area, and any new surface condition should be checked for its fatigue effect.

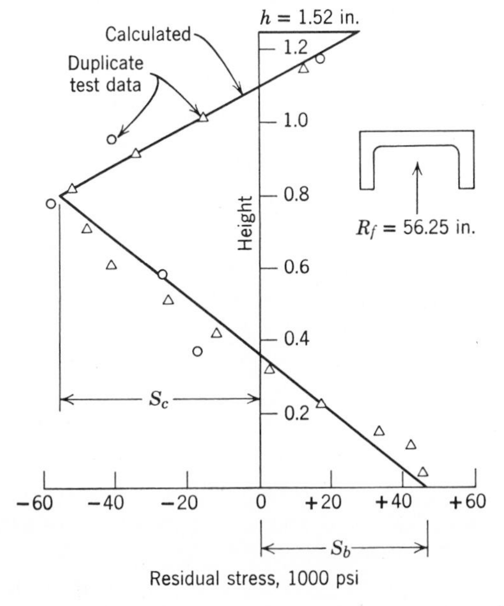

Figure 4.25 Residual stress gradient due to forming operations, 7075-T6 aluminum alloy.

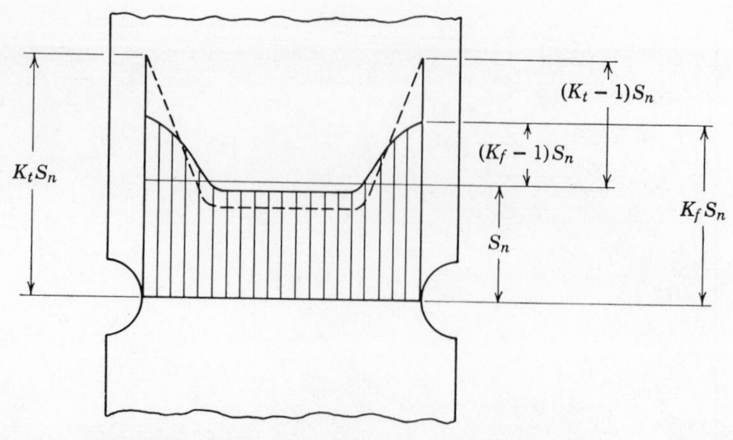

Figure 4.26 Lowering of peak stress by plastic action.

Sometimes internal defects and large second-phase particles can also introduce stress concentration at the interface [6].

To study the stress concentration effect on fatigue, we usually design and test specimens with notches or holes. The ratio of the peak stress in the notched specimen to that of a corresponding unnotched specimen is called the theoretical stress concentration factor K_t. The peak stress at the root of notch may be determined mathematically, photoelastically, or by X-ray measurement. The peak stress in the unnotched specimen is always calculated from an elementary stress formula such as $S = P/A$, $S = Mc/I$, or $S = Tc/J$.

Because the peak stress is raised by a factor of K_t due to the notch, it may be expected that the strength of the notched specimen will be reduced by the factor of K_t. In fact, the reduction of strength of notched specimens tends, indeed, to increase with the factor K_t but is usually not so large as K_t. The ratio of the nominal fatigue strength of an unnotched specimen to that of a notched specimen is called the strength reduction factor, the fatigue notch factor or the effect stress concentration factor and is denoted by K_f or K_e. The factor K_f is usually smaller than K_t.

To explain why K_f tends to be smaller than K_t it is claimed that the theoretical peak stress K_tS_n is lowered to K_fS_n by plastic flow (Figure 4.26), where S_n is the nominal stress in the corresponding unnotched specimen as found by an elementary formula. The increment of the plastic stress K_fS_n over the nominal stress S_n, divided by the increment of the theoretical or elastic stress K_tS_n over the nominal S_n, is defined as an index to the notch sensitivity q. Thus

$$q = \frac{K_fSn - Sn}{K_tSn - Sn} = \frac{(K_f - 1)Sn}{(K_t - 1)Sn} = \frac{K_f - 1}{K_t - 1} \tag{4.10}$$

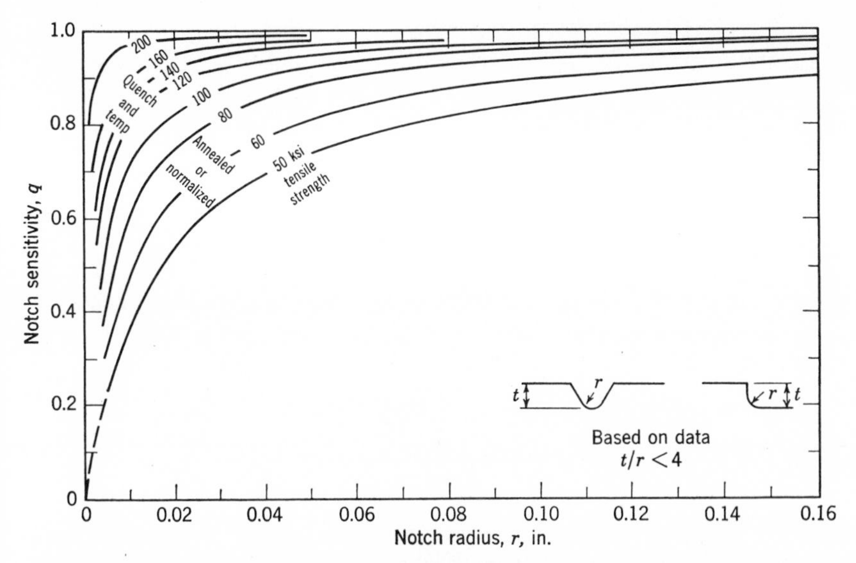

Figure 4.27 Curves q versus r for steels, bending or axial loading. (Permission granted from "Metal Fatigue," by Sines and Weisman, McGraw-Hill, 1959.)

For engineering design use Petersen relates notch sensitivity with notch radius and tensile strength for steels as shown in Figure 4.27. Yen and Dolan have made a thorough review of this subject in [20].

4.6 RANDOM LOADING AND CUMULATIVE DAMAGE

As pointed out earlier, the fatigue strengths or lives of many specimens or parts are always scattered over a wide band (Figures 4.10 to 4.13). Similarly, the service-load magnitudes also occur in several different values as shown in Figure 4.28 [23].

The aircraft service loads contributing to fatigue include many different situations:

1. Gust load in clear air and storm.
2. Maneuver loadings.
3. Landing impacts.
4. Taxiing and ground handling.
5. Ground-air cycle.
6. Buffeting, stalling, and supersonic shock wave instability.
7. Acoustical noise due to propeller tip or jet and aerodynamic boundary layer noise.

In these loadings the sequence of varying load magnitudes occurs more or

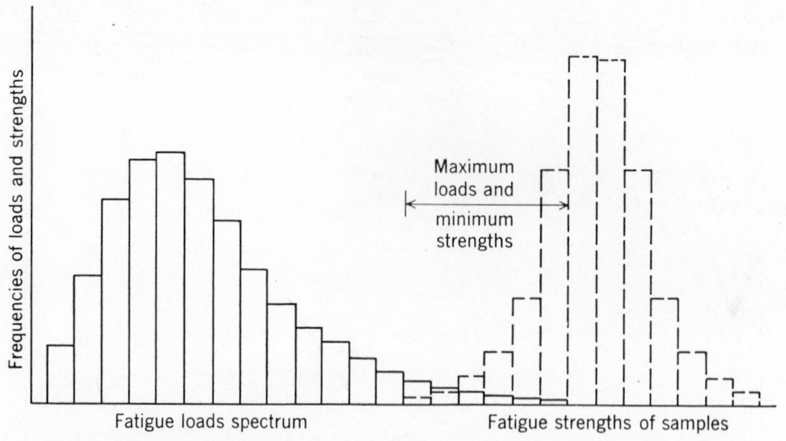

Figure 4.28 Histograms of fatigue loads and strengths of a structural component.

less in a random manner. As an example, the random sequence of the high and low flight gust loads may be represented in Figure 4.29 [7].

In the laboratory it is a standard practice to test each specimen at constant load amplitude and obtain an S-N curve, but in service the load on each part critical in fatigue varies a great deal. How should the constant amplitude laboratory data or the S-N curve be used to design the part to resist the service load with variable amplitude? A simple answer is to use the following equation proposed by Palmgren and later reprosposed by Miner.

$$\frac{n_1}{N_1} + \frac{n_2}{N_2} + \frac{n_3}{N_3} + \cdots + \frac{n_m}{N_m} = 1 \tag{4.11}$$

or simply

$$\Sigma \frac{n}{N} = 1, \tag{4.12}$$

where $n_1, n_2, n_3 \cdots n_m$ is the number of cycles applied at the different stress levels $S_1, S_2, S_3 \cdots S_m$, respectively. Each n causes partial damage and the

Figure 4.29 Random sequence of flight loads versus times.

total Σn causes complete failure; $N_1, N_2, N_3 \cdots N_m$ is the number of cycles in which each N causes complete failure at the corresponding stress levels S_1, $S_2, S_3, \ldots, S_m$. According to this equation, the damage done to a part at a given stress (S_1, S_2, or S_m) is directly proportional to the number of cycles at that stress.

This equation has been used by designers for many years, but at the same time it is under criticism by researchers. It is found that:

1. In many test results the summation of n/N is far from one.
2. The fatigue damage is not linearly proportional to the number of cycles or the cycle ratio n_1/N_1.
3. There is interaction in the fatigue damages between various stress levels which Miner neglected. In the interaction there is also a sequence effect which means that the fatigue damage resulting from the high load first, with the low load next, will be different from the damage resulting from the low load first, with the high load next.

Aware of its limitations, the designers still use Miner's equation as a preliminary guide. Cummings stated that for the design engineer there appears to be, for the present, no alternative to using the linear cumulative damage hypothesis (Miner's equation) as a starting point [23]. In another report [24] 20 different methods to assess cumulative damage were reviewed, compared, and evaluated. The first conclusion after the laborious evaluation is this: the linear cumulative damage procedure (Miner's equation) is recommended for its simplicity, versatility, and sufficient accuracy, commensurate with the data currently available for this type of analysis when satisfactory constant amplitude S-N type data can be provided for specific structure.

What are the consequences when Minner's equation is used for random loading conditions? Random load fatigue testing is time-consuming and costly, and the scant data available do not permit a general clear-cut answer to this question. According to Frendenthal [25], the sum $\sum n/N$ is always less than one. In some cases the sum is as low as 0.13 but it is mostly between 0.20 and 0.60. Head also reported [25] that Miner's equation overestimated the fatigue life in random loading by factors of 2 to $3\frac{1}{2}$. The primary reason for this discrepancy, according to Freudenthal, is the interaction of fatigue damages between various load amplitudes.

4.7 EFFECT OF TEMPERATURE AND FREQUENCY

The temperature of the materials and structures and the frequency (defined as time rate of the repetitions of load cycles) both have some influence on the

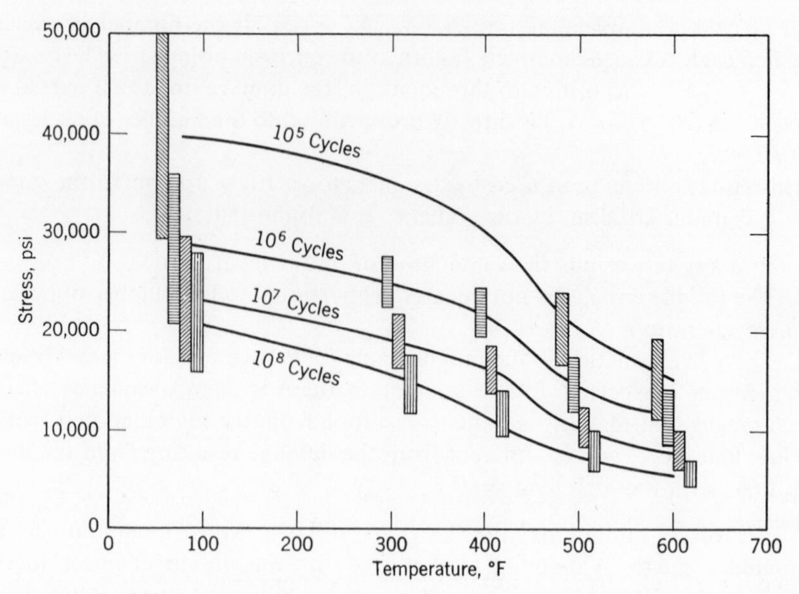

Figure 4.30 Fatigue strength of wrought aluminum alloys at various temperatures. ASTM [26].

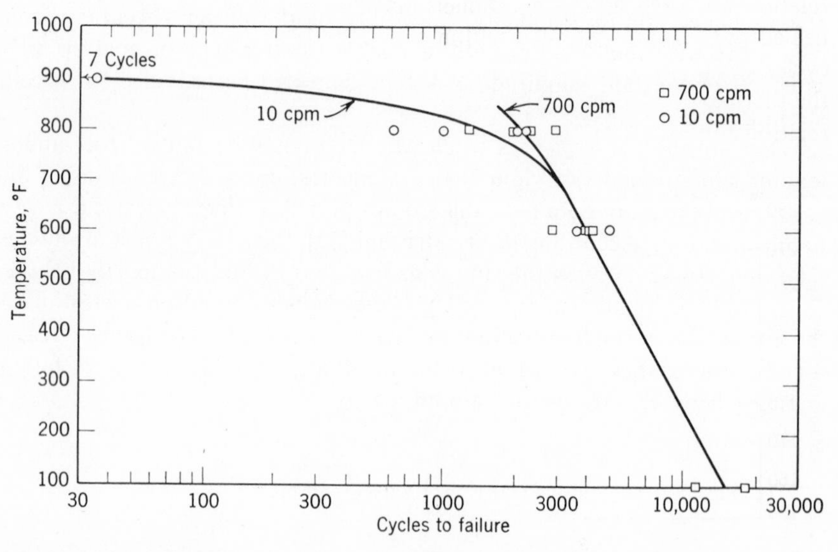

Figure 4.31 Fatigue life versus temperature, PH15-7Mo RH950: $K_t = 2.33$; $\sigma_{\max} = 166.8$ ksi; $\sigma_{\min} = 66.7$ ksi.

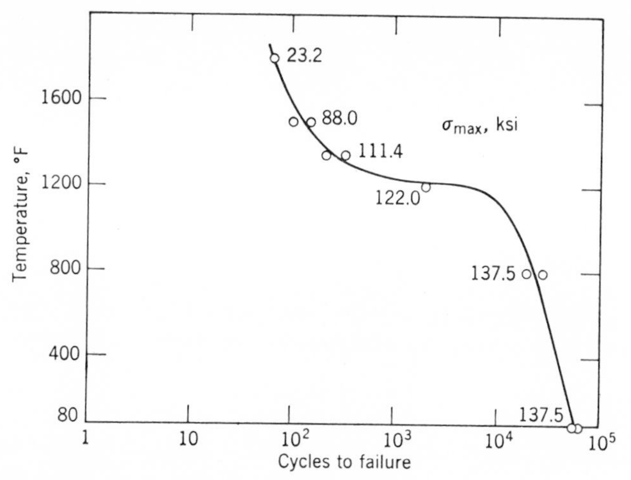

Figure 4.32 Fatigue life versus temperature, René 41: $K_t = 2.33$; $\sigma_{\max} = 0.7\,\sigma_u$ at temperature; $R = 0.54$.

fatigue life. Generally speaking, the higher the temperature or the lower the frequency, the lower the fatigue life or strength (Figures 4.30 to 4.33. [13], [26] and [27]).

From the viewpoint of thermal activation on the properties of materials the temperature effect and the frequency effect are closely inter-related. The relation was brought out by Daniels and Dorn in their fatigue study of annealed pure aluminum [2]. Within the scope of their tests, the number of cycles to failure N is related to temperature T, frequency ν, and stress σ by the thermal activation energy ΔH and the gas constant R according to this relationship,

$$N = f(\nu e^{\Delta H/RT}, \sigma) \tag{4.13}$$

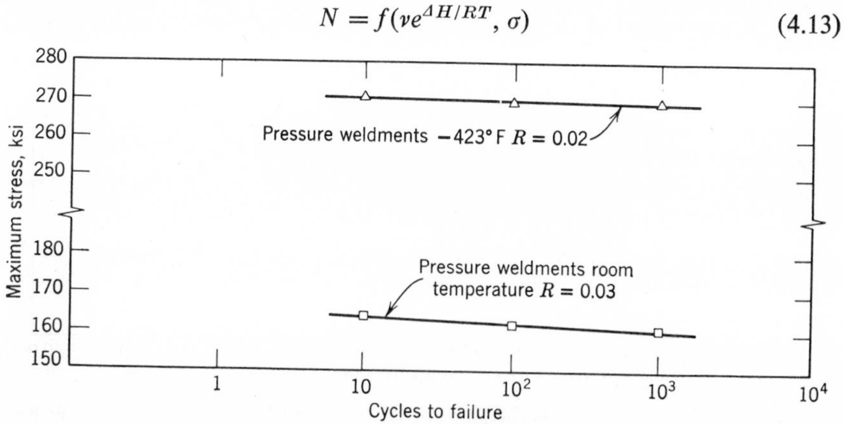

Figure 4.33 Effect of temperature on fatigue strength of Ti-6Al-4V weld. ASTM [31].

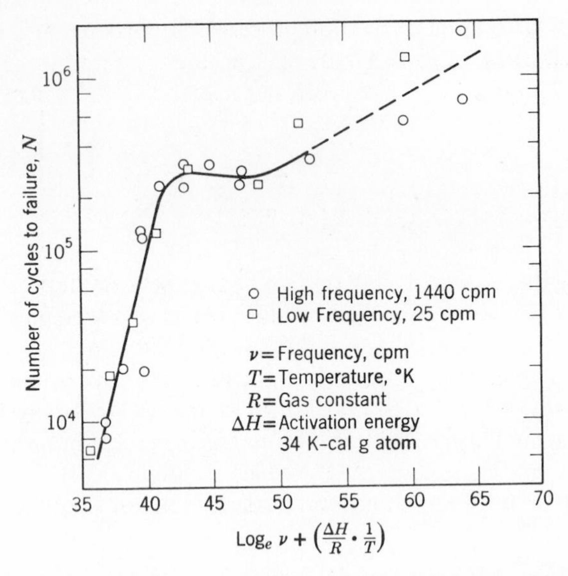

Figure 4.34 Composite N-temperature diagram (2450 psi). ASTM [2].

where f is the experimentally determinable function shown in Figure 4.34. Thus for a given stress the same number of cycles can be endured at the same value of $\nu e^{\Delta H/RT}$ and higher frequencies of testing are equivalent to lower temperatures.

From a metallurgical point of view, higher temperatures may cause instability in the microstructure of a metal, depending on the alloy, the temperature, the time of exposure, and the situation. Some of the phenomena are recovery, phase change, grain coarsening, recrystallization, reprecipitation, reversion, diffusion, aging, overaging, sigma phase formation, graphitization, decarburization, annealing, oxidation, and other corrosion and chemical changes. Many of these phenomena are known or are suspected of being factors affecting fatigue properties.

From a macroscopic point of view it should be noted that higher temperatures often hasten crack initiation but may sometimes retard crack propagation and increase critical crack length. Lower temperatures, on the contrary, often retard crack initiation, hasten crack propagation, and reduce critical crack length.

In high-temperature fatigue the failure or the initiation of failure is sometimes excessive deformation or creep instead of pure progressive fracture often encountered in low-temperature fatigue testing. Hence there is a close relationship between high-temperature fatigue and creep that should be considered when the thermal activation process is important. This consideration is supported by (4.7.1) and Figure 4.34, where the parameter $\nu e^{\Delta H/RT}$ or $\log_e \nu + \Delta H/RT$ is similar to the creep parameter $te^{-\Delta H/RT}$, where t is the time under stress [28].

4.8 METAL FATIGUE PROBLEMS IN SUPERSONIC TRANSPORT

Before closing this chapter I shall outline briefly the metal fatigue problems and suggest the fatigue data needed for the design of supersonic transport (SST).

To design and construct efficient fatigue-resistant supersonic transport (SST) structures reliable for 50,000 hours in the severe environment encountered in Mach 3 flight is a considerable challenge. The successful operation of aircraft in this environment, which is different from any other experience, may be described in general as depending on the following three conditions:

1. Temperature variation of about −70 to about 630°F. (The temperature of the aircraft structure during cruising is shown in Figure 4.35.)
2. A combination of mechanical and thermal stresses.
3. Up to 50,000 hours of operating life under the first two conditions.

As a first step in providing optimum fatigue resistance as well as many other functional requirements, data on service loads, temperature, and atmospheric conditions as functions of time are needed. After an analysis of environmental factors, the problem of converting these factors into laboratory testing parameters must be solved in order to evaluate the fatigue resistance of the structures and materials.

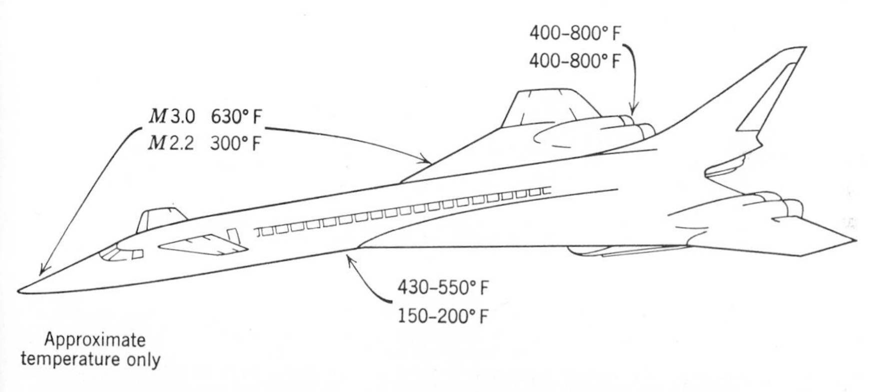

Figure 4.35 M2.2-M3.0 transports, temperature at cruise Mach number.

In addition to these general problems, several examples of more specific metal fatigue data are involved in the design of SST:

1. Conventional fatigue data at about −70°F, 70°F, and 630°F. These data include three items:
   a. *S-N* curves or a modified Goodman diagram (e.g., Figures 4.36 to 4.38).
   b. Crack propagation rate per cycle.
   c. Measurement of remaining static strength of fatigue-cracked specimens.
2. Extrapolation of short- and long-term fatigue data. (Since 30,000 hours is 3½ years and 50,000 hours is 5¾ years, it is impossible or impractical to obtain data on a real time (5¾ years) basis.)
3. Simultaneous load and temperature spectrum effects.
4. High temperature fatigue data of structural components on which little information is available.
5. Fail-safe designs, crack arresters, and redundant structures for high reliability.
6. At cruising altitude (70,000 feet) the ozone in the atmosphere may impair the fatigue resistance of metals. The effect should be determined.
7. Because the wing skin acts more like a two-dimensional plate than a one-dimensional beam surface and the fuel tanks and fuselage are internally pressurized, there are biaxial stresses. Fatigue data under biaxial stresses are required.
8. Prevention of explosive failure of pressurized fuselage cabins as a result of fatigue cracking under two-dimensional loading and thermal cycles.

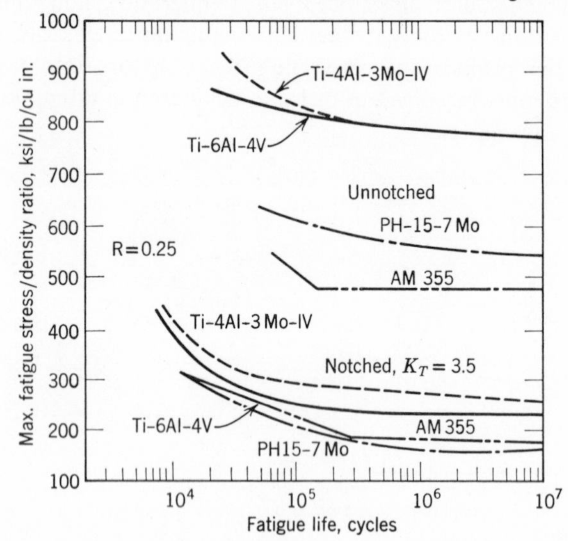

Figure 4.36 Comparative tension-tension fatigue curves for Ti-6Al-4V, TI-4AI-3Mo-IV, PH15-7Mo, and AM 355.

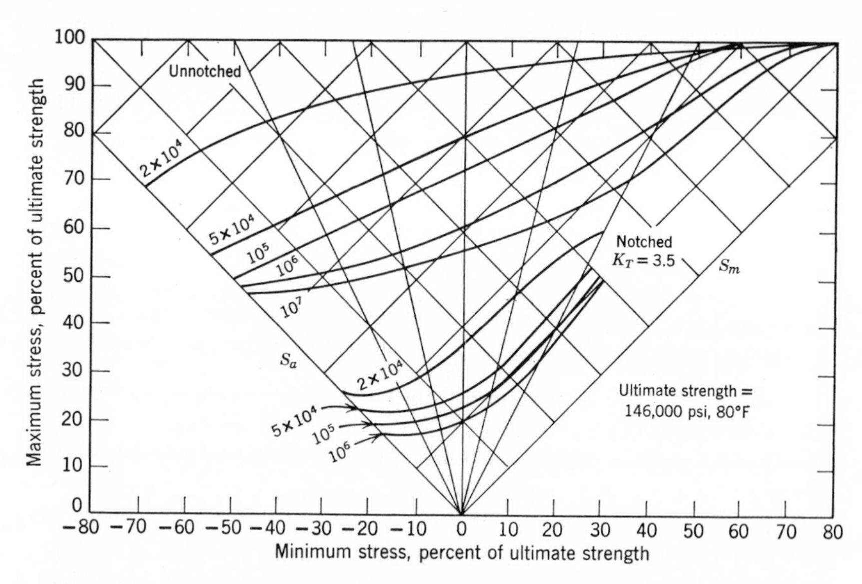

Figure 4.37 Modified Goodman diagram for 8Al-1Mo-1V (80°F) annealed titanium alloy.

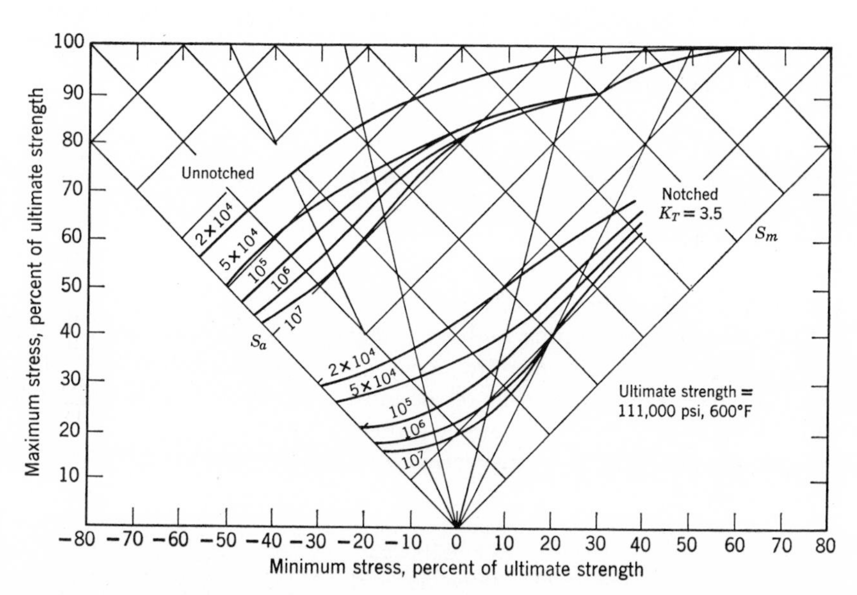

Figure 4.38 Modified Goodman diagram for 8Al-1Mo-1V (600°) annealed titanium alloy.

9. Control of cracking due to sonic fatigue in wing and fuselage skin, especially in honeycomb panels.
10. Prediction of cumulative damage as a result of load and thermal cycles.

Present knowledge of metal fatigue, as discussed in this and other chapters, may help to provide a general approach to the solution of these problems, but the solutions themselves still depend on large amount of research and development work.

4.9 SUMMARY

To summarize our interpretation of fatigue data, we have touched on several important subjects and provided some insight into them.

We have described the typical shapes of *S-N* curves. In addition to *S* and *N*, a third quantity relating to the scatter of the probability of survival or the minimum curve is also important but quite expensive to obtain. The entire *S-N* curve may be divided into two parts: low-cycle and high-cycle fatigue, which operate in the metal microstructures with different basic mechanisms.

The stress that causes fatigue damage always has two components, such as mean stress and amplitude or maximum stress and ratio *R*. Stresses can combine in two- or three-dimensional directions into different states, but the most important stress is the maximum shear or the octahedral shearing stress. We have also used Goodman's, Gerber's, and Sines' equations to express the effects of mean stress.

Practical engineers say that fatigue failures are tension failures, stress concentration failures, and surface failures. They also say that fatigue damage is only skin deep. These statements are based on experience and are usually true.

Because load amplitude is not constant, a method of assessing cumulative damage is necessary. For this purpose the industry has been using Miner's equation extensively in spite of criticism.

The effects of temperature and load frequency on mechanical properties appear to be related. Higher frequency may be equivalent to lower temperature. This effect is also one of the most critical problems in supersonic transport fatigue. Many SST fatigue problems have been reviewed. Though difficult and extensive, they are certainly not insurmountable.

REFERENCES

[1] H. F. Hardrath et al., NACA-TN 3107, 1953.
[2] G. Sachs et al., ASTM STP 196, pp. 83, 94, 1956.

[3] G. H. Stevens, U.S. Forest Product Lab. Report, FPL-027, 1964.
[4] W. Weibull, *Fatigue Testing and Analysis of Results*, 174, 1961.
[5] S. R. Valluri, "A United Engineering Theory of High Level Fatigue," GAL CIT SM61-1, 1961.
[6] J. L. Waisman, L. Soffa, P. W. Kloeris, and C. S. Yen, Effect of Internal Flaws, Non-destructive Testing, November 1958, 477.
[7] W. J. Crichlow, ASD TR61-434, p. 274, 1961.
[8] *A.I.A. Aircraft Fatigue Handbook*, Aerospace Industry Association, 1958, pp. 3.34–5.
[9] G. M. Sinclair and T. J. Dolan, *ASME Trans*, **75,** 867–873 (1953).
[10] C. S. Yen and T. J. Dolan, University of Illinois, 12th Progress Report to the Office of Naval Research, 1949.
[11] ASTM STP 91-A, 2nd ed., 1963.
[12] J. T. Yao and W. H. Munse, *Welding Res. Suppl.*, 182s (April, 1962).
[13] ASTM STP 338, pp. 25, 62, 1963.
[14] L. F. Coffin Jr., and J. F. Tavernelli, *Trans AIME,* **215,** 794–807 (1959).
[15] Wood *Fracture*, M.I.T., Wiley, New York, 1959, pp. 374 and 412.
[16] F. M. Howell and J. L. Miller, ASTM Proceedings, 1955.
[17] G. Sines and J. L. Waisman, *Metal Fatigue*, McGraw-Hill, New York, 1959 p. 145.
[18] C. S. Yen, "Calculating Residual Stresses in Some Formed Parts," ASME Aviation Conference, 1961.
[19] J. L. Waisman and C. S. Yen, "The Effect of Forming," ASTM STP 203 or ASTM STP 196, 1956.
[20] C. C. S. Yen and T. J. Dolan, "A Critical Review of the Criteria for Notch-Sensitivity in Fatigue of Metals," University of Illinois Engineering Experimental Station, Bulletin 398, 1952.
[21] C. S. Yen and M. Toth, "Fatigue Life After Removal of Crack," Douglas Aircraft Company, Long Beach, Calif., Lab. Report MP 6521, 1957.
[22] J. P. Butler, ASTM STP 203, p. 29, 1956.
[23] H. N. Cummings et al., WADC TR59-230 pp. 29, 174, 1959.
[24] W. J. Chrichlow et al., WADC TR61-434, 1961.
[25] IME and ASME, International Conference on Fatigue of Metals, Session 3, Papers 4 and 9, 1956.
[26] R. L. Templin, "Fatigue of Aluminum," ASTM Gillett Memorial lecture, 1954.
[27] WADC TR60-410 Part I, p. 66, Part II, p. 75, 1960.
[28] *Mechanical Behavior of Materials at Elevated Temperatures*, John E. Dorn, ed., McGraw-Hill, New York, 1961.
[29] J. C. Grosskeutz, AFML TR, 64-415.
[30] M. S. Hunter and W. G. Fricke, Jr., ASTM Proceedings, 1955.
[31] R. R. Hilsen, C. S. Yen and B. V. Whiteson, ASTM STP 338, 1963.

5

Fatigue Statistical Analysis

CHARLES S. YEN

As we know from our daily experience, many events in the world cannot be predicted with certainty. When I toss a coin, we do not know with certainty whether the head will show up. The measure of the likelihood or the chance that a certain event will happen is called probability. The probability that the head will turn up when I toss a coin is 0.5, or 50%.

When we test many metal specimens or components, their strengths are never exactly the same but are scattered over a wide range. The probability that a random specimen from the population† will have a strength higher than the median‡ strength is 50%. The probability that a random specimen will have a strength somewhat higher than the median strength is less than 50%. The exact value depends on the shape of frequently distribution§ of strengths.

In the study of statistics the terms survival, failure, reliability, confidence level, significance level, and tolerance level are all probabilistic events or concepts and are measured by the probability scale graduated from 0 to 1.

When we use the theories of probability and statistics, we can predict the property of the entire population (and the confidence level associated with the prediction) from the data of a portion of the population. Not only that, the theories also help to interpret the failure mechanism of materials, especially brittle materials. In the following discussion the failure mechanism is

† Population (or universe) is the hypothetical collection of a large number of all possible specimens or members prepared in the specified way from the material under consideration.

‡ Median is the middlemost number in a group of numbers arranged in order of magnitude; for example, the median of 1,3, and 4 is 3 and the median of 1, 3, 4 and 8 is $3\frac{1}{2}$.

§ Frequency distribution is the way in which the frequencies of occurrence of members of a population or a sample are distributed according to the values of the variable under consideration (e.g., the strength).

interpreted first. Next, special methods of long-life fatigue testing are described in which each method produces its special type of statistically meaningful data. The detailed procedure to analyze these data is not included here but is referred to another book in which the procedure has been already clearly described. The general methods of statistical analysis of fatigue data are then introduced. Finally, the concept of the reliability of structures is discussed.

The subsequent sections are intended primarily to provide an introduction or foundation for the understanding of the current state of the art in the application of statistics to fatigue phenomena. Examples to illustrate some established statistical methods are given. Relatively important and essential concepts are included and emphasized, but many details are necessarily omitted because of the nature of this book, which is primarily a survey.

5.1 BASIC STATISTICAL MECHANISM OF MATERIALS FAILURE

Theories based on statistical concepts to interpret the material fractures introduced here are derived for brittle materials that have no plastic deformation or strain hardening to complicate the fracture mechanism. In practice, however, they are often used for materials that are not very brittle.

5.1.1 Static Failure

It is known that the observed strength of materials are considerably less than the theoretical strength, which is calculated from atomic cohesive forces or energy. The reason for this discrepancy has been attributed to the fact that real materials contain defects, flaws, or weak points. For metals these defects in fundamental forms include dislocations, foreign particles, vacancies, and grain boundaries. For brittle materials like ceramics these defects or flaws are cracks and voids. In Weibull's statistical theory of failure [1] he assumes that these flaws are randomly distributed and are of random severity. The weakest link theory was adopted as a criterion of failure, which states that a component or a sample will fail when the stress at any flaw becomes larger than the ability of surrounding material to resist the imposed stress. In developing the theory, Weibull proposes that the probability of fracture† S is related to the observed fracture strength σ in this form:

$$S = 1 - \exp\left[-V\left(\frac{\sigma - \sigma_\mu}{\sigma_0}\right)^m\right],$$

† The probability of fracture is a cumulative frequency function and is also the area under the frequency distribution curve.

where V = the volume of the component subjected to tensile stress,
σ_μ = the lowest limiting stress below which the fracture cannot occur, also called "zero fracture probability stress" or briefly "zero strength,"
σ_0 = the strength of a "flawless" specimen,
m = flaw density per unit volume in the body.

In this equation, if the fracture strength σ varies from σ_μ to infinity, the probability of fracture will vary from 0 to 1. This equation may explain the size effect. Since larger components have a higher probability of having serious flaws, they are weaker. The relation between the volume of the component and the strength may be derived by putting S as a constant; then

$$V(\sigma - \sigma_\mu)^m = \text{constant.}$$

Sometimes σ_μ may be assumed to be zero, as Weibull originally did; then we have

$$\frac{\sigma_1}{\sigma_2} = \left(\frac{V_2}{V_1}\right)^m.$$

It has been found that $m = 24$ for a phosphorus steel. This theory appears to be valid for some materials but not for others. Sometimes, instead of the volume, the surface area is more important. With glass and similar materials not even the assumption of a proportionality between the number of flaws and the surface area is likely to be correct. Experiments have shown that for glass and silica tubes (and perhaps also for fibers and rods) the number of flaws is probably proportional to the length of the tube rather than to its surface area [2].

5.1.2 *Fatigue Failure*

Both Freudenthal and Aphanasiev used statistical approaches to explain fatigue phenomena and the notch effect. Freudenthal [3] considered the fatigue phenomenon as an expression of the repetitive action of an external load. This progressive destruction had the typical features of a mass phenomenon; both the cohesive bonds in the material and the load repetitions were treated as collective in a statistical sense. To avoid complications the plastic deformation was neglected in the development of theory and only the finite fatigue life rather than the endurance limit (which is related to some plastic deformation typical of some metals) was considered.

By assuming a statistical distribution of cohesive bonds and applying the basic rules of the theory of probability, Freudenthal reached the conclusion that the fatigue strength S is equal to the static tensile strength minus $k \log N$, where k is a constant and N is the number of cycles representing the fatigue life at fatigue strength S. Thus

$$S = (\mathrm{TS})\, k - \log N.$$

This relation agrees generally with the S-N curves of several metallic and nonmetallic materials.

If a nonuniform stress distribution due to a notch or to flexural action in a beam is approximated by a discontinuous step function with two intensities, it can be assumed that a percentage i of the total Q bonds is subjected to the high stress intensity in the vicinity of the notch, whereas the majority of bonds are in the field of comparatively low homogeneous stress. If p is the probability of destruction of a bond in the field of low stress intensity and p_1 is this probability in the immediate vicinity of the stress concentration, it is shown that the probability P_2 of rupture of all Q bonds under nonuniform stress distribution for N load repetitions is

$$P_2 = 1 - (1 - P)e^{-iQN(p_1 - p)},$$

in which e is the base of natural logarithms and P is the probability of rupture of all Q bonds under the homogeneous low stress, without stress concentration.

According to this equation, the probability P_2 increases rapidly with increasing value of the exponent of e; that is (a) with increasing number of bonds iQ in the volume affected by the stress concentration, (b) with increasing load repetitions N, and (c) with intensity of stress concentration $(p_1 - p)$. These three influences are of the same order of importance; a similar increase in the probability of rupture will be obtained either by increasing the volume affected by stress concentration or by increasing the number of load repetitions. Therefore the effective strength reduction due to notch should increase considerably at the lower stress levels (for which the number of load repetitions is large). The notch effect $(p_1 - p)$ is multiplied by the factor N simply because the load is repeated N times. This agrees with Bennett's observation that for *SAE X*4130 steel fatigue-strength reduction factors based on the endurance limits were about the same as the relative slopes of the log S log N curves; it also agrees with Almen's suggestion that the slopes of an S-N curve are a relative measure of the strength-reduction factor [4].

Aphanasiev [4,5] also developed a statistical theory of fatigue and assumed that the metal was an aggregate of crystal grains regarded as elementary structural units with identical yield limits and cohesive strength in the direction of the external force. These grains, however, were subjected to different stresses due to the inhomogeneity of the material, porosity, inclusion, influence of the grain boundaries, and so on.

Thus the frequency of the occurrence of any particular stress value acting on an individual grain might be expressed as a function of the value of the imposed stress.

For fatigue loading the critical condition or criterion for the value of the endurance limit was formulated by assuming that no grain was subjected to a stress exceeding the cohesive strength. The parameters of the probability function in the equations were evaluated from static tension test curves for the metal, which also included the effect of strain-hardening.

For the purpose of determining the fatigue strength reduction or notch factor the general solution presented considerable difficulties. Finally a tentative expression for the strength-reduction factor was proposed and made to agree with test data.

5.2 SPECIAL TESTS FOR FATIGUE LIMITS AT LONG LIFE

In addition to the use of statistical analysis to explain the basic fracture phenomena, as shown above, statistics may also be used to analyze some fatigue test data to provide more and better information. In order that the test data may be analyzed to provide the most and the best information, the test method and the number of specimens should be chosen or designed to fit the original objective and to produce statistically meaningful data.

The following discusses some special testing methods and the number of specimens (i.e., group size or sample size) needed for each type of test.

In ordinary fatigue testing at least one of two kinds of data is often required:

1. *S-N* curve.
2. The endurance limit or the fatigue limit at long life.

When the objective is to determine an *S-N* curve, the conventional constant amplitude-test method is generally used, in which one or a group of specimens is tested at each stress level. To determine long-life fatigue limit, as is often done by the power engine manufacturers, there are two types of testing method, the constant-amplitude response test or survival test and the increasing amplitude test. These two types are described here [6].

5.2.1 Response Test

There are two methods: probit† and staircase.

1. Probit method is one in which several specimens are tested for a given number of cycles at each of four or five stress levels near the stress of interest.

† Probit = probability unit.

First, the desired life is determined; for example, 10^7 cycles. Next, from preliminary or previous data four or five stress levels around 10^7 cycles and the expected per cent survival associated with each stress level are chosen. For example, for a certain test the following stress levels are selected to determine the fatigue limit at 10^7 cycles.

TABLE 5.1

PROBIT TEST PLAN (EXAMPLE)

Stress level	68	70	72	74	76	
Expected per cent Survival (first approximation)	90	80	50	30	5	
Relative group size	2	1½	1	1	3	
Number of specimens	12	9	6	6	18	Total 51

To facilitate statistical analysis and to assign equal weight to each observed percentage survival value the relative group size for each stress level is planned according to Table 5.2. In probit tests a group should consist of not less than five specimens and the total tested at all stress levels should be at least 50. In the preceding example the group size is six and the total is 51, as shown in Table 5.1.

TABLE 5.2

ALLOCATION OF TEST SPECIMENS FOR PROBIT TEST

Expected Per Cent Survival	Relative Group Size
25 to 75	1
15–20, 80–85	1½
10, 90	2
5, 95	3
2, 98	5

Figure 5.1 presents data that might be obtained in a probit test planned in Table 5.1. A properly designed probit test will give more useful fatigue data than any of the other response or increasing amplitude tests.

2. In the staircase method the first specimen is tested at the estimated fatigue strength at say 10^7 cycles for the prescribed number of cycles (10^7 in this case) or until it fails, if it fails prematurely (less than 10^7 cycles in this case). If this specimen fails, the next specimen is tested at a stress level that is one increment below the first stress level. If the first specimen does not fail, however, the second specimen is tested at a stress level that is one increment above the first stress level.

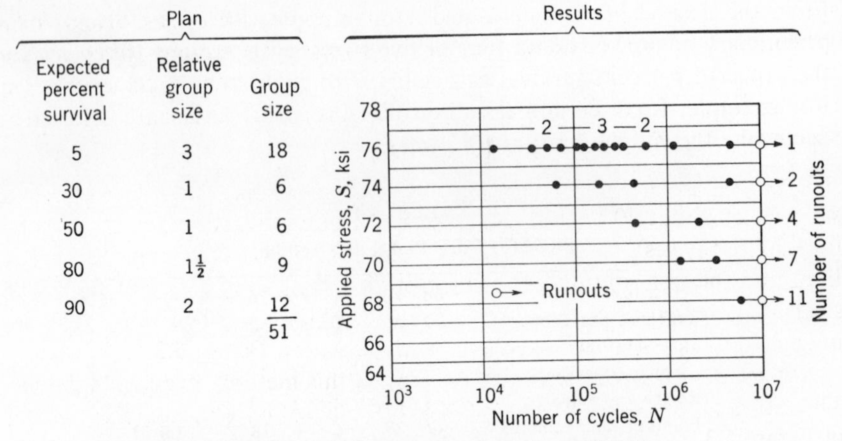

Plan		
Expected percent survival	Relative group size	Group size
5	3	18
30	1	6
50	1	6
80	$1\frac{1}{2}$	9
90	2	12
		51

Figure 5.1 Probit test plan and results. ASTM [6].

The third specimen is tested at a stress higher or lower than the stress for second specimen, depending on whether the second specimen survives or fails. The data are recorded as shown in Figure 5.2. In this figure specimens are numbered in chronological order and the number of cycles for each test is constant unless failure occurs beforehand.

When the total number of specimens is large, it may be divided into several groups, and each group is tested on one of several machines as a separate staircase program. These machines are checked to prevent them from giving significantly different results. Thus several machines may be used simultaneously and the time required to complete the test is reduced. This is called the modified staircase method.

5.2.2 *Increasing Amplitude Test*

When more than 50 specimens are available to determine the fatigue limit at a long life, the probit method is the best. When the number of the available

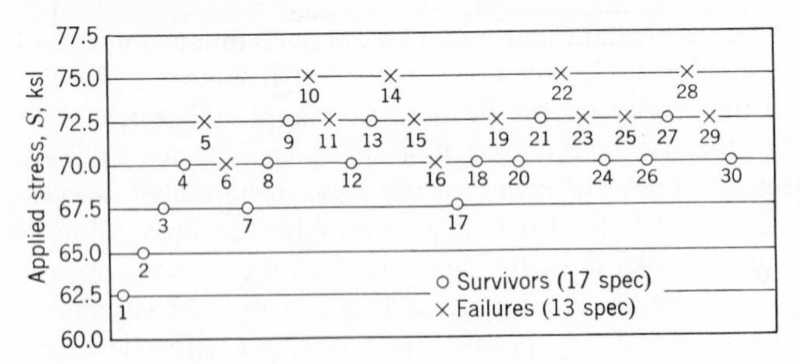

Figure 5.2 Staircase test results. ASTM [6].

specimens is about 20 to 50, the staircase method is often the most suitable. When the number is less than 50 and the material has no coaxing or understressing effect, the increasing amplitude test may also be used. There are two methods of increasing the stress amplitude in fatigue testing each specimen:

1. *Step method.* First, test a specimen at a stress level corresponding to a survival of about 70 to 90%. If the specimen survives the prescribed long life (say 10^7 cycles), the stress is increased about 5% of the estimated fatigue limit and the test continues for another 10^7 cycles if it does not fail. The stress is increased after each survival run until the specimen fails. The sequence of testing for each specimen is illustrated in Figure 5.3.

Any number of specimens may be used in this method, even a single one. The estimates of fatigue strength from the result of one specimen are shown in Figure 5.3, but more specimens will give more precise estimates, as in any other tests.

2. *Prot method.* In the Prot method test a number of specimens in the same way, starting from a low stress of about 60 or 70%, say, of its estimated fatigue limit. The stress is increased continuously in each cycle at a constant rate until fracture. Then test a number of specimens at a different rate of increase in loading, starting from the same low stress. At least three rates of increase in loading are used. The loading rates, etc., are expressed in pounds per square inch per cycle. Figure 5.4 is an illustration of the Prot test data. The fracture stress may also be plotted against the abscissa α^n, where α is the loading rate and the value of n may be taken initially as 0.5. A straight line may be obtained. If it is not straight, the value of n may be empirically adjusted until the curve becomes straight. Then the intercept of the straight

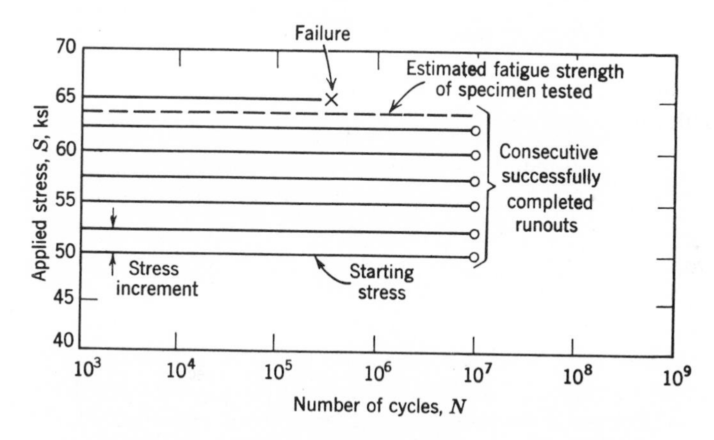

Figure 5.3 Step method result of a single specimen. ASTM [6].

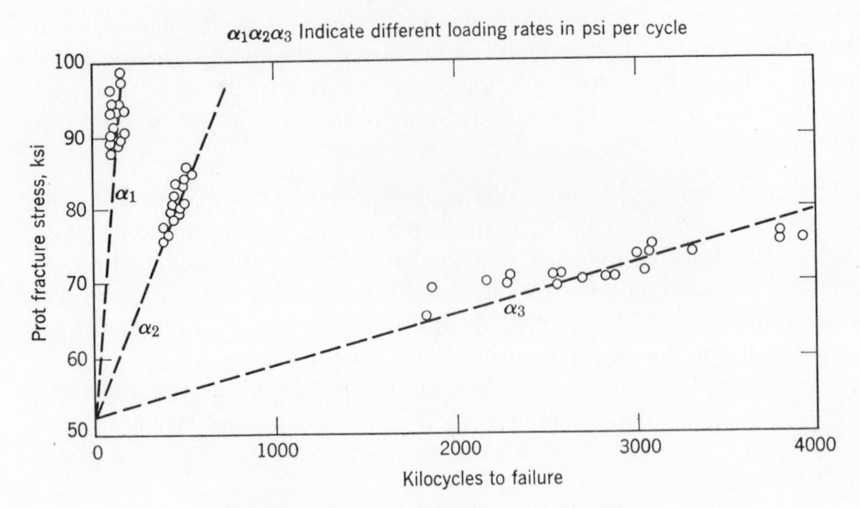

Figure 5.4 Probit test results. ASTM [6].

line with the fracture stress axis is considered as the fatigue or the endurance limit.

The Prot test method usually requires less time than the step method. Because of the wide scatter of Prot data, it is suggested that at least 20 specimens be used to obtain the data for the Prot analysis.

We have introduced four special test methods to obtain the fatigue limit at long life. For more complete procedures to analyze the data obtained from these special test methods, [6] is recommended. Other statistical methods for analyzing more general test data, such as a portion of or the entire *S-N* curve data, are discussed in the next section.

5.3 STATISTICAL ANALYSIS OF FATIGUE DATA

It is well known that fatigue-test data scatter widely. To describe the scatter a frequency distribution curve is often used. Such a curve may be plotted from the raw test data or from the treated or transformed data; that is, either from values of cycles N or strength S or from a function of N or S, such as the values of $\log N$, $\log S$, $\log \log N$, and $N^{1/2}$.

The purpose of statistical analysis is usually to estimate the property of the entire population from the data of a portion of the population. The appropriate method to be used in the estimation depends on the knowledge or assumption of the shape of the frequency distribution of the population. If the distribution shape is not known or conservatism is preferred, the methods of estimating the property of the population described in the first part of this section may be used. If the test data warrant, however, the distribution curve may be assumed to have a definite shape and the method used may be standardized and relatively convenient. A brief discussion is given later in this section for cases in which the distribution curve is assumed to be normal.

5.3.1 *Frequency Distribution Shape Not Assumed*

When the shape of the distribution curve of fatigue life or strength cannot be estimated or assumed, the following techniques may be used to give conservative results (usually more conservative than if the shape were assumed to be normal).

P-S-N Curves (probability-stress-cycle curves). The first step in the analysis of fatigue data is to draw the *S-N* curve for 50% survival; it is the curve fitted to the median of the groups tested at several stress levels.

If the group size is greater than one, that is, if more than one specimen is tested at each stress level, the *S-N* curve for other than 50% survival may be estimated or faired. To accomplish this the life values of specimens in each group are arranged in order of magnitude: N_1 is the minimum cycle life or the first-order statistic, N_2 is the second minimum life or the second-order statistic; and so forth. Then, depending on the sample size or group size n, the estimated percent survival for the population at cycle life N_1 and N_2, respectively, may be found in Table 5.3, [6].

TABLE 5.3

PERCENTAGE OF SURVIVAL

1 Sample Size, n	2 At the Lowest Value, N_1	3 At the Next Lowest Value, N_2
1	50	...
2	70	30
3	79	50
4	84	61
5	87	69
6	89	73
7	90	77
8	91	80
9	92	82
10	93	84
11	94	85
12	94	86
13	95	87
14	95	88
15	95	89
16	96	90
20	97	92

The percentage of survival at the maximum value of the group N_n is 100 minus (percent for N_1), the survival at N_{n-1} is 100 minus (percent for N_2), etc. The following information should be noted in connection with the use of Table 5.3.

1. The 50% survival curve may be estimated from the median of any sample size.

2. If three specimens are tested at each stress level, the 79, 50, and 21% survival curves may be estimated from the entries in Table 5.3 and their complements.

3. If five specimens are tested at each stress level, the 90, 77, 50, 23, and 10% survival curves may be estimated from Table 5.3, its complement, and the group median.

4. In practice, values of percent survival less than 50 usually are not wanted. Hence the great values of fatigue lives larger than the median may not be wanted and tests of these specimens may be stopped.

5. The percentages of survival listed in Table 5.3 are median for the population. They are median values because they are based on a confidence level of 50%. In other words, the true (population) percentage is higher than the sample median for half the time when the median is determined and smaller for the other half.

6. The percentages of survival in Table 5.3 are close to but usually not equal to the *expected* percentage of survival, which is equal to $1 - i/(n + 1)$, where i is the number of the order statistic and n is the sample size. The confidence level associated with *expected* percentages varies with the sample size, whereas it is constant (50%) for (median) percentages† in Table 5.4.

7. Percent survival values corresponding to higher confidence levels such as 95 or 99% are available in [6].

Estimates of Parameters at a Single Stress Level. Assume that a sample of 10 specimens is tested at a particular stress level and that the observed values of fatigue life in kilocycles are 201, 224, 226, 230, 232, 238, 240, 244, 245, and 248. What properties of the population may be predicted from the data of the sample? The answers are as follows:

1. Point estimate of median. A point estimate of the population median is the sample median, the average of the two middle-most values $(232 + 238)/2 = 235$ kc.

2. Confidence interval estimate of median. The population median may be above or below the sample median 235 kc, but the chances are at least 95 in 100 that the statement "the median lies between $N_2 = 224$ and $N_9 = 245$ kilocycles" is correct if the sample came from one population. The interval

† They are called median percentages because of the 50% confidence level.

between N_2 and N_9 is the confidence interval for the median at confidence level ≥ 0.95 (as found from Table 9 in [6]).

3. Percent survival for a given fatigue life. Let us calculate in the previous test data of the 10 specimens what the percent of survivors of the population is for 230 kc. Since there were 7 out of 10 specimens that survived 230 kc, a point estimate of the percentage of the population that has fatigue life equal to or above 230 kc is 70%. When statistical tables are available [6], we find that the percent survival of population for 230 kc may be above or below 70%, but the chances are at least 95 in 100 that the population survival is between 34 to 94% for 230 kc.

4. Fatigue life for a given percent survival. To determine the fatigue life for a given percent survival, Table 5.3 or corresponding information [6] may be used. For 93% survival fatigue life is $N_1 = 201$ kc; for 84% it is $N_2 = 224$ kc; for 50%, 235 kc. For percentages between 93 and 84 the fatigue life may be interpolated.

Another method of determining the fatigue life for a given percent survival is to obtain it from the cumulative frequency distribution of the observed values. In general these two methods do not give exactly the same results.

Tests of Significance. When two or more groups of specimens are tested, in the same condition, differences usually appear in the observed values between the groups. Sometimes the differences are due to chance of sampling from the same population; the groups are then considered not significantly different. At other times the difference is not due to chance alone but to some differences in the population from which group samples were drawn; for example, the observed differences may arise from differences in material lots or in the testing techniques. In this case the groups and the observed values are said to be significantly different.

In many practical situations the question of the significance of such differences cannot be answered with a clear-cut yes or no. Statistical techniques are available to determine the level of the significance graduated numerically from 0 to 1. When the level of significance is low, say 1%, the observed differences may be considered insignificant and due to sampling chance effects from the same population, but if the level of significance is high it is likely that the observed differences between different samples are due to population difference. These significance test techniques are described in detail in many books on statistics or fatigue; for example, [6], [7], [11], or [12].

5.3.2 *Frequency Distribution Shape Assumed Normal*

Normal Distribution. When the frequency distribution curve of fatigue data has a bell shape as defined by the following equation $f(x)$ and shown in

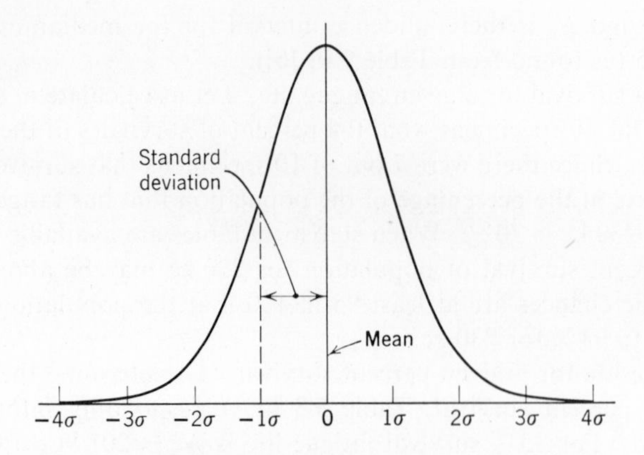

Figure 5.5 Normal distribution curve.

Figure 5.5, the data are said to have normal or Gaussian distribution.

$$f(x) = \frac{1}{\sigma\sqrt{2\pi}} \exp\left[-\frac{1}{2}\left(\frac{x-\mu}{\sigma}\right)^2\right],$$

where x = the observed random variable, for example, the number of cycles at a given stress,
μ = the population mean,
σ = the population standard deviation, a measure of dispersion or scatter,
$f(x)$ = the ordinate of the curve.

When a group of specimens is tested to failure at a single stress level, the test results often show that the logarithm of the fatigue life ($\log n$) gives a frequency distribution that appears normal [6,8]. There is evidence also to show that in some cases the fatigue strengths at any given life follow a normal distribution [9]. Once there is agreement between the test data and the normal distribution all the results of normal distribution theory are available for data analysis. The information from the sample may also be used to predict the information for the population. The mean (arithmetic average) of the normal population, which is equal to the median of the population, may be estimated either by the sample mean or by the sample median. The sample median is the less efficient of the two; that is, it experiences a larger variation from experiment to experiment but has the advantage of being simpler to determine. The standard deviation, a measure of the dispersion of the individuals about the center of distribution, is estimated by the sample standard deviation. Knowing the mean and the standard deviation, we can determine everything about the whole distribution, which is presumed as normal. The sample mean μ and the standard deviation σ are calculated from the following equations, where x_i is the ith specimen value from a sample of size n:

$$\mu = \frac{1}{n}\Sigma x_i$$

$$\sigma = \left[\frac{\Sigma(x_i - \mu)^2}{n-1}\right]^{1/2} = \left[\frac{n\Sigma x_i^{\,2} - (\Sigma x_i)^2}{n(n-1)}\right]^{1/2},$$

where Σ is the sum from 1 to n.

Life Normal or Strength Normal. There are two different assumptions regarding the relations between fatigue test data and normal distribution. One states that the cyclic life (or some transformation of cyclic life such as $\log n$) at a given stress level is normally distributed. The other assumes that the fatigue strength (or transformed or normalized strength) of a given life is normally distributed. Based on the former assumption, statistical methods to determine *P-S-N* curves and to make significance tests are developed in [6]. Based on the latter assumption, an analysis that estimates the fatigue limit at long life of the population is given in [6], and a simple and quick method to determine 50 and 95% *P-S-N* curves is given in detail in Section 5.5.

5.3.3 Extreme Value Theory and Weibull Distribution

Extreme Value Theory. Suppose we bought a large batch of nominally identical bolts to be installed on airplanes. The mean strength of these bolts can be estimated from the test results of some specimens. Very often, however, what we are interested in is not really mean strength but the minimum strength of the batch of bolts. In engineering design procedures, hardware specifications, or guarantee contracts the minimum quality, not the average, is often emphasized. The statistical theory that deals with the property of the lowest or extreme values is called the extreme value theory. It is useful not only in the study of fatigue strength or life but also in many other natural phenomena such as gust load, duration of human and other lives, radioactive emission, and rainfall [13,14].

Assume that the fatigue strength of a large number of bolts is normally distributed and that 100 groups of specimens are tested to determine the lowest fatigue strength in each group. After we complete the tests, we will have the data of 100 lowest strengths. When the frequency distribution of the 100 lowest strengths is plotted, we will find that the distribution is no longer normal (Figure 5.6). It is not even symmetrical, for it has a long tail to the left and a short tail to the right.

The exact distribution of the lowest strengths depends on the sample size (the number of specimens in each group). If the sample size is increased, it is

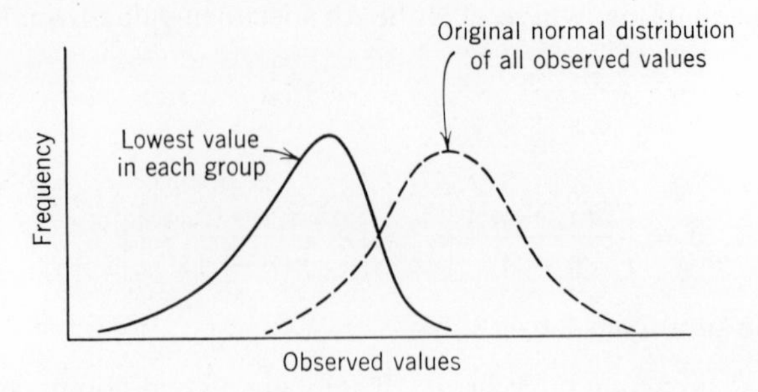

Figure 5.6 Lowest value distribution versus normal distribution.

conceivable that we will tend to obtain a lower least value in each sample and the mean of the smallest value distribution will shift to the left with increasing sample size. Theory shows that this shift of the lowest value distribution is accomplished by a reduction in its spread. Hence the larger the sample, the lower and more stable its lowest value. This is shown in Table 5.4.

TABLE 5.4

MEAN AND STANDARD DEVIATION OF LOWEST VALUES FOR VARIOUS SAMPLE SIZES

Sample Size	Mean of Lowest Value[a]	Standard Deviation of Lowest Value[b]
2	−0.564	0.826
3	−0.846	0.748
4	−1.029	0.701
5	−1.163	0.669
6	−1.267	0.645
7	−1.352	0.626
8	−1.424	0.611
9	−1.485	0.598
10	−1.539	0.587
15	−1.736	0.549
20	−1.867	0.525
100	−2.508	0.429
500	−3.037	0.370
1000	−3.241	0.351

[a] In units of standard deviation of the original distribution and measured from the mean of the original distribution.
[b] In units of standard deviation of the original distribution.

In the earlier discussion of the basic statistical mechanism of materials failure a structural component or a specimen is considered as something like a chain that consists of many small links or units, such as crystal grains, cohesive bonds, or the sound material between the tiny flaws. Because a chain is no stronger than its weakest link, the breaking strength will tend to decrease as the physical size of the specimen increases. The effect of the specimen size here is similar to the effect of sample size already discussed, and the law that governs the strength or failure of materials and components is intimately related to the distribution of extreme values.

Weibull Frequency Distribution. One of the types of extreme-value distribution is the Weibull frequency distribution [6,12]. Assuming that fatigue failures are initiated at the "weakest link," the fatigue lives of a group of specimens tested under a given set of conditions may be represented by the Weibull frequency distribution function:

$$f(N) = \frac{b}{N_a - N_0}\left(\frac{N - N_0}{N_a - N_0}\right)^{b-1} \exp\left[-\left(\frac{N - N_0}{N_a - N_0}\right)^{b}\right],$$

where N = specimen life,
N_0 = minimum life ≥ 0,
N_a = characteristic life at 36.8% survival of the population, 36.8% = $1/e$, $e = 2.718$;
b = shape parameter of the Weibull distribution curve
= slope of the Weibull cumulative distribution curve.

To simplify the appearance of this equation, let

$$x = \frac{N - N_0}{N_a - N_0}.$$

Then

$$f(N) = \frac{b}{N_a - N_0}\, x^{b-1} \frac{1}{e^{x^b}}.$$

If $b = 1$,

$$f(N) = \frac{1}{N_a - N_0} \cdot \frac{1}{e^x}.$$

This is a simple exponential distribution function and sometimes represents the structural reliability or value in a time interval. If $b = 2$,

$$f(N) = \frac{2}{N_a - N_0} \frac{x}{e^{x^2}}.$$

This is a Rayleigh distribution function, used often in accoustic fatigue analysis. If $b = 3.57$,

$$f(N) = \frac{3.57}{N_a - N_0} \cdot x^{2.57} \cdot e^{-x^{3.57}}$$

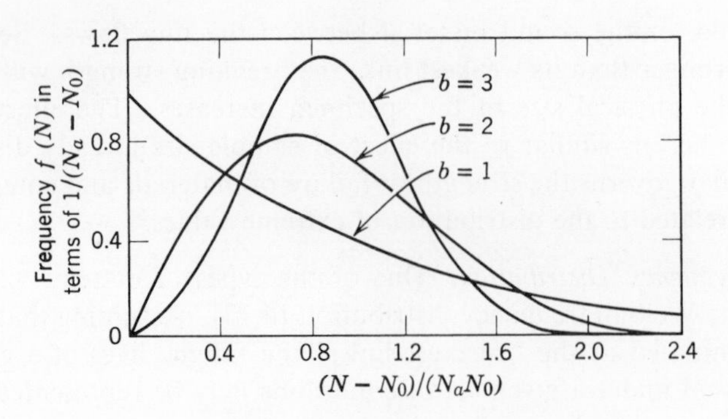

Figure 5.7 Weibull distributions.

This is a good approximation of the normal distribution function. The mean and the median values are equal.

The curves representing the Weibull function when $b = 1$, 2, and 3 are shown in Figure 5.7.

In fatigue testing the data are usually plotted or analyzed cumulatively to determine the percent survival or percent failure. The Weibull cumulative function for the fraction of population failed at life N is an integration of $f(N)$. It is

$$F(N) = 1 - \exp\left[-\left(\frac{N - N_0}{N_a - N_0}\right)^b\right].$$

This equation is similar to the Weibull equation for interpreting the statistical nature of static material failure in the early part of this chapter. This function can be transformed into the straight-line relationship by taking the logarithm of the logarithm of the equation:

$$\log\log\left[\frac{1}{1 - F(N)}\right] = b\log(N - N_0) - 0.36222 - b\log(N_a - N_0).$$

Plotting this equation as a linear relation between $\log[1/1 - F(N)]$ and $(N - N_0)$ on log–log paper allows a simple graphical method for fitting the Weibull cumulation distribution to the data and a graphical estimation of the parameters b, N_0, and N_a.

To plot the fatigue test data of a group of specimens tested at the same stress on a log–log chart, the value of N_0 is assumed zero first. The horizontal plotting position is the values of $1/1 - F(N)$, where $F(N)$ is equal to $q/(n + 1)$ or similar estimates; q is the order number when specimens are arranged according to the magnitude of individual life and n is the sample size.

If the lower end of the plotted points does not lie in a straight line, the value of N_0 will be estimated by trial and error and the value of $(N - N_0)$ will be plotted instead of N until a straight line is obtained.

The parameter b is equal to the slope of the line. The parameter N_a is evaluated on the chart at the point at which the percentage failed $F(N)$ or $F(N - N_0)$ is 63.2%.

When the distribution of population is thus estimated, other estimates and significance tests may be performed [15].

Since Weibull distribution has three parameters (N_0, N_a, and b) to be adjusted, it fits the test data closer or more accurately than the normal distribution, in which only two parameters (mean and standard deviation) are available. In other words, the Weibull distribution function is more general than the normal distribution which may be considered a special case of Weibull distribution. The general nature of the Weibull distribution accounts for its popularity today.

5.4 STRUCTURAL RELIABILITY

Reliability, a relatively new subject, is a methodology. Even though several books have been written on this subject [16,17,18], its scope is not completely defined. This brief section introduces some elementary concepts, especially in relation to the structural fatigue.

As technology progresses, systems like air vehicles, space vehicles, computers, communications, and weapons become more complex. Because a system contains a large number of subsystems and components, the failure of a single part may cause the failure of the entire system, and each detail may influence the reliability of the entire system. Hence there is a slogan: reliability is everybody's business—from management, designer, manufacturer, and inspector to the user. Highly reliable systems are certainly not easily obtained and are the result of conscious effort by all concerned [16].

Reliability may be defined as a probability of success. The percent survival of fatigue specimens already discussed is the reliability of single components. The statistical concepts and analyses presented in the preceding sections are also useful in reliability calculations.

To be more elaborate, structural (or system, device, etc.) reliability may be defined as the probability that a structure will perform its function under specified conditions of operation, inspection, maintenance, and environment for a certain length of time.

5.4.1 Failures

When a structure ceases to perform its function, the condition is called unreliability, malfunction, or failure. For the convenience of reliability theory and practice failures may be generally classified as early, wearout, and chance failures. Early failures occur early in the life of a structure during the debugging or development test period. Wearout failures occur near the end of the expected life. Between the early and the late periods of the life of the structure there are chance failures that occur randomly and unexpectedly. This is a rough classification according to the age of the structure, not according to any definite physical causes. Fatigue failures may occur in any one of the three periods. Although they may appear in the early period, such as low-cycle fatigue, they are often expected in the wearout period.

Other physical phenomena identified as structural failures are excessive plastic deformation, creep, wear, corrosion, and embrittlement. The failure and reliability information of the structures should be collected as the structure is developed and goes into service so that its real reliability may be determined.

5.4.2 *Factors Affecting Structural Reliability*

The real reliability of a large system is difficult to predict before the initial testing, for too many factors are involved; For example, in fatigue loaded structures the following random variability or uncertainty, including unavoidable human errors, will affect reliability [19].

1. Design. The variability specified in design drawings and specifications, such as (a) dimensional tolerances, (b) material properties tolerances, and (c) processing tolerances.
2. Production. (a) The variability in the interpretation of the design properties by the production and inspection systems, and (b) the variability of measuring systems and standards used to control the specified design attributes.
3. Operation. (a) The variability in the specification and in the interpretation of operating procedures, and (b) the variability of the operating environment (temperature, moisture, atmospheric pressure, gusts, natural and man-induced).
4. Maintenance. The variability in the specifications and interpretation of maintenance and inspection procedures.

5.4.3 *Theoretical Guidance*

Even though the real reliability is difficult to predict, theoretical guidance to reach a reliability goal may be obtained in the design stage; for example, when the reliabilities of simple components are known, the reliability of a system of components may be calculated. Two simple systems, that is, series and parallel systems, are discussed.

If the reliabilities of individual components are $r_1, r_2, r_3 \cdots r_n$, respectively, the unreliabilities of these components will be $1 - r_1, 1 - r_2, 1 - r_3 \cdots 1 - r_n$, respectively. Now these components are connected (or disconnected) to a system in any way. In this system the probability that all components will survive is the product

$$R_s = r_1 r_2 r_3 \cdots r_n.$$

The probability that all components will fail is the product

$$(1 - r_1)(1 - r_2) \cdots (1 - r_n)$$

and

$$= (1 - r)^n, \quad \text{if} \quad r_1 = r_2 = \cdots = r_n = r.$$

The probability that not all components will fail (i.e., at least one component survives) is

$$R_p = 1 - (1 - r_1)(1 - r_2) \cdots (1 - r_n);$$

R_s in the above is called the reliability of a series system of components, which means that if any one component fails the system will have failed. All complex systems belong in this category, with components connected in a series.

The reliability of a parallel system of components is called R_p, which means that the system will not fail so long as at least one of these components remains operative. From the above equation it can be seen that R_s decreases with the number of components and R_p increases. In theory, a parallel system is more reliable than a series system, but practically it is limited by cost, weight, and resources.

5.4.4 *Safety Margin*

In engineering design a factor of safety or a safety margin (interval between strength and stress) is often used to take care of the uncertainties of the applied stress and structural strength. It seems that, other things being equal, the safety margin is proportional to the safety or reliability of structures. The following simple analysis, however, shows that the safety margin and reliability are not always proportional.

Assume that we are designing a simple component. The applied external stress which the component has to resist is known to vary over a range and the frequency distribution of the applied stress is normal (Figure 5.8).

The strength of the component, as determined from the strength tests, is also normally distributed. When the two normal distribution curves are plotted as in Figure 5.8, a portion of the areas under the curves is overlapped, as shaded, and indicates the possibility of failure. The area under the

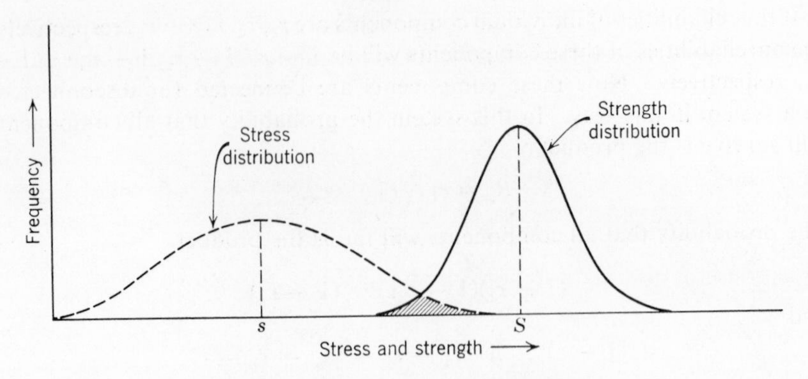

Figure 5.8 Stress versus strength distribution.

component strength distribution which is not covered under the stress distribution is the reliability. The ratio of the mean strength S to the mean stress s is defined as factor of safety in may books on structural or machine design.

When the difference between the strength and the stress increases, the shaded area or the failure rate will decrease and the reliability and the factor of safety will increase. When the reliability is plotted against the difference between the mean strength S and the mean stress s, we obtain the curve

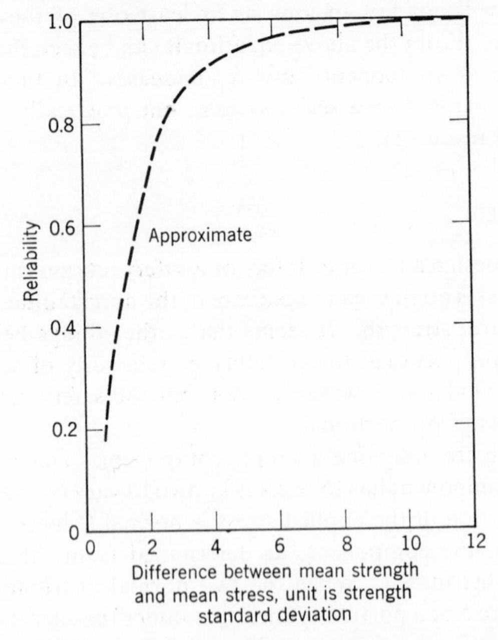

Figure 5.9 Reliability versus the difference between mean strength and stress.

shown in Figure 5.9 [20]. The precise position of the curve depends on the values of both standard deviations of stress and strength, and the curve shown in Figure 5.9 is only approximate. It does, however, show clearly that when the difference between mean strength and mean stress becomes very high the increment of reliability becomes increasingly smaller as the curve approaches the limit asymptotically. This small increase in reliability, however, will add rapidly to the weight and cost of the component, as shown in Figure 5.10 [20]. This conclusion is another example how theoretical analysis supplies useful information on reliability.

5.5 ESTABLISHING OF MEAN AND 95% SURVIVAL S-N CURVES†

As an extension of the discussion in Section 5.3.2, this section describes a relatively simple statistical method in detail that provides for the establishing of the mean and 95% survival *S-N* curves. Although the method is intended for use with *S-N* data with a limited number of points, it cannot be employed if too small a number is available, since in such cases the mean curve can be influenced to include almost all points. No scatter would then be evident. Use of the statistical technique is not recommended for zones of *S-N* data in which less than five points are available with a 10-to-1 life ratio.

When using the statistical technique, it should be borne in mind that it is merely a mathematical tool that assumes that the data analyzed represent chance variation of the same variables that would be operating for the "population" from which the data are drawn. It is important to use judgment in determining whether this condition is satisfied; for example, an

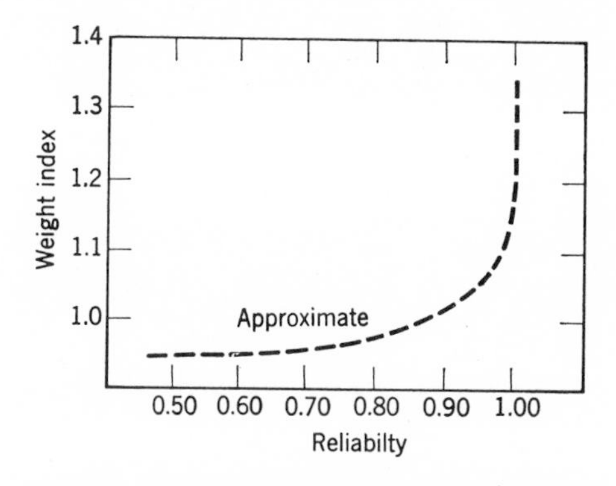

Figure 5.10 Reliability versus weight of structures.

† The method in this section was originially developed by J. L. Waisman, R. Keating, and Charles C. S. Yen at Douglas Aircraft Co. for the *A.I.A. Aircraft Fatigue Handbook* [10].

S-N curve for thin sheet material will not give information on heavy sheet, bar, or lot-to-lot variations. Such data should be labeled with their limitations.

Another assumption made in applying these methods is that fatigue data follow a normal distribution on the stress scale [9].

5.5.1 Calculation of the Mean Curve

To illustrate the calculation of the mean curve and a curve above which 95% of points will fall (95% survival limit) a typical set of experimental points is used.

An estimation mean curve is first drawn by "inspection" for the data plotted on the usual *semilogarithmic* scale, as in line 1, Figure 5.11.

In Figure 5.12 we see how the data may be divided into groups, each of which includes five or more test points within the 10-to-1 life ratio. Any one of these groups now constitutes a valid group of data to analyze. Do not, however, include in your calculations any data representing specimens that had a fatigue life equal to or greater than the shortest life of an unbroken

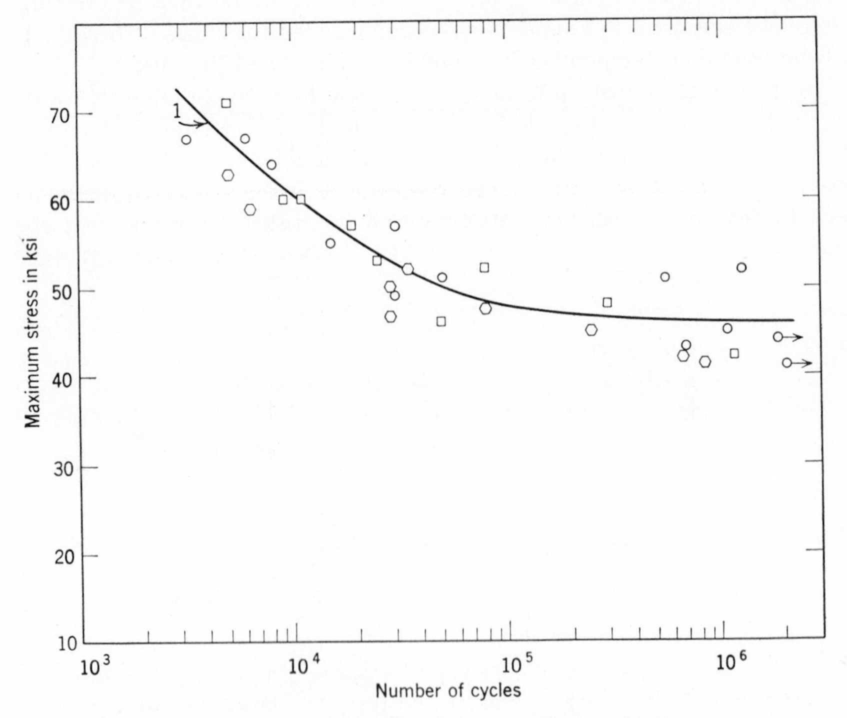

Figure 5.11 Data and first "mean curve" approximation.

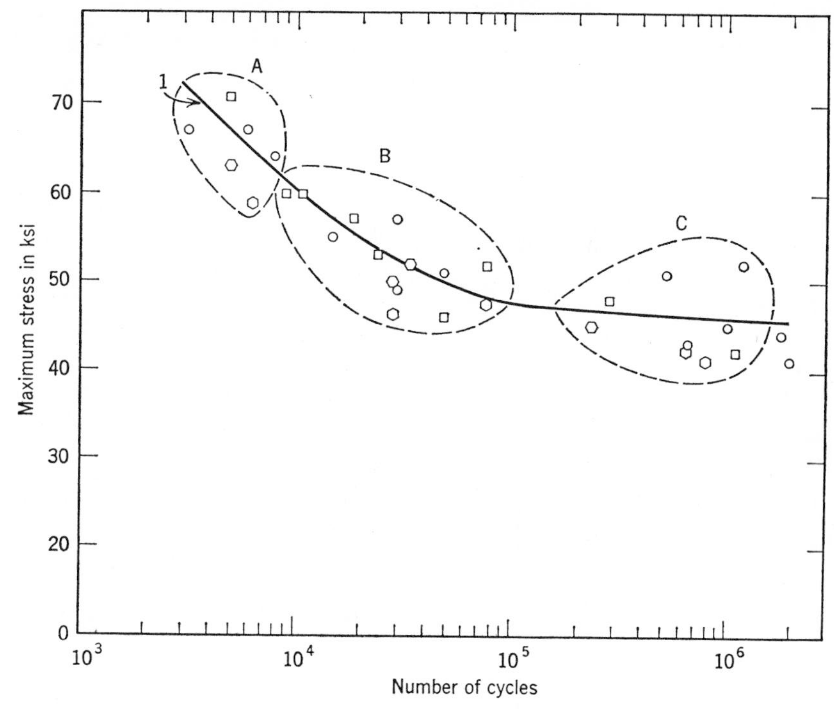

Figure 5.12 Division into groups.

sample. If portions of a curve cannot be included within any group, represent that portion by estimated dashed lines.

Figures 5.13 and 5.14 show how to average the data within a group to arrive at a more accurate mean curve than the "eye-estimated" curve. A vertical line is drawn through the center of the group, *S-T* in Figure 5.13. This is the mean-log lifeline.

Then measure with a ksi scale the stress by which each test point deviates from the eye-estimated curve. Points below the curve are negative, those above, positive. To average these deviations compute their algebraic sum and mean. As an example, in Group B, Figure 5.13, the average is $-\frac{1}{2}$ ksi. In Figure 5.14 half of $-\frac{1}{2}$ ksi is plotted on the mean-log lifetime *S-T* at point *W*. If the sign is negative, *W* will fall below the eye-estimated curve; if positive, it will lie above.

Repeat this averaging for each group. Figure 5.15 shows how line 2 through the "*W*" points would represent a conversion from the eye-estimated mean *S-N* curve to a calculated mean curve. If the calculated line deviates from the original estimate by more than 2 stress percent, repeat the calculation and revise the location of the calculated line accordingly. For that portion

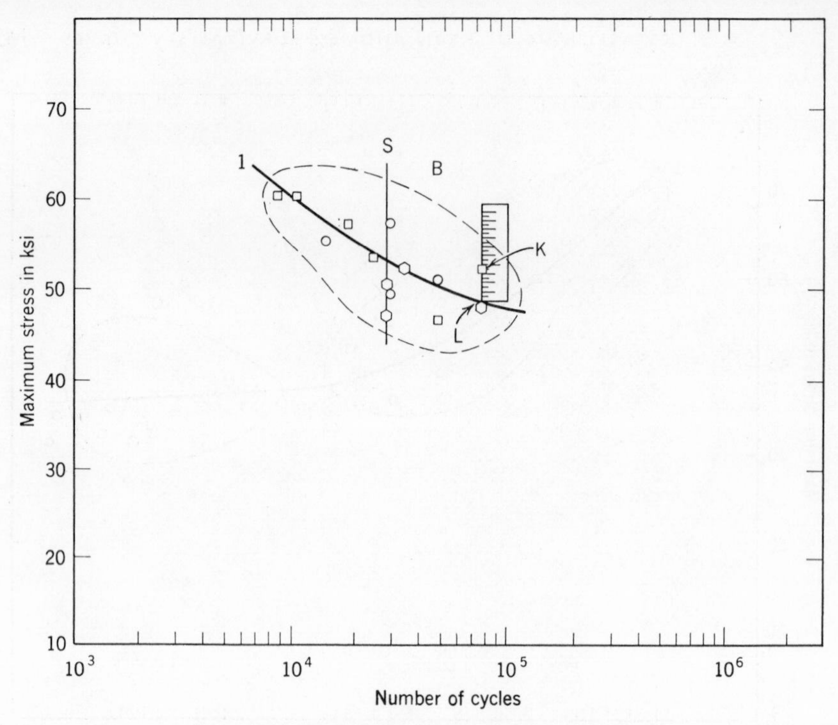

Figure 5.13 Calculating "mean curve" position.

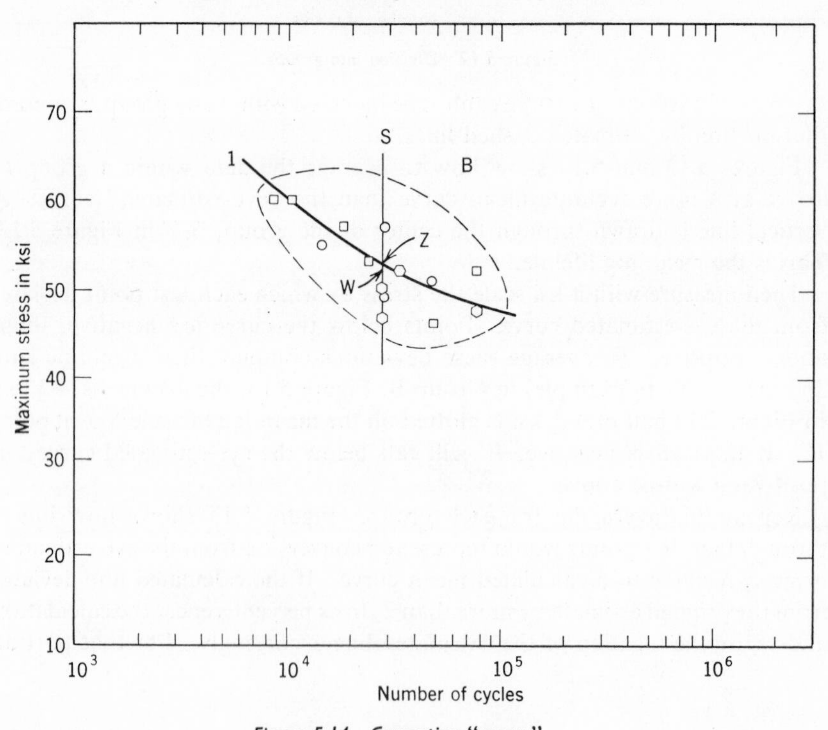

Figure 5.14 Correcting "mean."

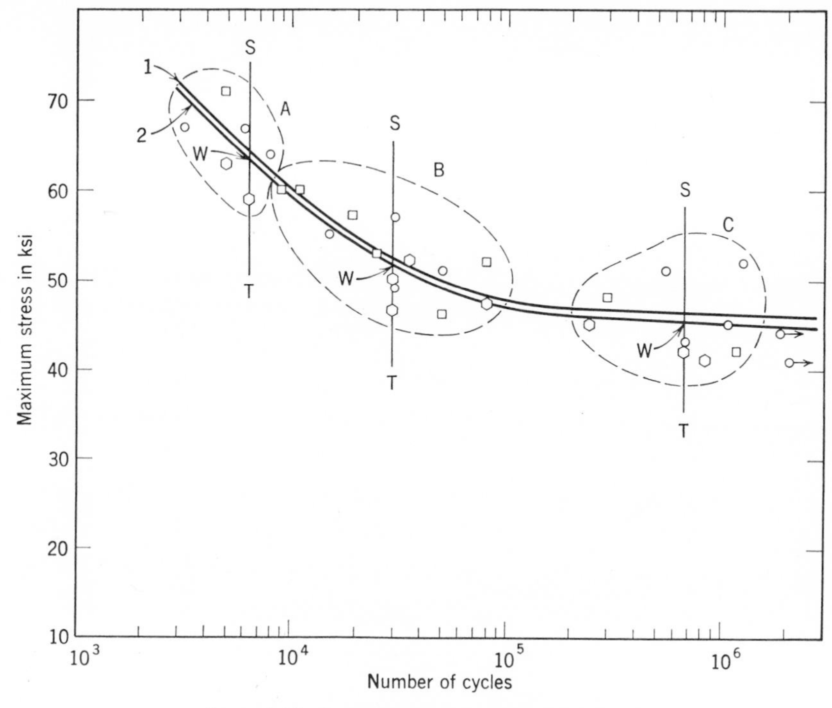

Figure 5.15 Second approximation, mean curve.

of the S-N curve in which the group does not include five or more test points within a 10-to-1 life ratio, an eye-estimated dashed line should be used for the mean curve.

5.5.2 *Establishing the 95% Survival Curve*

With our mean S-N curve established, we may now compute the 95% survival curve with the following steps:

Within each group of test points on the S-N curve establish the standard deviation from the corrected mean curve. To do this measure each plus or minus deviation in ksi and enter in column $(X_i - \bar{X})$, Table 5.5. In column $(X_i - \bar{X})^2$ each of these values is squared. Table 5.5 also shows how these squared deviations for the 14 test points in Group B, Figures 5.12 to 5.16, are averaged and how the standard deviation (σ) for the group is computed.

This standard deviation is for the actual test points in the group only. It is a measure of the group's scatter. To convert it to a deviation corrected for the 95% survival limit of an *unlimited number* of tests refer to Table 5.6.

TABLE 5.5
CALCULATION OF STANDARD DEVIATION

N	$(X_i - \bar{X})$ ksi	$(X_i - \bar{X})^2$
1	0.50	0.25
2	1.25	1.56
3	1.50	2.25
4	2.25	5.06
5	0.25	0.06
6	5.25	27.56
7	2.00	4.00
8	2.75	7.56
9	5.50	30.25
10	1.00	1.00
11	2.00	4.00
12	3.25	10.56
13	0.00	0.00
14	4.50	20.25
Total		115.0
$\sigma^2 = \dfrac{115}{14} =$		8.2
σ = Standard deviation =		2.9

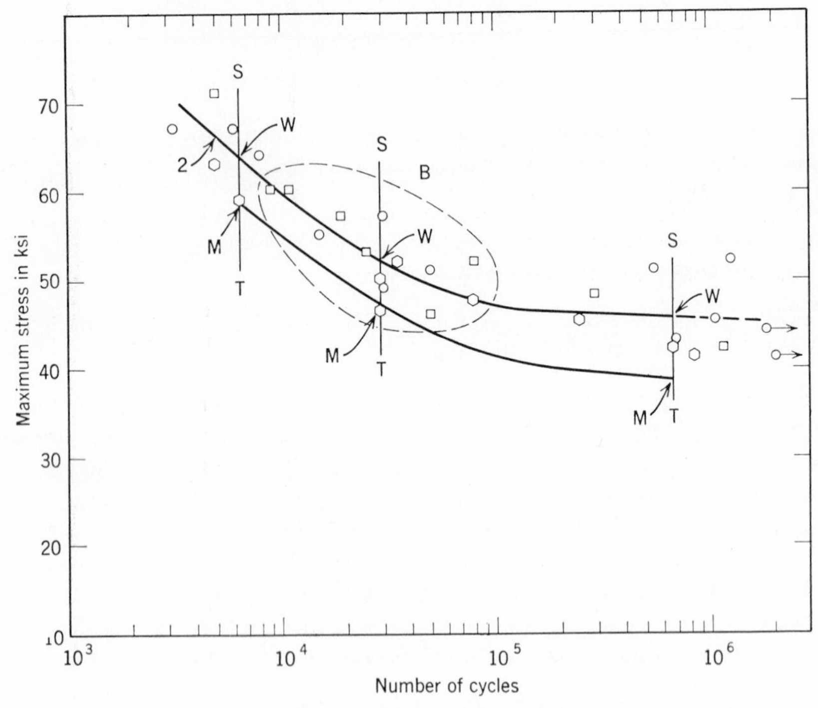

Figure 5.16 95 per cent survival curve.

TABLE 5.6

CALCULATING 95% SURVIVAL LIMIT

N	S	N	S
Number of Samples	Stress Deviation of 95% Survival Limit From Mean	Number of Samples	Stress Deviation of 95% Survival Limit From Mean
1	—	21	−1.707
2	−2.916	22	−1.704
3	−2.273	23	−1.701
4	−2.062	24	−1.699
5	−1.957	25	−1.696
6	−1.894	30	−1.688
7	−1.852	35	−1.681
8	−1.822	40	−1.677
9	−1.800	45	−1.673
10	−1.783	50	−1.670
11	−1.769	55	−1.668
12	−1.758	60	−1.666
13	−1.748	65	−1.664
14	−1.740	70	−1.663
15	−1.733	75	−1.662
16	−1.727	80	−1.661
17	−1.722	85	−1.660
18	−1.718	90	−1.659
19	−1.714	95	−1.658
20	−1.710	100	−1.657

The number of test points in a group are in column N. The computed factors in column S are to be used in determining the 95% survival limit. These S factors are corrected for infinite sample size and each is computed for the number of test samples involved, found in column N [7].

Multiply the standard deviation (σ) already determined by its proper factor. Subtract this product from the mean to get the position of the 95% survival limit.

As an example, there were 14 test points in Group B. The standard deviation was found to be 2.9 ksi in Table 5.5. The factor in Table 5.6 for 14 tests is -1.740; $(2.9) \times (-1.740) = -5.0$ ksi. Plot a point 5 ksi below the mean W to give the point M, in Figure 5.16.

Similar points established for the other groups of tests on the curve permits the drawing of a curve through them to establish the 95% survival curve.

This procedure is used to draw curves of 50 and 95% probability of survival. The two curves are the most important and useful. When curves for other probabilities of survival are required, two additional calculations are necessary:

1. Correct the *group* standard deviation to the standard deviation for *unlimited number* of tests.
2. Determine the probability of survival in terms of the units of the corrected standard deviation.

5.6 SUMMARY OF FATIGUE STATISTICAL ANALYSIS

In Section 5.1 we discussed how the theories of probability and statistics may be used to explain the size effect in static failures and to explain the size effect, notch effect, and the slope of *S-N* curve in fatigue failures.

In Section 5.2 four long-life test methods were described to determine the endurance limit or the fatigue limit. Each method gives a particular set of data to be analyzed statistically. They are constant amplitude tests (probit method and staircase method) and increasing amplitude tests (step method and Prot method).

Section 5.3 concerns the statistical analysis of the fatigue data of a sample to estimate the properties of the population. Before the analysis the shape of the frequency distribution of the fatigue strength or life data should be determined. Often the strength, life, or transformation of strength or life like log N is determined or assumed to be normally distributed. Recently the Weibull distribution, which is more general than normal distribution, has become more popular. When no decision or assumption can be made about the shape of distribution of a fatigue property, estimates of the mean value and *P-S-N* curves of the population can still be made.

In Section 5.4 the concept of reliability is introduced. Fatigue failure is often a cause of structural unreliability. To achieve a high reliability is not an easy matter, but it is a necessary effort to be made by all concerned. Factors affecting reliability of a complex system are too many to determine the real reliability, but the theories of probability and statistics, some of which have been discussed here, can help us to reach a reliability goal.

In the last section a detailed procedure for establishing the 50 and 95% survival *S-N* curves is described.

REFERENCES

[1] W. Wiebull, "A Statistical Theory of the Strength of Materials," *Ingen. Vetenskaps Akad. Handl. Proc.* **151,** 1–45 (1939).
[2] E. N. Andrade and L. C. Tsien, *Proc. Roy. Soc.* (*London*) **A159,** 346 (1937).

[3] A. M. Freudenthal, "The Statistical Aspect of Fatigue and Metals Proceedings," *Proc. Roy. Soc.* (*London*) **187,** 416 (1946).
[4] C. S. Yen and T. J. Dolan, "A Critical Review of the Notch-Sensitivity in Fatigue of Metals," University of Illinois Bulletin 398, 1952.
[5] N. N. Aphanasiev, *Z. Tech. Fiz.*, **4,** 349, and **19,** 1553 (1940–1941).
[6] A.S.T.M. STP 91-A, 2nd Ed., "Guide for Fatigue Testing and Statistical Analysis," 1963.
[7] L. E. Simon, *An Engineers' Manual of Statistical Methods*, Wiley, New York, 1941.
[8] A. M. Freudenthal, C. S. Yen, and G. M. Sinclair, "The Effect of Thermal Activation on the Fatigue Life of Metals," University of Illinois Project NR-031-005, 1948.
[9] A. Bender and A. Hamm, "The Application of Probability Paper to Life or Fatigue Testing," Delco Remy Division, General Motors Corp., 1957.
[10] A.I.A. Aircraft Fatigue Handbook, Aerospace Industry Association ARTC, pp. 3.34–5, 1958.
[11] E. J. Moroney, *Facts from Figures*, Penguin Books, New York, 1951.
[12] W. Weibull, *Fatigue Testing and Analysis of Results*, Pergamon, New York, 1961.
[13] E. J. Gumbel, "Statistical Theory of Extreme Values and Some Practical Applications," *Nat. Bur. Std.* (*U.S.*). Applied Mathematics Series, **33,** 1954.
[14] J. Mandel, "The Theory of Extreme Values," A.S.T.M. Bulletin, February 1959, pp. 29–30.
[15] A.S.T.M. Committee E-9, "The Weibull Distribution Function for Fatigue Life," *Mater. Res. Std.*, Vol. 2, No. 5, 405–411 (1962).
[16] M. Zelen, "Statistical Theory of Reliability," The University of Wisconsin Press, Madison, 1963.
[17] I. Bazovsky, *Reliability Theory and Practice*, Prentice-Hall, Englewood Cliffs, N.J., 1961.
[18] D. K. Lloyd and M. Lipow, *Reliability*, Prentice-Hall, Englewood Cliffs, N.J., 1962.
[19] H. T. Jensen, *The Application of Reliability Concepts to Fatigue Loaded Helicopter Structures*, American Helicopter Society Forum, Washington, D.C. 1962.
[20] L. H. Abraham, *Structural Design of Missiles and Spacecraft*, McGraw-Hill, New York, 1962.

6

Cumulative Damage Theories

HOWARD L. LEVE

Fatigue prediction is concerned with estimating the time length that a material can serve its intended design function when it is subjected to fluctuating stress environments. Because of the multitude of possible stress patterns, it does not appear, in general, that fatigue life results can be compiled for complex stress patterns similar to that accumulated for pure sinusoidal stress histories. This indicates that for complex stress histories a certain amount of analysis must be resorted to in order to overcome the expected deficiency in test data directly relating to particular histories.

As yet no fatigue prediction approach has been developed for complex fluctuating stress histories that does not require some sort of experimental support. Recognizing this need, it has been the hope that no more than pure sinusoidal fatigue data would be necessary for fatigue prediction associated with any arbitrary stress environment. It appears, however, after many years of effort, that the consideration of sinusoidal data as the cornerstone for more complex environmental situations does not lead to a fatigue prediction approach which is applicable to a wide range of materials and environments.

Although lacking universality, fatigue prediction theories for complex fluctuating stress environments have been useful in evaluating fatigue lives for specific designs. In the subsequent sections a set of fatigue prediction theories is brought together, analyzed, compared, and discussed with the intent of giving the person engaged in fatigue design a flexible background of possible approaches when confronted with the problem of complex environmental fatigue life determination. These theories, usually referred to as "cumulative fatigue damage theories," are not restricted solely to those

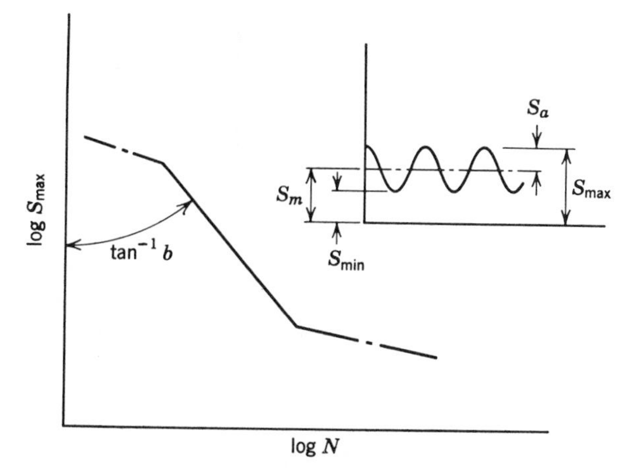

Figure 6.1 S-N diagram.

based upon pure sinusoidal test data but include some approaches that require more complex test support. An attempt has been made in the ensuing discussion to introduce some logic behind the construction of the various cumulative damage theories.

6.1 FATIGUE UNDER PURE SINUSOIDAL STRESSES

Since most cumulative damage theories rely on pure sinusoidal fatigue data, it would be worthwhile to describe briefly the form in which this data will be used in these theories. The fatigue life N of a material subjected to a pure sinusoidal stress history depends on the mean value of the stress S_m and the stress amplitude S_a. The fatigue life can also be referred to the maximum and minimum stresses S_{max} and S_{min} of the sinusoidal history or any combination of two of the four descriptive quantities mentioned. A typical fatigue plot is shown in Figure 6.1 which relates fatigue life to the maximum stress of the sinusoidal history. This plot is referred to as an S-N diagram.

The magnitudes indicated in Figure 6.1 depend especially on the specific material tested, the mean applied stress level, and the temperature and surface finish of the material. Frequency and oscillatory shape (i.e., constant amplitude oscillations but not necessarily sinusoidal) have insignificant effects on the S-N diagram except at high temperatures. It is seen that the plot can be depicted by three straight-line regions. The cumulative damage theories discussed are primarily concerned with stress magnitudes that fall within and below the central portion of the diagram. The functional relation for the central portion of the diagram can be written as

$$\frac{N}{N_r} = \left(\frac{S_r}{S}\right)^b. \tag{6.1}$$

In (6.1) N is the number of cycles to failure associated with the maximum stress S (for brevity, the "max" notation is omitted from the subscript of the maximum stress when it is evident that clarity is not being sacrificed) and N_r is the number of cycles to failure associated with a reference maximum stress S_r. The exponent b is seen in Figure 6.1 to be the slope of the central portion of the diagram with respect to the vertical axis (i.e., log $S_{\max}$ axis).

It should be pointed out with regard to Figure 6.1 that repeated determinations under the same sinusoidal environment will produce widely dispersed values for the fatigue life. The individual value N for the number of cycles to failure associated with each maximum stress level in Figure 6.1 represents a magnitude about which the dispersed fatigue life data tends to centralize. It is therefore to be understood in the following discussions that the application of the S-N diagram of Figure 6.1 in the cumulative damage theories will result only in a prediction of some "central measure" of fatigue life.

6.2 STRESS HISTORIES

There exist many dynamic situations in which the fluctuating stress environment induced into a structural material cannot be described by a pure sinusoid imposed on a constant mean stress level. This indicates that more general descriptions of stress environments must be developed and fatigue prediction techniques constructed for these environments. The most complex stress environment that can be anticipated is one consisting of a sequence of excursions of differing magnitudes and proportions ordered in an unrecognizable or obscure manner, as shown in Figure 6.2.

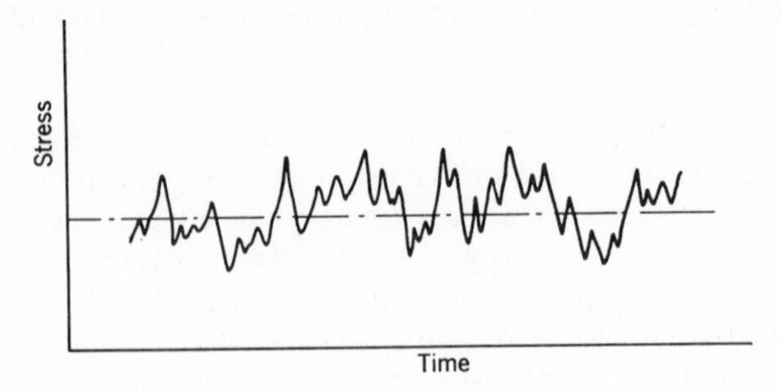

Figure 6.2 Complex stress pattern.

The type of stress pattern represented in Figure 6.2 commonly occurs in structural elements and is referred to as a "random" history. The random stress history precludes any representation by an analytic function of time due to the lack of any describable pattern. A description of this type of history is offered in the general case by the use of time-sensitive statistical measures.

6.2.1 *Multilevel Sinusoidal History*

Stress patterns of all degrees of complexity which are functionally describable can be produced. One of the simplest in this category is the multilevel sinusoidal history shown in Figure 6.3.

The multilevel sinusoidal history, as seen in Figure 6.3, is composed of a sequence of sinusoids in grouped arrangements in which each group is characterized by the magnitude and number of sinusoids contained in it. The set of distinct groups composing the history is arranged in a pattern that is repeated as many times as desired in an invariant fashion, as indicated in Figure 6.3. A complex case of the multilevel sinusoidal history occurs when every group of an extremely large number of different groups contains one sinusoidal cycle. A mixture between this multilevel sinusoidal pattern and a random history can be produced by randomly combining in sequence the set of distinct sinusoidal groups.

For the purpose of providing a quantitative description of the multilevel sinusoidal history the sinusoidal groups are characterized by their severity according to the number of cycles to failure associated with the group as found in pure sinusoidal testing. The mildest applied group of sinusoids is the one that is associated with the largest number of cycles to failure. The maximum stress of the sinusoids in this group is designated as S_1 and the

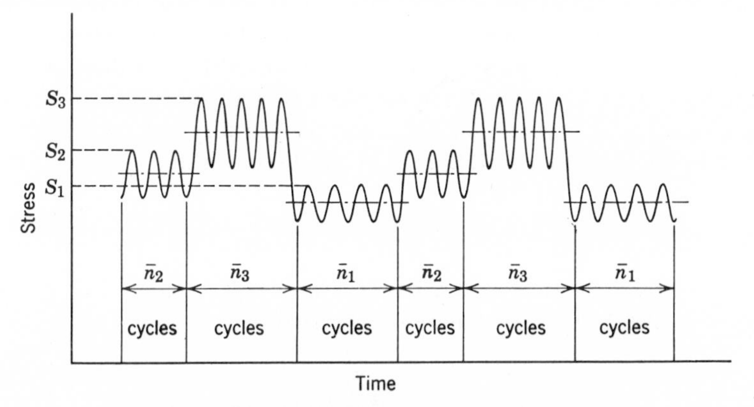

Figure 6.3 Multilevel sinusoidal stress history.

number of cycles to failure, as N_1. The group with the next level of severity has a maximum stress S_2 and an associated number of cycles to failure N_2, where $N_2 < N_1$. The designations continue in order to the group with the greatest severity, hence the lowest associated number of cycles to failure. The maximum stress for this group is denoted by S_h and the number of cycles to failure, by N_h. From these definitions it follows that

$$N_h < N_{h-1} < N_{h-2} \cdots < N_2 < N_1.$$

Each distinct group of sinuoids, in the subsequent discussion, is referred to as a "stress condition." Stress condition S_q is that group of sinusoids for which the associated number of cycles to failure is N_q. The number of cycles in a group at stress condition S_q is designated as $\bar{n}_q$, as shown in Figure 6.3. These descriptions indicate that there are h distinct sinusoidal groups, hence a corresponding number of stress conditions. The h sinusoidal groups are arranged consecutively in a prescribed pattern and the pattern is then rigidly repeated, as indicated in Figure 6.3. The combination of the h sinusoidal groups is referred to as a "block" in subsequent discussions and the total number of repetitions of the block is designated by n_B. From this notation it follows that the total applied stress history contains n_B groups at stress condition S_q, where $q = 1, 2, \ldots, h$. If n_q denotes the number of cycles in the total multilevel sinusoidal history at stress condition S_q, the above definitions will produce the relationship

$$n_q = n_B \bar{n}_q. \tag{6.2}$$

A description of the multilevel sinusoidal history thus entails the values S_q and N_q for each stress condition, the number of cycles $\bar{n}_q$ in the group at each stress condition S_q, the order of the groups in the block pattern as designated by an arrangement of the q values (i.e., an arrangement of the numbers 1 through h), and the number of repetitions n_B of the block pattern.

The number of repetitive blocks in a failure history is denoted by n_{BF} when the total number of cycles in the block is small in relation to the number of cycles required by the multilevel sinusoidal history to cause failure. For this case the number of cycles n_{qF} in the failure history applied at stress condition S_q is found from

$$n_{qF} = n_{BF} \bar{n}_q. \tag{6.3}$$

It follows that the total number of cycles N_F required by the multilevel sinusoidal history to cause failure is expressed by

$$N_F = \sum_{q=1}^{h} n_{qF}. \tag{6.4}$$

If α_q is the percentage of cycles in the total history applied at stress condition S_q, then for the case in which the history causes failure

$$n_{qF} = \alpha_q N_F. \tag{6.5}$$

The multilevel sinusoidal stress history has been used extensively as an equivalent representation for diverse types of complex patterns even though the correspondence may be quite vague. Because of the sinusoidal character and extensive usage of the multilevel sinusoidal stress history, it has been the predominant stress pattern for testing the validity of cumulative damage theories based on sinusoidal test data, hence is much referred to in subsequent discussions.

6.3 FAILURE CRITERIA BASED UPON TEST DATA

The *S-N* diagram, previously discussed, is usually obtained by testing small size specimens of materials formed as cylindrical shapes in push-pull tests, as wires in rotating beam tests, and as flat bars in cantilever bending tests. Failure of these specimens occurs when a crack of the order of the specimen's cross section has been produced. In most cases the specimen has been completely split into two parts at failure.

A crack of the size of a typical specimen's cross section is usually not sufficient to cause failure of many structural shapes that are composed of an extensive amount of material. These shapes have a redundancy of stress paths and lack the regular or homogeneous distribution of internal stresses that is applied to the simple fatigue specimens. Fatigue failure of extensive structural shapes should be dictated by a lack of functional performance. The application of the *S-N* diagram or any approach based on specimen testing implies, however, that a failure must be defined when a crack of the order of a specimen's cross sectional dimension appears. This is a conservative criterion for many structural shapes, but its application provides some margin against fatigue-life variability.

6.4 CYCLE RATIO AND LINEAR DAMAGE

In the sections that follow various cumulative damage theories that provide an interesting contrast of viewpoints are discussed. The simplest and most used of these theories is that propounded by Miner [1] and proposed earlier by Palmgren [2]. This theory is quite often referred to as the linear cumulative damage rule and utilizes the simple cycle ratio as its basic measure of damage.

To specify analytically the concept of damage referred to above consider the application of n cycles of a pure sinusoidal history with maximum stress S_{max} and mean stress S_m. The linear damage rule proposes that these n cycles will cause an amount of damage D given by

$$D = \frac{n}{N}, \tag{6.6}$$

where N is the number of cycles to failure (i.e., a central measure for the number of cycles to failure) under the specified sinusoidal history. The quantity (n/N) is termed the "cycle ratio" and has a value less than or equal to unity. It is implied for the pure sinusoidal history that failure occurs when $n = N$; that is, when $D = 1$. The damage concept is stated as being linear because of the linear relationship to the cycle ratio expressed in (6.6).

6.5 MINER'S THEORY

The basis of Miner's theory of fatigue prediction is the linear damage concept discussed in Section 6.4. If a multilevel sinusoidal stress history, as given in Figure 6.3, is applied to a structural material, it is hypothesized in Miner's theory that (a) each group of sinusoids contributes an amount of damage given by the linear cycle ratio for the group; (b) the damage accruing from any group of sinusoids is not dependent on the group's location in the stress history, and (c) the total applied damage is equal to the sum of the damages contributed by each sinusoidal group.

From the definitions in Section 6.2.1 this discussion can be put into quantitative terms. Thus, first, consider the group of $\bar{n}_q$ sinusoids in the multilevel sinusoidal stress pattern whose maximum stress level is equal to S_q. The number of sinusoids at this stress condition S_q needed to produce a failure is given by a central or representative value denoted by N_q. The damage $\bar{D}_q$ resulting from this group of sinusoids, according to Miner's criterion, is obtained as

$$\bar{D}_q = \frac{\bar{n}_q}{N_q}. \tag{6.7}$$

Miner's theory further specifies that the damage D_B produced by the block of h distinct sinusoidal groups is found from

$$D_B = \sum_{q=1}^{h} \bar{D}_q = \sum_{q=1}^{h} \frac{\bar{n}_q}{N_q}. \tag{6.8}$$

For the complete stress history which contains n_B repetitions of the basic block of h sinusoidal groups, the total damage D is given by

$$D = n_B D_B = n_B \sum_{q=1}^{h} \frac{\bar{n}_q}{N_q}. \tag{6.9}$$

From (6.2) the expression (6.9), in terms of the total number of cycles n_q in the history at stress condition S_q, takes the form

$$D = \sum_{q=1}^{h} \frac{n_q}{N_q}. \tag{6.10}$$

Miner's theory now requires that the amount of damage at failure D_F, caused by the multilevel sinusoidal stress history, is equal to unity. Hence Miner's failure criterion becomes

$$\sum_{q=1}^{h} \frac{n_{qF}}{N_q} = 1. \tag{6.11}$$

In (6.11) $n_q = n_{qF}$ signifies the number of cycles applied at stress condition S_q in a total history that causes failure of the acted on material. The result in (6.11) has been interpreted as indicating that each distinct stress condition in the complex history "removes an amount of life" given by its associated cycle ratio. The total number of cycles N_F needed by the multilevel sinusoidal history to cause failure (i.e., a central or representative value for the number of cycles to failure) can now be found from (6.11). Substituting (6.5) into (6.11), we find that

$$N_F = \frac{1}{\sum_{q=1}^{h} \frac{\alpha_q}{N_q}}. \tag{6.12}$$

6.5.1 *Structure of Miner's Theory*

To establish a perspective for the more complex fatigue prediction theories to be discussed subsequently, it is worth looking at the Miner fatigue failure criterion in more detail. For this purpose consider the two-level sinusoidal stress history shown in Figure 6.4 in which the two sinusoidal stress conditions, denoted by their maximum stress levels S_1 and S_2, are applied to a material alternately in groups of $\bar{n}_1$ and $\bar{n}_2$ cycles, respectively.

The number of cycles to failure at stress conditions S_1 and S_2 is represented by N_1 and N_2, respectively. It is presumed that S_2 is the severer stress condition, hence $N_2 < N_1$. From (6.6) and the discussion in Section 6.5, it is seen that the damage due to a pure sinusoidal history is linearly proportional

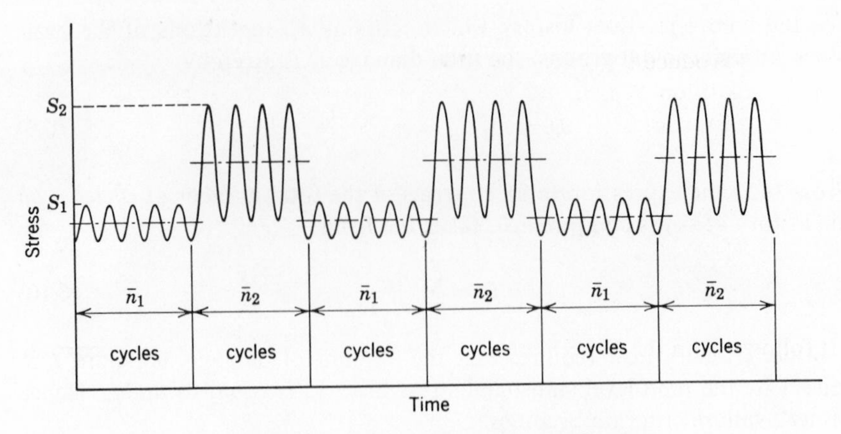

Figure 6.4 Two-level sinusoidal stress history.

to the number of applied cycles n. Thus a representation of the type shown in Figure 6.5 can be given for Miner's damage rule.

From (6.6) the equations for the straight lines through the origin in Figure 6.5 are given by $D = (n/N_1)$ and $D = (n/N_2)$ for stress conditions S_1 and S_2, respectively. From these equations or Figure 6.5 the number of cycles $\bar{n}_{21}$ applied at stress condition S_2 that would produce the same amount of damage $\bar{D}_1$ as the $\bar{n}_1$ cycles applied at stress condition S_1 is found from

$$\bar{D}_1 = \frac{\bar{n}_1}{N_1} = \frac{\bar{n}_{21}}{N_2},$$

or (6.13)

$$\bar{n}_{21} = \frac{N_2}{N_1}\bar{n}_1.$$

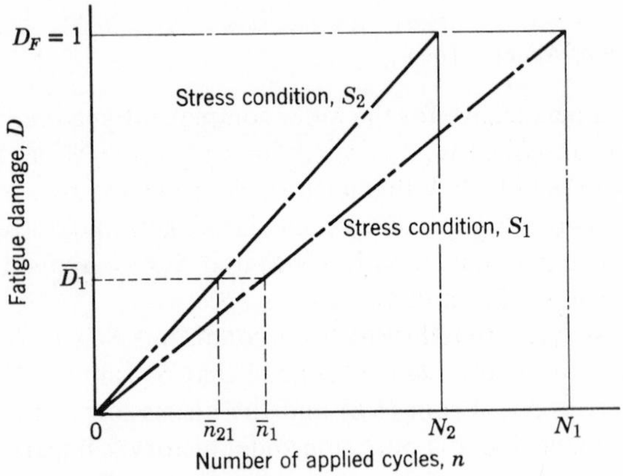

Figure 6.5 Damage representation for Miner's theory.

By a similar analysis the number of cycles $\bar{n}_{12}$ applied at stress condition S_1 that would produce the same amount of damage $\bar{D}_2$ as the $\bar{n}_2$ cycles applied at stress condition S_2 is found again by equating the cycle ratios associated with the two histories. Thus

$$\bar{D}_2 = \frac{\bar{n}_2}{N_2} = \frac{\bar{n}_{12}}{N_1},$$

or

$$\bar{n}_{12} = \frac{N_1}{N_2}\,\bar{n}_2. \tag{6.14}$$

It follows from the underlying assumptions specified for Miner's theory in Section 6.5 that the damage $\bar{D}_1$ caused by the group of $\bar{n}_1$ cycles at stress condition S_1 does not depend on the group's location in the stress pattern. Corresponding remarks can be stated in reference to the damage $\bar{D}_2$. Now the number of cycles $\bar{n}_{2B}$ at stress condition S_2 that yields the same amount of damage as that caused by the block containing $\bar{n}_1$ cycles at stress condition S_1, plus $\bar{n}_2$ cycles at stress condition S_2, is obtained as follows:

$$\bar{n}_{2B} = \bar{n}_{21} + \bar{n}_2. \tag{6.15}$$

Utilizing (6.13),

$$\bar{n}_{2B} = N_2\left(\frac{\bar{n}_1}{N_1} + \frac{\bar{n}_2}{N_2}\right). \tag{6.16}$$

Considering $\bar{n}_{2B}$ to be relatively small compared with the number of cycles to failure N_2 associated with stress condition S_2, the number of repetitive blocks to failure n_{BF} is obtained with sufficient accuracy from

$$n_{BF} = \frac{N_2}{\bar{n}_{2B}}. \tag{6.17}$$

Substituting (6.16) into (6.17), we find that

$$n_{BF}\left(\frac{\bar{n}_1}{N_1} + \frac{\bar{n}_2}{N_2}\right) = 1. \tag{6.18}$$

Using (6.3) in (6.18), we obtain the result

$$\sum_{q=1}^{2} \frac{n_{qF}}{N_q} = 1. \tag{6.19}$$

The derived equation (6.19) is a special case of Miner's failure criterion given in (6.11). The determination of the more general expression in (6.11) can be obtained by the same procedure with respect to the analysis of a multilevel history which contains an arbitrary number h of distinct sinusoidal groups.

Miner's theory (6.11) or (6.12) can be simply expressed in terms of the maximum stress levels S_q of the sinusoidal groups if the stress conditions S_q are associated with the same *S-N* diagram. This is the case, for instance, when each of the distinct sinusoidal groups has the same mean level or the same stress ratio ($S_{\min}/S_{\max}$). To proceed to the desired result, (6.1) for the *S-N* diagram is expressed as

$$\frac{N_q}{N_r} = \left(\frac{S_r}{S_q}\right)^b. \tag{6.20}$$

By solving (6.20) for N_q and substituting the result into (6.11) and (6.12) Miner's theory becomes

$$\sum_{q=1}^{h} \frac{n_{qF}}{N_r}\left(\frac{S_q}{S_r}\right)^b = 1, \tag{6.21}$$

or

$$N_F = \frac{1}{\sum_{q=1}^{h} \frac{\alpha_q}{N_r}\left(\frac{S_q}{S_r}\right)^b}. \tag{6.22}$$

The reference maximum stress level S_r appearing in these equations need not coincide with any of the maximum stress levels S_q that describe the multilevel sinusoidal stress history. In many representations the level S_h is selected for the reference maximum stress level S_r and, in this case, N_r is replaced by N_h. Of course, the predictions from Miner's theory do not depend on the particular level chosen for the reference maximum stress, as can be seen from the basic equations (6.11) and (6.12).

6.6 STRUCTURE OF CUMULATIVE DAMAGE THEORIES

Miner's theory was developed in Section 6.5.1 by using the hypotheses specified in Section 6.5. The approach applied in Section 6.5.1 to establish Miner's theory is referred to as an "equivalent cycles" approach. This terminology stems from the consideration that the number of cycles in a sinusoidal group at stress condition S_q is transformed into an equivalent number of cycles at reference stress condition S_r such that the original and transformed groups produced the same damage.

Another procedure that can be used to determine Miner's failure criterion is referred to as an "equivalent stress" approach, which directly considers a history that produces failure and establishes the contribution of each stress condition to the total damage according to the number of cycles in the total history applied at the stress condition. For the composite of stress conditions total damage is again given by the sum of the contributions from each stress condition. In the equivalent stress approach damage resulting from each stress condition in the failure history is not obtained by accumulating the damage contributed by each sinusoidal group of the given stress condition, as is the case in the equivalent cycles approach. The equivalent stress approach requires, first, the accumulation of the total number of cycles applied at the particular stress condition in the history and then associating the damage with respect to this total number of cycles. This distinction between the manner in which the two approaches establish the damage arising from each stress condition in a total history will not produce a difference in damage accumulation when damage is specified by the simple linear cycle ratio of Section 6.4. In other damage specifications to be discussed the two approaches can lead to different results.

The equivalent stress approach derives its name from the consideration that there exists a stress condition which will cause failure in the same total number of cycles as that needed by the complex history. To illustrate the equivalent stress approach consider that damage is defined by the linear cycle ratio described in Section 6.4. From this discussion the equivalent stress approach specifies, in this case, that the total damage at failure D_F, associated with a multilevel sinusoidal history containing h distinct sinusoidal groups, is given as

$$D_F = \sum_{q=1}^{h} \frac{n_{qF}}{N_q}. \tag{6.23}$$

In (6.23) n_{qF} is the number of cycles in the failure history at stress condition S_q. The total number of cycles required by the multilevel sinusoidal history to cause failure is designated by N_F in Section 6.2.1. Now the equivalent stress approach specifies that a stress condition S_e would also produce failure in N_F cycles. If D_{Fe} denotes the damage at failure associated with the stress condition S_e, a fatigue failure criterion can be established by using the result in (6.23) and setting

$$D_F = D_{Fe}. \tag{6.24}$$

If N_e is the number of cycles to failure at stress condition S_e, as obtained from an S-N diagram, then, from the above description, $N_e = N_F$ or

$$\frac{N_F}{N_e} = 1. \tag{6.25}$$

Now, since damage is defined by the simple linear cycle ratio of Section 6.4 and (6.25) expresses the cycle ratio at failure for stress condition S_e, it follows that

$$D_{Fe} = \frac{N_F}{N_e} = 1. \tag{6.26}$$

From (6.23) and (6.24),

$$\sum_{q=1}^{h} \frac{n_{qF}}{N_q} = 1.$$

Thus the equivalent stress approach also yields Miner's fatigue failure criterion when damage is expressed by the simple linear cycle ratio.

6.6.1 Modifying Miner's Theory

It has been found from testing under multilevel sinusoidal histories that Miner's theory predicts a longer life (i.e., a greater central or representative number of cycles to failure) for many metallic materials than that actually witnessed. It would thus be of interest to modify Miner's theory in order to produce more conservative fatigue-life predictions. For this purpose consider that the damage D resulting from n cycles applied at a stress condition S with an associated number of cycles to failure N is defined as

$$D = \left(\frac{n}{N}\right)^x, \tag{6.27}$$

where x is a positive constant. This damage concept, it is noted, is specified as a nonlinear relationship of the cycle ratio for which the linear concept used in Miner's theory is a special case. As in the case of linear damage, the damage specification in (6.27) implies that failure occurs for the pure sinusoidal history when $n = N$; that is, when $D = 1$.

From the two-level sinusoidal history in Figure 6.4 and the equivalent cycles approach in Section 6.5.1 it can be determined whether the definition of damage given by (6.27) will lead to a useful modification of Miner's theory.

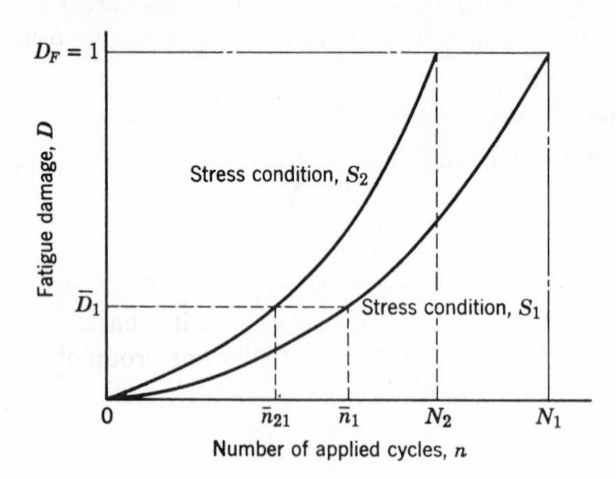

Figure 6.6 Nonlinear damage representation.

To pursue this determination a plot parallel to Figure 6.5 is shown in Figure 6.6 for the damage specification in (6.27).

From (6.27) the nonlinear plots in Figure 6.6 are expressed as $D = (n/N_1)^x$ and $D = (n/N_2)^x$ for stress conditions S_1 and S_2, respectively. The curvature of the plots in Figure 6.6 are based on the presumption that the damage accelerates with increasing numbers of applied cycles; that is, (dD/dn) becomes larger as n increases or $(d^2D/dn^2) > 0$ for each stress condition and all values of the number of applied cycles n. Differentiating (6.27), we find that

$$\frac{dD}{dn} = \frac{x}{N}\left(\frac{n}{N}\right)^{x-1}. \qquad (6.28)$$

From (6.28) it is seen that for (dD/dn) to increase as n increases requires $x > 1$. It does not appear plausible to consider $x < 1$, since this implies that the rate of damage decreases as the number of applied cycles increases.

From these expressions for the plots in Figure 6.6 the number of cycles $\bar{n}_{21}$ applied at stress condition S_2 that would produce the same amount of damage $\bar{D}_1$ as the $\bar{n}_1$ cycles applied at stress condition S_1 is found from

$$\bar{D}_1 = \left(\frac{\bar{n}_1}{N_1}\right)^x = \left(\frac{\bar{n}_{21}}{N_2}\right)^x,$$

or

$$\bar{n}_{21} = \frac{N_2}{N_1}\bar{n}_1.$$

This last result is the same as that obtained in (6.13) for the linear damage specification. From this result (6.15) through (6.19) would again follow. Thus the equivalent cycles approach produces the Miner's failure criteria for the damage specification in (6.27), regardless of the value of the exponent x.

This result is not an accident but forms a special case of broader consideration systematized by Kaechele [3]. Kaechele refers to damage specifications such as (6.6) and (6.27) as stress-independent, implying that a plot of damage against the cycle ratio produces the single curve shown in Figure 6.7, regardless or independent of the stress condition that causes the damage. A cumulative damage theory based upon a stress-independent damage specification is termed in [3] a stress-independent theory.

Under the circumstances in which the sinusoidal groups have small numbers of cycles with respect to the total number applied it is basic to the equivalent cycles approach that the damage contributed by any group of sinusoids is the same, regardless of its location in the history. This hypothesis implies that the damage perpetrated by any stress group does not depend on the previously acquired damage, thus ruling out the possibility of damage interaction resulting from the manner in which the sinusoidal stress groups are sequenced. A

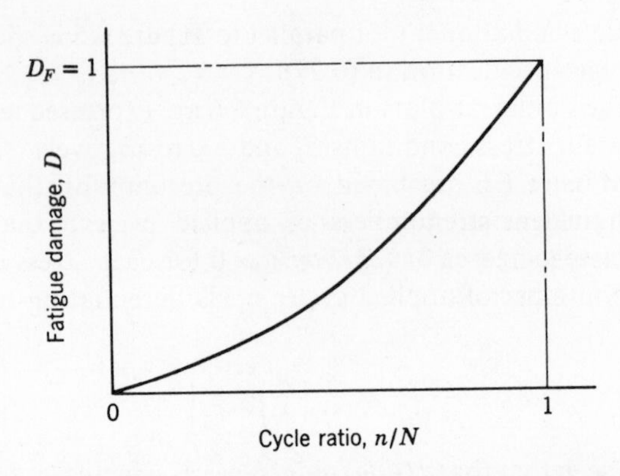

Figure 6.7 Stress independent damage representation.

cumulative damage theory arrived at by using this hypothesis is termed, in [3], an interaction-free theory. It will be noticed from these descriptions that Miner's theory is a stress-independent, interaction-free theory.

A derivation in [3] shows that the use of the equivalent cycles approach will always lead to Miner's failure criterion when a stress independent damage criterion is specified and the damage accrued by the sinusoidal stress groups is taken as interaction free. This result can be anticipated from the previous discussion by use of the plot in Figure 6.7 which represents stress independent damage. This plot can be expressed in functional form as

$$D = f\left(\frac{n}{N}\right). \tag{6.29}$$

Specifically, when $\bar{n}_1$ cycles are applied at stress condition S_1, the resulting damage $D = \bar{D}_1$ is found from

$$\bar{D}_1 = f\left(\frac{\bar{n}_1}{N_1}\right). \tag{6.30}$$

The number of cycles $\bar{n}_{21}$ at stress condition S_2, which cause the same amount of damage $\bar{D}_1$ as the $\bar{n}_1$ cycles at stress condition S_1, is obtained from (6.29) as

$$\bar{D}_1 = f\left(\frac{\bar{n}_{21}}{N_2}\right). \tag{6.31}$$

In Figure 6.7 it is seen that damage and cycle ratio are related by a one-to-one correspondence. From this relationship and (6.30) and (6.31) it follows that

$$\frac{\bar{n}_1}{N_1} = \frac{\bar{n}_{21}}{N_2},$$

or

$$\bar{n}_{21} = \frac{N_2}{N_1}\bar{n}_1.$$

This is the same result found in (6.13) for the linear damage specification. As shown in Section 6.5.1, this result leads directly to Miner's failure criterion.

Before leaving the considerations of this section, it would be of interest to determine whether the equivalent stress approach can provide a significant modification to Miner's failure criterion when damage is specified by (6.27). Following the equivalent stress approach as discussed in Section 6.6 and using the damage specification in (6.27), the total damage at failure, D_F, associated with the two-level sinusoidal history depicted in Figure 6.4 is given as

$$D_F = \left(\frac{n_{1F}}{N_1}\right)^x + \left(\frac{n_{2F}}{N_2}\right)^x = \sum_{q=1}^{2} \left(\frac{n_{qF}}{N_q}\right)^x. \tag{6.32}$$

In (6.32) n_{qF} denotes the total number of cycles in the failure history applied at stress condition S_q. The total number of cycles N_F required by the two-level history to cause failure is found from (6.4):

$$N_F = n_{1F} + n_{2F} = \sum_{q=1}^{2} n_{qF}. \tag{6.33}$$

An equivalent stress condition S_e can now be specified which also produces failure in N_F cycles. The results in (6.24) and (6.25) directly follow and it is seen from (6.25) that

$$\left(\frac{N_F}{N_e}\right)^x = 1. \tag{6.34}$$

The left-hand side of (6.34) expresses the damage at failure D_{Fe} for the stress condition S_e, as obtained from the damage specification in (6.27). The result in (6.34) is based on the equality between N_F and the number of cycles to failure N_e from an *S-N* diagram at stress condition S_e. Thus

$$D_{Fe} = \left(\frac{N_f}{N_e}\right)^x = 1, \tag{6.35}$$

and from (6.24) and (6.32)

$$\sum_{q=1}^{2} \left(\frac{n_{qF}}{N_q}\right)^x = 1. \tag{6.36}$$

Using the damage specification in (6.27), the equivalent stress approach produces the modification to Miner's failure criterion obtained in (6.36). Substituting (6.5) into (6.36), the following result is obtained for N_F:

$$N_F = \left[\left(\frac{\alpha_1}{N_1}\right)^x + \left(\frac{\alpha_2}{N_2}\right)^x\right]^{-1/x}. \tag{6.37}$$

Requiring that $x > 1$, as discussed above, we see that

$$\left[\left(\frac{\alpha_1}{N_1}\right)^x + \left(\frac{\alpha_2}{N_2}\right)^x\right]^{1/x} < \left[\left(\frac{\alpha_1}{N_1} + \frac{\alpha_2}{N_2}\right)^x\right]^{1/x} = \left(\frac{\alpha_1}{N_1} + \frac{\alpha_2}{N_2}\right). \tag{6.38}$$

From (6.38) and (6.12) it follows that the modification arrived at in (6.37) by using the damage specification in (6.27) is contrary to the desire to obtain a theory that yields more conservative (i.e., shorter) fatigue-life predictions than found from Miner's theory. Thus neither the equivalent cycles approach nor the equivalent stress approach produces a satisfactory variation to Miner's theory by using the damage specification in (6.27) with $x > 1$.

6.7 GROVER'S THEORY

An interesting variation to Miner's theory has been proposed by Grover [4]. Grover's theory considers that the fatigue life of a material subjected to a complex stress history is composed of two stages: (a) an initial number of cycles, N'_F, required to nucleate the failure producing crack and (b) a terminal number of cycles, N''_F, needed to propagate this crack to the point at which the material no longer performs its required function. The total number of cycles N_F required to fail the material is thus given by

$$N_F = N'_F + N''_F. \tag{6.39}$$

We now consider a multilevel sinusoidal failure history with h distinct sinusoidal stress conditions

$$N'_F = \sum_{q=1}^{h} n'_{qF}, \tag{6.40}$$

and

$$N''_F = \sum_{q=1}^{h} n''_{qF}. \tag{6.41}$$

In (6.40) and (6.41) n'_{qF} and n''_{qF} designate, respectively, the number of cycles at stress condition S_q applied during the crack nucleation and crack propagation stages. If each stress group has a relatively small number of cycles, the percentage of cycles α_q in the total history at stress condition S_q also designates the percentage of cycles in the crack nucleation and propagation stages applied at stress condition S_q. Thus

$$n'_{qF} = \alpha_q N'_F, \tag{6.42}$$

and

$$n''_{qF} = \alpha_q N''_F. \tag{6.43}$$

The special case in which the applied failure history is a pure sinusoid at stress condition S_q can now be considered. For this case denote the number of cycles in the initial stage required to nucleate the failure-producing crack by N'_q and the number of cycles in the terminal stage needed to propagate this crack to failure by N''_q. If N_q is the central or representative value for the total number of cycles to failure at stress condition S_q, then

$$N_q = N'_q + N''_q, \tag{6.44}$$

or

$$\frac{N'_q}{N_q} + \frac{N''_q}{N_q} = 1. \tag{6.45}$$

Take

$$a_q = \frac{N'_q}{N_q}. \tag{6.46}$$

From (6.45) and (6.46) it is seen that

$$1 - a_q = \frac{N''_q}{N_q}. \tag{6.47}$$

Grover's theory now utilizes Miner's theory separately for the nucleation stage and for the propagation-to-failure stage. The fatigue criteria prescribed by Grover thus takes the form of the following two expressions:

$$\sum_{q=1}^{h} \frac{n'_{qF}}{N'_q} = 1, \tag{6.48}$$

and

$$\sum_{q=1}^{h} \frac{n''_{qF}}{N''_q} = 1. \tag{6.49}$$

Equation 6.48 determines whether the duration of the stress history is sufficient to nucleate the critical crack and (6.49) determines whether any excess of stress history beyond the nucleation stage is sufficient to propagate the critical crack to the point at which failure occurs. Substituting (6.42) and (6.46) into (6.48), it will be found that

$$N'_F = \frac{1}{\displaystyle\sum_{q=1}^{h} \frac{\alpha_q}{a_q N_q}} \tag{6.50}$$

Putting (6.43) and (6.47) into (6.49), it follows that

$$N''_F = \frac{1}{\displaystyle\sum_{q=1}^{h} \frac{\alpha_q}{(1 - a_q) N_q}}. \tag{6.51}$$

The prediction of fatigue life can now be obtained for Grover's theory by substituting (6.50) and (6.51) into (6.39). For the case in which all the a_q values are the same (i.e., equal to a constant value) independent of the stress condition, it can readily be seen that Grover's theory coincides with Miner's theory, (6.12). When the a_q magnitudes depend on the stress condition, it is shown in [3] that Grover's theory gives shorter fatigue life predictions than Miner's theory.

Grover's theory is interaction free in that the value of a_q associated with stress condition S_q is not influenced by the application of other stress conditions. Unlike the previously discussed theories, it is shown in [3] that Grover's theory is stress-dependent, meaning in the general case that a single plot, as in Figure 6.7, will not suffice to represent the relationship between damage and the linear cycle ratio for all stress conditions. The following section delves further into stress-dependent damage specifications.

The use of Grover's theory is based upon being able to determine the magnitude of a_q as a function of the stress condition S_q. Since, at present, means do not exist for establishing this relationship accurately (i.e., accurately determining the number of cycles needed to nucleate a critical crack), Grover's theory, although intriguing, has not found useful application.

6.8 MARCO-STARKEY THEORY

A stress independent cumulative damage theory is typified by a damage specification that relates damage to the linear cycle ratio by a single curve for all stress conditions, as shown in Figure 6.7. In contrast, a single curve will not suffice to show the damage specification for a stress-dependent theory. The theory proposed by Marco and Starkey [5] aptly illustrates the nature of a stress-dependent theory while featuring a damage specification that is a logical modification of the damage criterion in (6.27). The Marco-Starkey specification for the damage D arising from n cycles applied at stress condition S with an associated number of cycles to failure N is given by

$$D = \left(\frac{n}{N}\right)^{x_v}. \tag{6.52}$$

In (6.52) the exponent x_v is a variable quantity whose magnitude is dependent on the applied stress condition. When $x_v = x$ independent of the stress condition (i.e., a constant value for all stress conditions), the damage specification in (6.52) reduces to (6.27). For a pure sinusoidal history (6.52) implies that failure occurs when $n = N$; that is, when $D = 1$.

A plot of damage against the linear cycle ratio for the damage specification in (6.52) is shown in Figure 6.8. Since the exponent x_v is a function of the

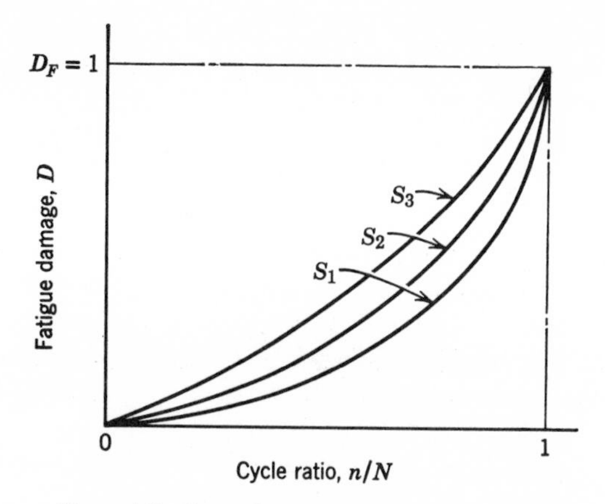

Figure 6.8 Stress-dependent damage representation.

stress condition, each stress condition, as shown in Figure 6.8, requires a separate plot. Marco and Starkey consider that x_v has a magnitude greater than one and approaches a value of one as the stress condition becomes severer. This characteristic is indicated in Figure 6.8, in which S_3 is shown as the severest stress condition and S_1 as the mildest.

Application of the equivalent cycles approach to the damage specification in (6.52) can now be considered. Suppose for analysis that the two-level sinusoidal history shown in Figure 6.4 is applied. Designate x_q as the value of the exponent x_v associated with stress condition S_q. Then from the damage specification in (6.52) the damage $\bar{D}'$ due to $\bar{n}_1$ cycles applied at stress condition S_1 is given by

$$\bar{D}' = \left(\frac{\bar{n}_1}{N_1}\right)^{x_1}. \tag{6.53}$$

The geometric representation of the result in (6.53) can be seen in Figure 6.9.

From (6.52) the number of cycles $\bar{n}_{21}$ applied at stress condition S_2 that would produce the same amount of damage $\bar{D}'$ as the $\bar{n}_1$ cycles applied at stress condition S_1 is obtained as

$$\bar{D}' = \left(\frac{\bar{n}_1}{N_1}\right)^{x_1} = \left(\frac{\bar{n}_{21}}{N_2}\right)^{x_2}$$

or

$$\frac{\bar{n}_{21}}{N_2} = \left(\frac{\bar{n}_1}{N_1}\right)^{x_1/x_2}. \tag{6.54}$$

The number of cycles $\bar{n}_{2B}$ at stress condition S_2 which causes the same amount of damage as the block of $\bar{n}_1$ cycles at stress condition S_1 and $\bar{n}_2$ cycles

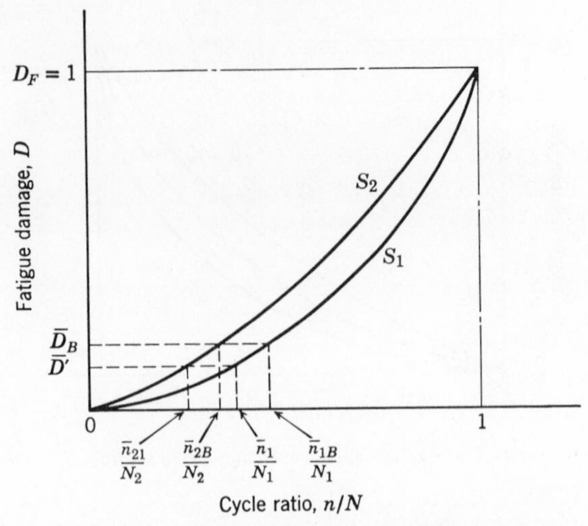

Figure 6.9 Damage representation for Marco-Starkey theory.

at stress condition S_2 is given by (6.15). From (6.15) and (6.54) it is found that

$$\bar{n}_{2B} = N_2\left[\left(\frac{\bar{n}_1}{N_1}\right)^{x_1/x_2} + \frac{\bar{n}_2}{N_2}\right]. \tag{6.55}$$

The damage $\bar{D}_B$ resulting from the first block of cycles is obtained from (6.52) and (6.55) as

$$\bar{D}_B = \left(\frac{\bar{n}_{2B}}{N_2}\right)^{x_2} = \left[\left(\frac{\bar{n}_1}{N_1}\right)^{x_1/x_2} + \frac{\bar{n}_2}{N_2}\right]^{x_2}. \tag{6.56}$$

The result in (6.56) is shown in Figure 6.9. The number of cycles $\bar{n}_{1B}$ applied solely at stress condition S_1 which causes an amount of damage $\bar{D}_B$ can be found from (6.52) and (6.56) as follows:

$$\bar{D}_B = \left(\frac{\bar{n}_{1B}}{N_1}\right)^{x_1} = \left(\frac{\bar{n}_{2B}}{N_2}\right)^{x_2} = \left[\left(\frac{\bar{n}_1}{N_1}\right)^{x_1/x_2} + \frac{\bar{n}_2}{N_2}\right]^{x_2}$$

or

$$\bar{n}_{1B} = N_1\left[\left(\frac{\bar{n}_1}{N_1}\right)^{x_1/x_2} + \frac{\bar{n}_2}{N_2}\right]^{x_2/x_1}. \tag{6.57}$$

The geometric basis of this result can be seen in Figure 6.9. The result in (6.57) permits the continued accumulation of damage, since the next group of $\bar{n}_1$ cycles is applied at stress condition S_1. Thus the number of cycles $\bar{n}'_{1B}$ at stress condition S_1 which causes the same amount of damage $\bar{D}'_B$ as the first block of cycles followed by the first group of $\bar{n}_1$ cycles in the second block is given by

$$\bar{n}'_{1B} = \bar{n}_{1B} + \bar{n}_1. \tag{6.58}$$

From (6.52), (6.58), and (6.57),

$$\bar{D}'_B = \left(\frac{\bar{n}'_{1B}}{N_1}\right)^{x_1} = \left\{\left[\left(\frac{\bar{n}_1}{N_1}\right)^{x_1/x_2} + \frac{\bar{n}_2}{N_2}\right]^{x_2/x_1} + \frac{\bar{n}_1}{N_1}\right\}^{x_1}. \tag{6.59}$$

Since the next group of cycles is applied at stress condition S_2, the number of cycles $\bar{n}'_{2B}$ at stress condition S_2 which causes an amount of damage equal to $\bar{D}'_B$ must now be determined. Hence from (6.52) and (6.59)

$$\bar{D}'_B = \left(\frac{\bar{n}'_{2B}}{N_2}\right)^{x_2} = \left(\frac{\bar{n}'_{1B}}{N_1}\right)^{x_1}. \tag{6.60}$$

Using the right-hand side of (6.59), we find from (6.60) that

$$\bar{n}'_{2B} = N_2\left\{\left[\left(\frac{\bar{n}_1}{N_1}\right)^{x_1/x_2} + \frac{\bar{n}_2}{N_2}\right]^{x_2/x_1} + \frac{\bar{n}_1}{N_1}\right\}^{x_1/x_2}. \tag{6.61}$$

The number of cycles $\bar{\bar{n}}_{2B}$ at stress condition S_2 which causes the same amount of damage as that accumulated by the first two blocks of the two-level sinusoidal history is found from

$$\bar{\bar{n}}_{2B} = \bar{n}'_{2B} + \bar{n}_2. \tag{6.62}$$

Substituting (6.61) into (6.62), we obtain

$$\bar{\bar{n}}_{2B} = N_2\left(\left\{\left[\left(\frac{\bar{n}_1}{N_1}\right)^{x_1/x_2} + \frac{\bar{n}_2}{N_2}\right]^{x_2/x_1} + \frac{\bar{n}_1}{N_1}\right\}^{x_1/x_2} + \frac{\bar{n}_2}{N_2}\right). \tag{6.63}$$

The damage $\bar{\bar{D}}_B$ resulting from the first two blocks of the history is then given by

$$\bar{\bar{D}}_B = \left(\frac{\bar{\bar{n}}_{2B}}{N_2}\right)^{x_2}. \tag{6.64}$$

The expression for $\bar{\bar{n}}_{2B}$ can now be substituted into (6.64). The analysis can be continued in the step-by-step manner followed above until the total accumulated damage equals unity. The total number of cycles to failure N_F is then determined by summing the applied cycles in the history through the sinusoidal group at which the damage accumulation reached unity.

This analysis and discussion indicate the difficulty in determining the number of cycles required by a multilevel sinusoidal history to cause failure for a stress-dependent damage specification of the continuous, increasing type shown in Figure 6.8. Although not quantitatively useful, it follows from a derivation in [3] that a stress dependent, interaction free damage specification of the type shown in Figure 6.8 will lead by use of the equivalent cycles approach to more conservative (shorter) fatigue-life predictions than Miner's theory. This result will be valid for the Macro-Starkey theory, since the damage specification in (6.52) is a special case of that shown in Figure 6.8.

The results of this section permit a determination of the validity of the simplified procedures used to determine the failure criterion and total number of cycles required to cause failure associated with the damage specifications in (6.6) and (6.27). From (6.63) and (6.55) we see that

$$\bar{\bar{n}}_{2B} = 2\bar{n}_{2B} \quad \text{when} \quad x_1 = x_2 = x. \tag{6.65}$$

It can be inferred from (6.65) and the formation of the results for each succeeding step of the analysis described in this section that the number of cycles at stress condition S_2 which causes the same amount of damage as B blocks of the history will be given by $B \cdot \bar{n}_{2B}$. This conclusion gives credence to (6.17) for establishing Miner's failure criterion and justifies the analysis and discussion in Section 6.6.1 pertaining to the damage specification in (6.27).

6.9 SHANLEY'S THEORY

A modification of Miner's theory, referred to as Shanley's theory [6], which also provides more conservative fatigue-life predictions than Miner's theory and gives further insight into the equivalent stress approach, is discussed in this section. The damage specification for Shanley's theory can be expressed as follows:

$$D = cS^{kb}n. \tag{6.66}$$

Equation 6.66 gives the damage D resulting from n cycles applied at stress condition S. The quantity b in (6.66) is the slope of the central portion of the S-N diagram, as shown in Figure 6.1, presuming that the set of distinct stress conditions in the multilevel sinusoidal history under consideration can be associated with the same S-N diagram. The quantities c and k in (6.66) are constants, where $k > 1$. The expression for damage in (6.66) differs from those already specified in that it is a function of the number of applied cycles rather than the cycle ratio.

From (6.1) it is seen that the equation for the central portion of the S-N diagram can be put in the form of

$$N = \frac{1}{cS^b}. \tag{6.67}$$

The constant c in (6.67) is obtained from the prescribed reference quantities S_r and N_r in (6.1):

$$c = \frac{1}{S_r{}^b N_r}. \tag{6.68}$$

Associating the constant c in (6.67) with the corresponding quantity in (6.66), we obtain the following result by substituting (6.67) into (6.66) and taking $k = 1$:

$$D = \frac{n}{N}.$$

Thus the damage specification in (6.66) reduces to the linear cycle ratio used in Miner's theory when $k = 1$.

Now consider that a multilevel sinusoidal history containing h distinct sinusoidal groups is applied until failure occurs. From the damage specification in (6.66) the damage in the failure history D_{qF} resulting from the n_{qF} cycles applied at stress condition S_q is, according to the equivalent stress approach,

$$D_{qF} = cS_q^{kb} n_{qF}. \tag{6.69}$$

Thus for the total history the damage at failure D_F is given by

$$D_F = \sum_{q=1}^{h} D_{qF} = \sum_{q=1}^{h} cS_q^{kb} n_{qF}. \tag{6.70}$$

The total number of cycles N_F required by the multilevel sinusoidal history to cause failure is given by

$$N_F = \sum_{q=1}^{h} n_{qF}. \tag{6.71}$$

An equivalent stress condition S_e that produces failure in N_F cycles is now specified in accordance with the equivalent stress approach. The stress condition S_e is taken as having N_e cycles to failure from the same S-N diagram applicable to the h stress conditions in the multilevel sinusoidal history. From (6.67) it is seen that

$$N_e = \frac{1}{cS_e^{\,b}}. \tag{6.72}$$

The damage at failure D_{Fe} resulting from stress condition S_e can be obtained from (6.66):

$$D_{Fe} = cS_e^{kb} N_F. \tag{6.73}$$

Solving for S_e in (6.72) and substituting the result into (6.73), we obtain the following expression for D_{Fe} by considering the equality between N_F and N_e:

$$D_{Fe} = \frac{1}{c^{k-1} N_F^{k-1}}. \tag{6.74}$$

Now, again from (6.67),

$$N_q = \frac{1}{cS_q^{\,b}}. \tag{6.75}$$

Solving (6.75) for S_q and substituting into (6.70),

$$D_F = \sum_{q=1}^{h} \frac{n_{qF}}{c^{k-1} N_q^{\,k}}. \tag{6.76}$$

From (6.74) and (6.76) the equivalent stress approach gives, by setting $D_{Fe} = D_F$,

$$\sum_{q=1}^{h} \frac{n_{qF}}{N_q^{\,k}} = \frac{1}{N_F^{k-1}}. \tag{6.77}$$

Using (6.5) in (6.77), we find that

$$\sum_{q=1}^{h} \alpha_q \left(\frac{N_F}{N_q}\right)^k = 1. \tag{6.78}$$

Solving (6.78) for N_F, Shanley's theory yields the following result for a central or representative value of the fatigue life:

$$N_F = \left[\left(\sum_{q=1}^{h} \frac{\alpha_q}{N_q^{\,k}}\right)^{1/k}\right]^{-1}. \tag{6.79}$$

As previously indicated, it is seen from (6.79) and (6.12) that Shanley's theory reduces to Miner's theory for $k = 1$. The value of $k = 2$ is commonly used for Shanley's theory. It is shown in [3] for this value of k that Shanley's theory yields shorter fatigue-life predictions than Miner's theory. It must be remarked, however, that even though Shanley's theory produces more conservative predictions than Miner's theory this does not signify that the fatigue life of a material will always be longer than the prediction found from Shanley's theory. Shanley's theory, as well as Miner's theory and the other theories discussed in this chapter, predicts only a central or representative value of fatigue life, hence gives no indication of the possible scatter or variability to be found in the fatigue lives of materials processed under and subjected to supposedly identical conditions.

6.10 CORTEN-DOLAN THEORY

The previously discussed theories were considered as interaction-free when accumulating the damage caused by the sequence of stress groups. In this section an interaction theory proposed by Corten and Dolan [7] is discussed. This theory begins with a damage specification similar to that utilized by Shanley in (6.66). For the Corten-Dolan theory the damage D in a material due to n cycles of a pure sinusoidal stress history can be expressed essentially as

$$D = rn^a. \tag{6.80}$$

In the simplest form of Corten and Dolan's theory r is a function of the stress condition and a is a constant independent of the stress condition. For stress condition S_q, r will take on the value r_q. It can be seen from (6.80) and (6.66) that Corten and Dolan's damage specification will coincide with Shanley's when

$$a = 1 \quad \text{and} \quad r = cS^{kb}. \tag{6.81}$$

Corten and Dolan use an equivalent cycles approach for determining a fatigue failure criterion based on the damage specification in (6.80). Considering the stress environment to be the two-level sinusoidal history in Figure 6.4, the damage $\bar{D}_1$ resulting from the first $\bar{n}_1$ cycles at stress condition S_1 is given on the basis of (6.80) as

$$\bar{D}_1 = r_1\bar{n}_1^{\,a}. \tag{6.82}$$

Now the number of cycles $\bar{n}_{21}$ at stress condition S_2 which causes the same amount of damage $\bar{D}_1$ as the $\bar{n}_1$ cycles at stress condition S_1 is found from

$$\bar{D}_1 = r_1\bar{n}_1^{\,a} = r_2\bar{n}_{21}^{\,a}.$$

Thus

$$\bar{n}_{21} = \left(\frac{r_1}{r_2}\right)^{1/a} \bar{n}_1. \tag{6.83}$$

The number of cycles $\bar{n}_{2B}$ at stress condition S_2 which causes the same amount of damage as the block of $\bar{n}_1$ cycles at stress condition S_1 and $\bar{n}_2$ cycles at stress condition S_2 is given by (6.15). Putting (6.83) into (6.15), we find that

$$\bar{n}_{2B} = \left(\frac{r_1}{r_2}\right)^{1/a} \bar{n}_1 + \bar{n}_2. \tag{6.84}$$

The number of repetitive blocks n_{BF} required to cause failure can now be obtained from (6.17). The use of (6.17) in this case is justified by the analysis and discussion in Section 6.8. It was shown in Section 6.8 that the result in (6.65) and the ensuing conclusions follow when the exponent in the Marco-Starkey damage specification is a constant, that is, when the damage specification (6.52) reduces to (6.27). The Corten-Dolan damage specification in (6.80) can be put into correspondence with (6.27) by setting

$$a = x \quad \text{and} \quad r = \left(\frac{1}{N}\right)^{x}. \tag{6.85}$$

Since (6.17) is appropriate to the determination of n_{BF}, the substitution of (6.84) into (6.17) with the use of (6.3) and some manipulation gives

$$\left(\frac{r_1}{r_2}\right)^{1/a} \frac{n_{1F}}{N_2} + \frac{n_{2F}}{N_2} = 1. \tag{6.86}$$

Using (6.5) in (6.86), we find that the total number of cycles N_F required by the two-level sinusoidal history to cause failure is

$$N_F = \left[\left(\frac{r_1}{r_2}\right)^{1/a}\frac{\alpha_1}{N_2} + \frac{\alpha_2}{N_2}\right]^{-1}. \tag{6.87}$$

This analysis can be readily generalized to the case of an applied multilevel sinusoidal history with h distinct stress groups. It will be found for this history that (6.86) and (6.87) become, respectively,

$$\sum_{q=1}^{h}\left(\frac{r_q}{r_h}\right)^{1/a}\frac{n_{qF}}{N_h} = 1, \tag{6.88}$$

and

$$N_F = \left[\sum_{q=1}^{h}\left(\frac{r_q}{r_h}\right)^{1/a}\frac{\alpha_q}{N_h}\right]^{-1}. \tag{6.89}$$

It can be seen from (6.89) that the determination of the total number of cycles to failure N_F requires evaluation of the factors $(r_q/r_h)^{1/a}$. By applying two-level stress histories with zero mean for each stress condition and using (6.87), Corten and Dolan [7] have established the following relationship empirically.

$$\left(\frac{r_q}{r_h}\right)^{1/a} = \left(\frac{S_q}{S_h}\right)^{d}. \tag{6.90}$$

In (6.90) d is a constant, depending on the material and is shown by Corten and Dolan to be relatively insensitive to the percentage of cycles in the repetitive blocks applied at each of the two stress conditions. Thus for the special case considered by Corten and Dolan (6.90) can be used in (6.88) and (6.89) to give, respectively,

$$\sum_{q=1}^{h}\left(\frac{S_q}{S_h}\right)^{d}\frac{n_{qF}}{N_h} = 1, \tag{6.91}$$

and

$$N_F = \left[\sum_{q=1}^{h}\left(\frac{S_q}{S_h}\right)^{d}\frac{\alpha_q}{N_h}\right]^{-1}. \tag{6.92}$$

It is now interesting to compare the results of the Corten-Dolan theory with the corresponding results obtained for Miner's theory in (6.21) and (6.22). Replacing the reference quantities S_r and N_r in Miner's theory with S_h and N_h, respectively, of the severest stress condition, we see that the Corten-Dolan theory can be visualized as utilizing Miner's equation in conjunction with a modified *S-N* diagram whose central portion is represented by a solid straight line described geometrically in Figure 6.10. Thus the Corten-Dolan theory

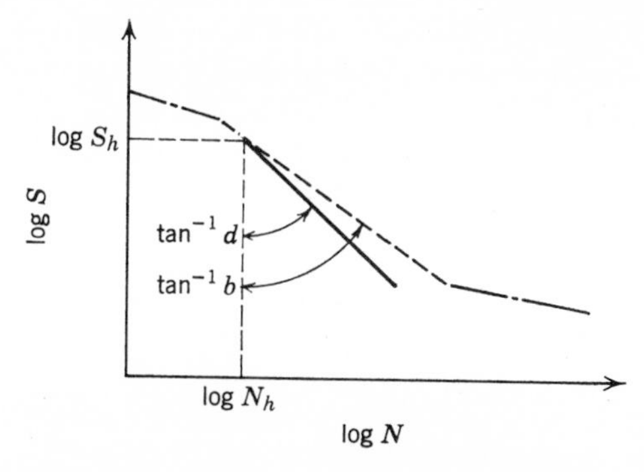

Figure 6.10 Modified S-N diagram for Corten-Dolan theory.

accounts for interaction between the stress groups in a history by the performance of two-level sinusoidal tests leading to the modification of the S-N diagram shown in Figure 6.10.

For any particular material the magnitude of d may be smaller or greater than b. By use of the severest stress condition S_h as the reference condition, the equation of the modified central portion of the S-N diagram is given parallel to (6.20) as

$$\frac{N_{qm}}{N_h} = \left(\frac{S_h}{S_q}\right)^d. \tag{6.93}$$

In (6.93) N_{qm} designates the central or representative number of cycles to failure on the modified S-N diagram associated with a pure sinusoidal stress history at stress condition S_q. It is seen from (6.93) and (6.20) that, for $d \neq b$, N_{qm} coincides with N_q only for the severest stress condition. Using (6.93) in (6.91), it is found that

$$\sum_{q=1}^{h} \frac{n_{qF}}{N_{qm}} = 1. \tag{6.94}$$

Although the appearance of the Corten-Dolan theory, as given by (6.94), is clearly as simple as Miner's theory, its application can be considerably more difficult. It can be suspected that the value of d for a particular material will change if the mean stresses are the same for all stress conditions in the history but at a value unequal to zero. Further, if the stress conditions have different means or, for instance, have no common stress ratio (S_{min}/S_{max}), it may occur that each stress condition S_q will have an uniquely associated value d_q. The Corten-Dolan approach can thus lead to a great deal of testing in order to make fatigue life predictions for complex stress histories. The circumstances described have not yet been explored to determine whether the Corten-Dolan approach is feasible for stress histories of a complex nature.

One other aspect has been pointed out by Schjelderup [8]. As developed, the Corten-Dolan theory depends on the severest stress condition being the reference condition for determining the damage accumulation in the application of the equivalent cycles approach. If a different stress condition were used for the reference condition, the result for the Corten-Dolan theory would be changed. Thus it can be seen for the random history in which stresses may become extremely large that a problem will occur in specifying some finite level for the maximum stress condition.

6.11 FREUDENTHAL-HELLER THEORY

Freudenthal and Heller [9] propose to consider interaction effects between stress groups in complex histories by the study and evaluation of results obtained from fatigue testing with histories closely simulating those which actually occur in the material in its operational environment. For the purpose of this consideration, Freudenthal and Heller postulate the fatigue failure criterion given in (6.94) in conjunction with a straight-line modification to the central portion of the *S-N* diagram resembling that obtained by Corten and Dolan. In terms of the representative, modified number of cycles to failure N_{qm} for stress condition S_q the modified *S-N* diagram used by Freudenthal and Heller can be expressed as

$$\frac{N_{qm}}{N_r^*} = \left(\frac{S_r^*}{S_q}\right)^{\delta}. \tag{6.95}$$

In (6.95) S_r^* is a defined reference stress condition that is unrelated to any applied stress history and N_r^* is the associated representative number of cycles to failure. Equation 6.95 is contrasted in Figure 6.11 with the following expression for the conventional *S-N* diagram:

$$\frac{N_q}{N_r^*} = \left(\frac{S_r^*}{S_q}\right)^{b}. \tag{6.96}$$

Expression (6.95) and (6.96) indicate that the modified and conventional *S-N* diagrams will be coincident only at stress condition $S_q = S_r^*$ for which $N_q = N_{qm} = N_r^*$. The reference stress condition S_r^* will usually have such intensity that N_r^* will fall in the range of 10^3 to 10^4 cycles. Specifying the reference condition S_r^*, as discussed above, eliminates the problem of the history related reference stress condition occurring in the Corten-Dolan theory.

From (6.94) and (6.5) the total number of cycles N_{Fm} required to cause failure of a material under a multilevel sinusoidal stress history is as follows,

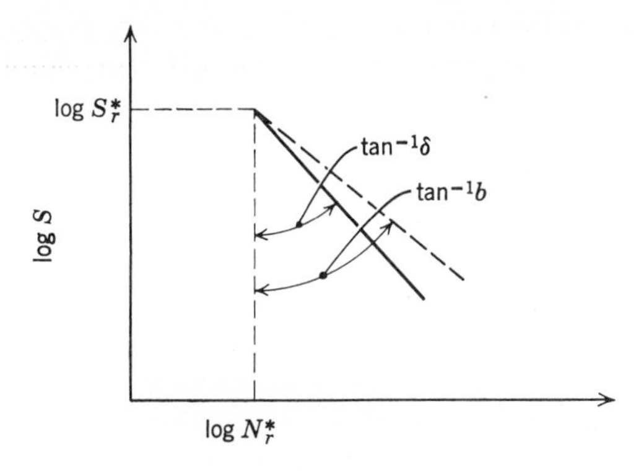

Figure 6.11 Modified S-N diagram for Freudenthal-Heller theory.

based on the modified *S-N* diagram:

$$N_{Fm} = \left(\sum_{q=1}^{h} \frac{\alpha_q}{N_{qm}} \right)^{-1}. \tag{6.97}$$

Using the result in (6.12) for the total number of cycles N_F required to cause failure, based on the conventional *S-N* diagram, we see that

$$\frac{N_{Fm}}{N_F} = \frac{\sum_{q=1}^{h} \alpha_q / N_q}{\sum_{q=1}^{h} \alpha_q / N_{qm}}. \tag{6.98}$$

Substituting (6.95) and (6.96) into (6.98), we find that

$$N_{Fm} = N_F \frac{\sum_{q=1}^{h} \alpha_q (S_q / S_r^*)^b}{\sum_{q=1}^{h} \alpha_q (S_q / S_r^*)^\delta}. \tag{6.99}$$

This last result considers that all stress conditions in the multilevel sinusoidal history under consideration have a common *S-N* diagram. For a particular multilevel sinusoidal history the magnitude of b associated with the stress conditions in the history is known from previous pure sinusoidal fatigue testing, and N_F is determined by (6.12). From the results of fatigue testing with the multilevel history N_{Fm} will be experimentally obtained, and thus the slope δ of the modified *S-N* diagram with respect to the vertical axis, shown in Figure 6.11, is the only remaining quantity to be determined. It can be seen that the value of δ associated with any experimental value of N_{Fm} can be found from (6.99).

The modified S-N diagram, which is used to account for interaction effects between stress groups, is obtained by Freudenthal and Heller by the performance of fatigue tests with multilevel sinusoidal histories that simulate the actual environment. This is in contrast to the Corten-Dolan approach which utilizes two-level sinusoidal histories for assessing interaction effects. Results from the Freudenthal-Heller approach indicate that the fatigue life of a material is sensitive to the percentage of the environmental stress history that occurs at each stress condition, whereas the Corten-Dolan two-level test results indicate a lack of sensitivity with respect to this characteristic. It is anticipated from the complex fatigue testing required in the Freudenthal-Heller approach that for any material the slope δ of the modified S-N diagram can be functionally related to the characteristics of the applied complex stress pattern. The availability of these relationships has not yet been sufficiently established to ascertain whether the approach proposed by Freudenthal and Heller can be feasibly utilized in a wide variety of circumstances.

6.12 SUMMARY AND REMARKS

The popular cumulative damage theories described, as well as other proposed theories, cannot be relied on to produce fatigue life predictions of sufficient accuracy for many encountered circumstances. Because there is no indication at present that any cumulative damage theory will give predictions significantly better than all other proposed theories, the use of Miner's theory in the majority of applications is still recommended. When experience dictates, it may be more desirable to use Miner's theory in the form of

$$\sum_{q=1}^{h} \frac{n_{qF}}{N_q} = K, \tag{6.100}$$

where $K \neq 1$. For conservative applications, as utilized by Mains [10], the value of $K = 0.3$ is used in (6.100).

Miner's theory depends on pure sinusoidal fatigue data, whereas the other theories presented require at least as much or more supporting test information. As mentioned at the beginning of this chapter and implied in the subsequent discussion, it has not been substantiated by experience that sinusoidal fatigue data are satisfactory as a basis for cumulative damage theories required to predict the fatigue lives of materials under complex fluctuating stress histories. Further, there is no assurance that complex stress patterns can be parametrically categorized in a simple manner to fill a role in cumulative damage approaches similar to that presently provided by the pure sinusoid. As an illustration of the complex nature of this problem, consider that an element of material is subjected to the random stress history shown in Figure 6.2. For the stress condition depicted in Figure 6.2, the stress levels are random quantities, hence have an associated probability distribution. Suppose that, for this illustration, the normal distribution represents the probability law of the randomly varying stress levels for a sufficiently long application of the stress condition. This signifies that the mean stress level and rms (root-mean-square) stress level completely describe the probability law underlying the randomly produced stress levels.

The stress condition shown in Figure 6.2 has some source or mechanism of generation. It is presumed that the repetitive generation of this stress condition will produce a set of histories with common probabilistic properties (i.e., a stationary process is assumed in this illustration). The appearance of the fluctuating portion of each of the stress histories, however, can be significantly different, as shown in Figure 6.12.

The differences in the sequential behavior of the stress-level excursions in the patterns of Figure 6.12 indicate that this group of histories may not have a common representative fatigue life, even though their underlying probability laws are the same. If random shaker fatigue tests indicate that the mean stress and rms levels are the primary parameters affecting the fatigue life of a material subjected to random, normally distributed stress histories, it is possible, for example, to establish a fatigue-life prediction theory similar to Miner's theory in which the random history assumes the role of the pure sinusoid. Applying normally distributed stress histories with common mean stress and varying rms levels to a set of specimens of a particular

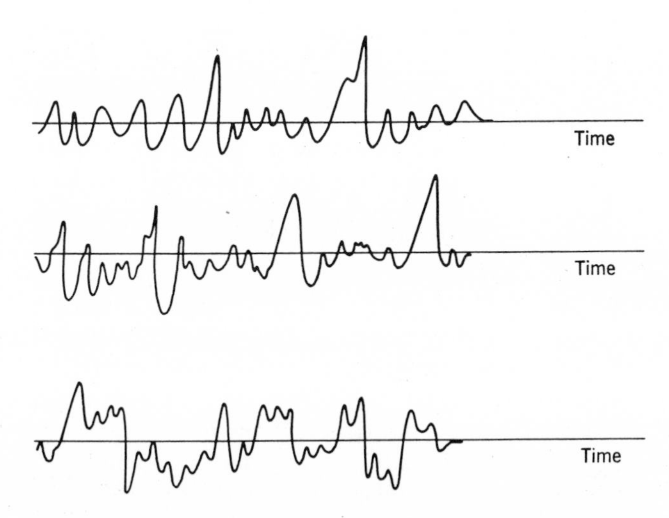

Figure 6.12 Repetitive stress patterns generated by a common source.

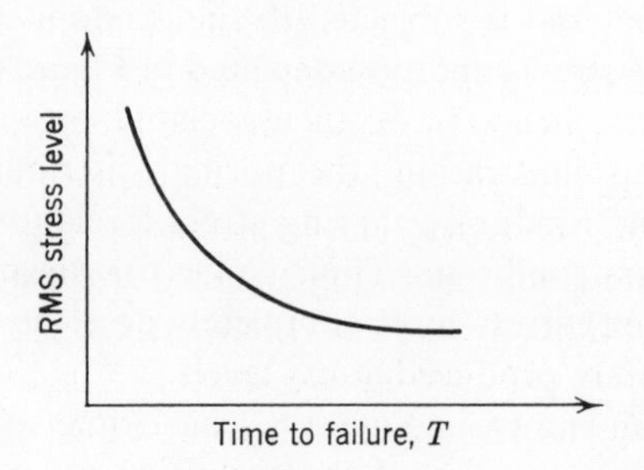

Figure 6.13 Diagram relating time to failure to RMS stress level.

material, a diagram of the relationship between the rms level of the applied history and the representative time to failure T can be determined. This diagram, which is analogous to the *S-N* diagram for pure sinusoidal histories, is displayed in Figure 6.13.

A set of diagrams of the type shown in Figure 6.13 can be obtained for various mean stress levels. Now consider that the stress history in Figure 6.14 is applied to a material.

For the random histories shown in Figure 6.14 with times of application t_1, t_2, and t_3 the corresponding representative failure times are designated T_1, T_2, and T_3. By analogy to Miner's theory the damage D resulting from the displayed history can be specified as

$$D = \sum_{q=1}^{3} \frac{t_q}{T_q}. \tag{6.101}$$

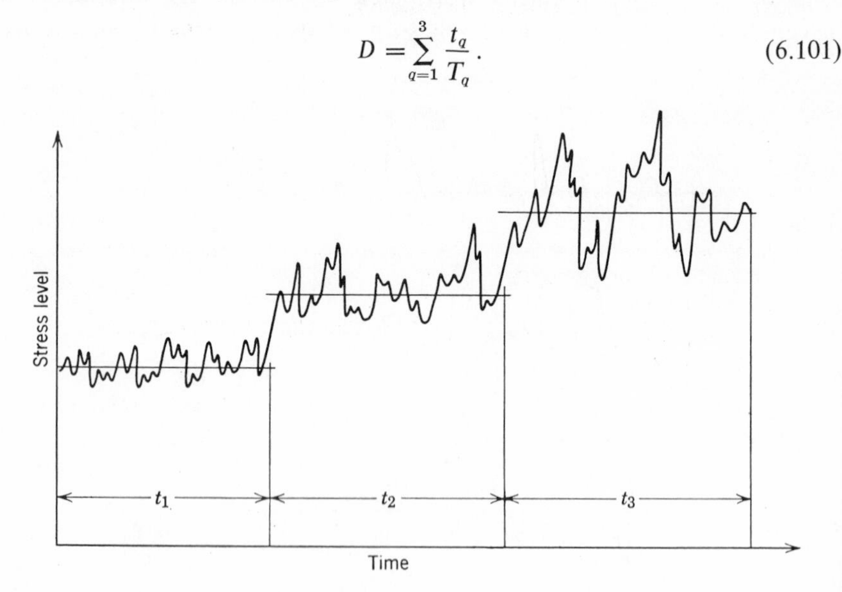

Figure 6.14 Complex stress history composed of random patterns.

This empirical approach to random histories has not been substantiated by tests. At present, purely empirical attempts have not resulted in a cumulative damage theory that will give satisfactory fatigue-life predictions for a wide range of metallic materials and environmental conditions. An increase in proficiency in these predictions will occur primarily when the physical processes involved with material deterioration under fluctuating stresses are more fully understood. It will then be the goal to describe these physical processes analytically and to evolve a cumulative damage theory with a minimal number, preferably none, of parameters that require time-consuming test evaluation. The most useful theory is one that not only predicts a representative fatigue life but also gives a probabilistic description of the random or variable nature of fatigue lives. It can be anticipated that greater emphasis will be given in the future to the incorporation of physical considerations into cumulative damage theories, but it can be further expected that progress will be slow because of the complex nature of the fatigue process.

REFERENCES

[1] M. A. Miner, "Cumulative Damage in Fatigue," *J. Appl. Mech.* (September 1945).

[2] A. Palmgren, "Die Lebensdauer von Kugellagern," *Z. Vereines Deutscher Ingenieure*, **68,** (1924).

[3] L. Kaechele, "Review and Analysis of Cumulative Fatigue-Damage Theories," The RAND Corporation, RM-3650-PR, August 1963.

[4] H. J. Grover, "An Observation Concerning the Cycle Ratio in Cumulative Damage," Symposium on Fatigue of Aircraft Structures, American Society for Testing Materials, Special Technical Publication No. 274, 1960.

[5] S. M. Marco and W. L. Starkey, "A Concept of Fatigue Damage," *Trans. Am. Soc. Mech. Eng.*, **76** (1954).

[6] F. R. Shanley, "A Theory of Fatigue Based on Unbonding During Reversed Slip," The RAND Corporation, P-350, November 11, 1952.

[7] H. T. Corten and T. J. Dolan, "Cumulative Fatigue Damage," *Proceedings of the International Conference on Fatigue of Metals*, Institution of Mechanical Engineers and American Society of Mechanical Engineers, 1956.

[8] H. C. Schjelderup, "Cumulative Fatigue Damage," *J. Aerospace Sci.*, **26** (June 1959).

[9] A. M. Freudenthal and R. A. Heller, "On Stress Interaction in Fatigue and A Cumulative Damage Rule," *J. Aerospace Sci*,, **26** (July 1959).

[10] R. M. Mains, "Mechanical Design for Random Loading," Chapter 12, *Random Vibrations*, S. H. Crandall, ed., Technology Press of the Massachusetts Institute of Technology, Cambridge, 1958.

7

Response to Random Loadings; Sonic Fatigue

PRITCHARD H. WHITE

The fatigue failure of aircraft and aerospace structures under intense acoustic loading has been a direct consequence of the large amount of acoustic power radiated by jet and rocket engines and the high pressures produced in the turbulent boundary layer during high speed flight. The failure of these elements is commonly said to be due to *sonic fatigue*, although in some cases the exciting pressure is not a true sound wave. In many instances the sonic fatigue criteria, and not the static strength criteria, determines the design of a structural element. Panels on the trailing edge of a wing or the rear of a fuselage of a jet airplane or section near the rear of a large rocket are examples.

The problem of sonic fatigue first became significant in the early 1950's when structural failures in panels and ribs in the vicinity of the jet engine exhaust were observed. Efforts to understand the problem led to the development of theories on the generation of sound by jets [1–4] and measurement of the acoustic radiation properties of jet and rocket engines [5–8]. These studies and later work showed that the sound generation and radiation is a complicated process. The sound is essentially random in time with a continuous frequency spectrum. The total power radiated is approximately proportional to the product of velocity to the eighth power and jet diameter squared for jet engines and to U^3d^2 to U^5d^2 for rockets. The strong directionality of the noise has its maximum between 30° and 45° from the jet exhaust axis. An excellent survey of the topic of jet noise has been given by Powell [9].

The response of structures to random pressure fields, and, in particular, jet noise, has also received attention at both the theoretical and experimental levels. Early theoretical work by Miles [10], Powell [11], and Thomson and Barton [12] provided methods by which the response of structures to jet noise and boundary layer turbulence could be obtained. A detailed and extensive examination of the entire problem of random vibration of structures is contained in two books edited by Crandall [13,14].

A great amount of experimental research on fatigue under random acoustic loading has been performed both by government agencies [15,16] and by the major airplane companies [17,18]. The results of these experiments have pointed out weaknesses both in the theories of response and the theories of fatigue under random loading. A comprehensive survey of the problems of sonic fatigue is contained in the proceedings of symposia on this topic sponsored by the U.S. Air Force [19,20], and the reader is urged to examine these works closely if he desires to become thoroughly competent in the field.

The object of this chapter on sonic fatigue is to introduce certain basic physical and mathematical concepts to the reader. Although one brief chapter cannot begin to cover all the many facets and complexities of the problem, it is intended to provide him with sufficient information to cope with many practical situations, and to serve as a background for reading and understanding the published literature on the subject. He is strongly urged to delve into the outside literature, for by seeing the various viewpoints and methods of analysis he can come closer to a true understanding of the problem and its solutions.

The material to be presented in this chapter may be logically divided into two parts: a description of the exciting pressure field, and presentation of methods for determining the response of structures to these pressure fields. The mathematics however kept to a minimum and emphasis has been placed on a physical interpretation of each aspect.

7.1 CONCEPTS OF ACOUSTICS

A study of acoustical fatigue begins most naturally with the fundamentals of acoustics. Although we are all familar with many aspects of sound, it is appropriate to develop here the basic physical and mathematical concepts of acoustics. In this way those ideas of most importance to sonic fatigue can be emphasized, and the other equally interesting but less pertinent topics such as psychoacoustics, architectural acoustics, and underwater acoustics can be dealt with in other books.

7.1.1 *The Acoustic Wave Equation*

The general differential equation for an acoustic wave may be derived by considering the flow of fluid into and the forces acting on a small element of fluid dV. Figure 7.1 shows the element and the appropriate coordinates. The rate of fluid entering the element from the rear face (normal to x_1) is $\rho u_1\, dx_2\, dx_3$, and that flowing out of the front face is

$$\left[\rho u_1 + \frac{\partial}{\partial x_1}(\rho u_1)\, dx_1\right] dx_2\, dx_3.$$

The net influx is thus

$$\left[\frac{\partial(\rho u_1)}{\partial x_1}\right] dx_1\, dx_2\, dx_3.$$

Taking the flow in and out of the other faces gives the net influx resulting from all directions to be

$$-\left[\frac{\partial(\rho u_1)}{\partial x_1} + \frac{\partial(\rho u_2)}{\partial x_2} + \frac{\partial(\rho u_3)}{\partial x_3}\right] dx_1\, dx_2\, dx_3. \tag{7.1}$$

This must clearly be equal to the rate-of-mass change in the element; thus

$$\frac{\partial \rho}{\partial t}\, dx_1\, dx_2\, dx_3 = -\left[\frac{\partial(\rho u_1)}{\partial x_1} + \frac{\partial(\rho u_2)}{\partial x_2} + \frac{\partial(\rho u_3)}{\partial x_3}\right] dx_1\, dx_2\, dx_3. \tag{7.2}$$

Dividing by $dx_1\, dx_2\, dx_3$ we find the equation of continuity†

$$\frac{\partial \rho}{\partial t} + \frac{\partial(\rho u_j)}{\partial x_j} = 0. \tag{7.3}$$

The condensation is a measure of the change of density of the fluid and is defined by

$$s = \frac{\rho - \rho_0}{\rho_0}, \tag{7.4}$$

where ρ_0 is the unperturbed density of the fluid. The continuity equation is now written as

$$\frac{\partial s}{\partial t} + (1 + s)\frac{\partial u_j}{\partial x_j} + u_j \frac{\partial s}{\partial x_j} = 0. \tag{7.5}$$

In order to arrive at a linear equation, it is necessary to apply restrictions to s

† The summation convention that repeated subscripts are to be summed gives a great simplification in the form of the equation, and is often useful in observing generalities. Thus

$$a_j x_j = a_1 x_1 + a_2 x_2 + a_3 x_3,$$

$$\frac{\partial(u_j)}{\partial x_j} = \frac{\partial u_1}{\partial x_1} + \frac{\partial u_2}{\partial x_2} + \frac{\partial u_3}{\partial x_3}.$$

This notation is used throughout this chapter.

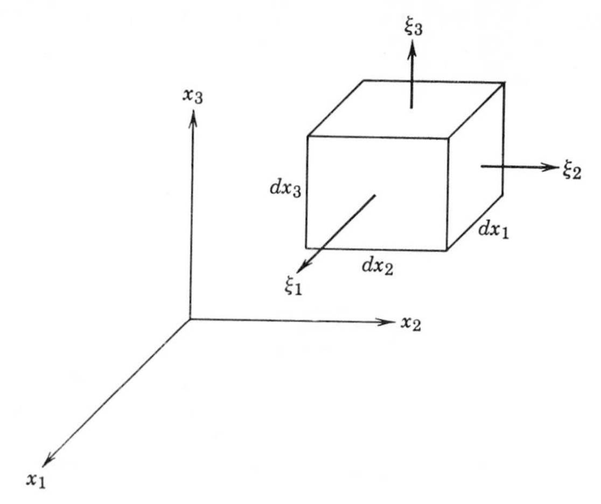

Figure 7.1 Elementary fluid element.

and u_j. By assuming that

$$|s| \ll 1 \tag{7.6}$$

and

$$\left|\frac{\partial u_j}{\partial x_j}\right| \gg \left|u_j \frac{\partial s}{\partial x_j}\right|. \tag{7.7}$$

The linear continuity equation

$$\frac{\partial s}{\partial t} + \frac{\partial u_j}{\partial x_j} = 0 \tag{7.8}$$

results. The implications of the assumptions are that the condensation be extremely small, and that the mean velocity of the fluid be small.

Taking the balance of forces on the element in the x_1 direction gives

$$\left[p - \left(p + \frac{\partial p}{\partial x_1} dx_1\right)\right] dx_2\, dx_3 = -\frac{\partial p}{\partial x_1} dx_1\, dx_2\, dx_3. \tag{7.9}$$

The rate of change of momentum of the fluid in the x_j direction in the element is given by

$$\frac{\partial}{\partial t}(\rho u_1) + u_j \frac{\partial(\rho u_1)}{\partial x_j}.$$

By again considering the condensation to be small, one arrives at the momentum equation

$$\frac{\partial}{\partial t} u_i + u_j \frac{\partial u_i}{\partial x_j} = \frac{-1}{\rho_0} \frac{\partial p}{\partial x_i}. \tag{7.10}$$

Were viscous stresses or some type of magnetic or gravitational force field present, then additional terms would appear on the right-hand side of

Equation (7.10). By requiring small amplitude fluctuations and

$$\left|\frac{\partial u_i}{\partial t}\right| \gg \left|u_j \frac{\partial u_i}{\partial x_j}\right|.$$

Equation 7.10 reduces to

$$\frac{\partial u_j}{\partial t} = -\frac{1}{\rho_0}\frac{\partial p}{\partial x_j}. \tag{7.11}$$

By differentiating (7.8) with respect to t and (7.11) with respect to x_j, u_j may be eliminated and the resulting second order equation is

$$\frac{\partial^2 s}{\partial t^2} = -\frac{1}{\rho_0}\frac{\partial^2 p}{\partial x_j^{\,2}}. \tag{7.12}$$

The compression and rarefaction of the fluid medium generally takes place so rapidly that there is not sufficient time for appreciable heat transfer, hence making the process adiabatic. The pressure and density changes are thus related by the equation [21,22]

$$\delta p = c_0^{\,2}\rho_0 s, \tag{7.13}$$

where c_0 has the dimensions of velocity. Making the substitution of (7.13) in (7.12) gives the *acoustic wave equation*

$$\frac{\partial^2 s}{\partial t^2} - c_0^{\,2}\frac{\partial^2 s}{\partial x_i^{\,2}} = 0. \tag{7.14}$$

Since

$$\begin{aligned}\frac{\partial^2 s}{\partial x_i^{\,2}} &= \frac{\partial^2 s}{\partial x_1^{\,2}} + \frac{\partial^2 s}{\partial x_2^{\,2}} + \frac{\partial^2 s}{\partial x_3^{\,2}},\\ &= \nabla^2 s,\end{aligned} \tag{7.15}$$

the wave equation may be written as

$$\frac{\partial^2 s}{\partial t^2} - c_0^{\,2}\,\nabla^2 s = 0, \tag{7.16}$$

which is applicable to all coordinate systems. In addition, all of the other variables such as pressure, displacement, and velocity also satisfy the wave equation.

$$\begin{aligned}\frac{\partial^2 p}{\partial t^2} - c_0^{\,2}\,\nabla^2 p &= 0\\ \frac{\partial^2 \xi}{\partial t^2} - c_0^{\,2}\,\nabla^2 \xi &= 0\\ \frac{\partial^2 u}{\partial t^2} - c_0^{\,2}\,\nabla^2 u &= 0.\end{aligned} \tag{7.17}$$

A velocity potential ϕ may be defined such that

$$u_i = \frac{-\partial\phi}{\partial x_i}, \tag{7.18}$$

and

$$p = \rho_0 \frac{\partial \phi}{\partial t}. \tag{7.19}$$

The velocity potential also satisfies the wave equation, thus by solving the one equation

$$\frac{\partial^2 \phi}{\partial t^2} - c_0^{\,2} \nabla^2 \phi = 0, \tag{7.20}$$

all the remaining variables may be easily found.

7.1.2 *Solution of the Wave Equation*

An understanding of the acoustic wave is best obtained by considering the simple case of a plane wave (motion in the x_2 and x_3 directions is zero). The solution to (7.20) is thus

$$\phi(x, t) = \phi_1(x - c_0 t) + \phi_2(x + c_0 t), \tag{7.21}$$

which represents a function ϕ_1 moving in the $+x$ direction with velocity c_0 and a function ϕ_2 moving in the $-x$ direction with velocity c_0. The boundary conditions determine the two functions. The most common case is a harmonic variation, which is expressed as

$$\phi(x, t) = A_1 e^{i(\omega t - kx)} + B e^{i(\omega t + kx)}, \tag{7.22}$$

where

$$k = \frac{\omega}{c_0}$$

and is called the wave number. This wave number is a measure of the waviness in space; since

$$k = \frac{2\pi}{\lambda}, \tag{7.23}$$

a small k denotes long waves and a large k implies a short wave length. The various acoustic variables of interest may now be determined by direct operation on (7.21);

$$\begin{aligned}
p &= \rho_0 \frac{\partial \phi}{\partial t} = i\omega[Ae^{i(\omega t - kx)} + Be^{i\omega + kx}], \\
s &= \frac{p}{\rho_0 c_0^{\,2}} = \frac{i\omega}{\rho_0 c_0^{\,2}} [Ae^{i(\omega t - kx)} + Be^{i(\omega t + kx)}], \\
u &= -\frac{\partial \phi}{\partial x} = ik[Ae^{i(\omega t - kx)} + Be^{i(\omega t + kx)}], \\
\xi &= \frac{u}{i\omega} = \frac{1}{c_0} [Ae^{i(\omega t - kx)} - Be^{i(\omega t + kx)}].
\end{aligned} \tag{7.24}$$

It may be seen that the pressure and the condensation are always in phase, and the particle displacement always lags the velocity by $\pi/2$ regardless of the direction of wave travel. The pressure and velocity are in phase or π out of phase for forward traveling or backwards traveling waves respectively.

A description of acoustic waves in two and three dimensions is considerably more complicated, and is beyond the scope of this chapter. This knowledge is helpful, but not absolutely necessary for a description of sonic fatigue. Complete and detailed descriptions of the three dimensional waves are given in standard acoustics texts [21,22,23].

7.1.3 Energy Density and Intensity

The energy density in an acoustic wave is a measure of the total amount of vibrational energy contained in a unit volume of the fluid. The total energy is made up of the sum of kinetic energy due to the velocity of the fluid, plus the potential energy due to its compression and rarefaction. The kinetic energy in a volume ΔV is

$$\Delta KE = \tfrac{1}{2}\rho u^2 \, \Delta V. \tag{7.25}$$

The potential energy due to compressing the volume ΔV to a smaller volume $\Delta V - dV$ is

$$\Delta PE = \int_{dV} p \, dV. \tag{7.26}$$

The excess pressure due to the compression is related to the condensation s by $\delta p = \rho_0 c_0^2 s$. In addition $dV = -\Delta V \, ds$, thus

$$\Delta PE = \int_0^s \rho c^2 s \, ds \, \Delta V = \tfrac{1}{2}\rho c^2 s^2 \, \Delta V. \tag{7.27}$$

The total instantaneous energy density W is thus the sum of (7.25) and (7.27), divided by ΔV,

$$W_{\text{inst}} = \tfrac{1}{2}\rho(u^2 + c^2 s^2), \tag{7.28}$$

or

$$W_{\text{inst}} = \tfrac{1}{2}p\left(u^2 + \frac{p^2}{(\rho c)^2}\right). \tag{7.29}$$

For a single plane wave, $p = u/\rho_0 c_0$, and the average energy density is

$$W = \rho\langle u^2\rangle_t = \frac{\langle p^2\rangle_t}{\rho_0 c_0^2}, \tag{7.30}$$

where $\langle u^2\rangle_t$ and $\langle p^2\rangle_t$ denote time averages of velocity and acoustic pressure These relations cannot be applied to cylindrical or spherical waves except at great distances from the source because the pressure, condensation, and velocity do not bear the same relationship to each other in these cases.

A plane wave transports energy as it progresses through the fluid. The amount of energy falling on a unit area normal to the direction of propagation is called *intensity*, and is usually measured in ergs/sec/cm² or watts/cm². The energy of a plane wave in a unit volume of fluid is transported through a unit area in a time $1/c_0$. Thus the intensity in a plane wave is the product of energy density times speed of propagation

$$I = W \times c_0 = \rho_0 c_0 \langle u^2 \rangle_t = \frac{\langle p^2 \rangle_t}{\rho_0 c_0}. \tag{7.31}$$

Alternately we may consider the average energy passing through a unit area as the product of the pressure and the component of velocity in phase with the pressure,

$$\mathrm{I} = \mathrm{Re}\,[\langle pu^* \rangle]. \tag{7.32}$$

7.1.4 Scale of Intensity and Pressure

The intensity of a sound wave is defined in terms of ergs/cm²/sec; however, for the range of values encountered in common usage it is more convenient to use a logarithmic scale. Furthermore, the human ear responds approximately in a logarithmic fashion and perceived changes in loudness correspond to logarithmic changes in intensity. The intensity level of a sound is defined by

$$\text{Level} = 10 \log_{10} \left(\frac{I}{I_{\text{ref}}} \right), \tag{7.33}$$

where the reference intensity, I_{ref}, is 10^{-16} W/cm². This reference intensity corresponds approximately to a 1000-cps signal just audible to the human ear.

Substitution of the reference intensity into (7.30) gives (at standard atmospheric conditions) a root mean square pressure of

$$\text{Pref} = 0.000204 \text{ dynes/cm}^2. \tag{7.34}$$

Most microphones do not measure intensity, but pressure instead. In addition, the relation between pressure and intensity is usually not simple for anything but a plane wave. Indeed, for many cases knowledge of the pressure field gives no indication of the intensity. It is thus common practice to consider the pressure as the acoustic variable of interest and define the *sound pressure level* by

$$\text{SPL} = 10 \log_{10} \left(\frac{\langle p^2 \rangle}{\langle p_{\text{ref}}^2 \rangle} \right), \tag{7.35}$$

where the reference pressure p_{ref} is 0.0002 dynes/cm². The sound pressure level and the intensity level are almost the same for plane waves. The sound pressure level for some common situations is shown in Table 7.1. The last two entries in the above table should be viewed as strictly mathematical

TABLE 7.1

SOUND PRESSURE LEVELS

Pressure	SPL
0.0002 dyne/cm^2	0 db
1.000 dyne/cm^2	74 db
1.0 lb/ft^2	128 db
1.0 lb/in^2	171 db
14.7 lb/in^2	194 db

examples. The amplitudes are so high that the linear wave equation does not hold true. Waves of this amplitude would be considered as shock waves and not sound waves. Sonic booms are examples of the latter type.

TABLE 7.2

TYPICAL AMPLITUDES OF ACOUSTIC VARIABLES

SPL	Pressure	Velocity	Displacement
70 db	63 dyne/cm^2	1.52×10^{-2} cm/sec	2.42×10^{-6} cm
100 db	20 dyne/cm^2	0.482 cm/sec	7.69×10^{-5} cm
120 db	200 dyne/cm^2	4.82 cm/sec	7.69×10^{-4} cm

Typical values for other acoustic variables are given in Table 7.2 for a 1000-cps signal. The particle velocity and displacement are found by using (7.24).

TABLE 7.3

NOISE LEVELS OF TYPICAL SITUATIONS

Description	SPL
Rocket engine	160 db
Threshold of pain	120 db
Loud orchestra	100 db
Elevated train	95 db
Busy street traffic	80 db
Ordinary conversation	70 db
Quiet automobile	50 db
Quiet radio in home	40 db
Average whisper	20 db
Rustle of leaves	10 db
Threshold of hearing	0 db

To aquaint the reader with the order of magnitude of different sounds, a list of typical sound pressure levels for common situations is presented in Table 7.3. It is an interesting exercise to listen carefully to everyday sounds and try to rank them in comparison with the sounds presented here.

7.2 RANDOM PRESSURES

One of the main features of pressure fields which excite aerospace structures and cause acoustic fatigue is their randomness in space and time. Only in very special cases is the pressure field a pure tone acoustic wave impinging on the surface at a fixed angle of incidence. In most instances the pressure field is not uniform over the surface, but varies from point to point, and the pressure at a given point is a random function of time.

In order to understand the properties of a random pressure field and how they come to bear on the subject of acoustical fatigue, it is necessary to examine the statistical properties of the pressure field. The quantities of interest discussed in this section are the expected values of certain products, the power spectral density and cross power spectral densities, and the probability distribution of the pressure.

7.2.1 *Expected Values*

A random process, or stochastic process, is an ensemble of time functions $\{x_k(t)\}$, $-\infty < t < \infty$, $k = 1, 2, \ldots$, such that the ensemble can be characterized by statistical properties [24]. Such a process may be represented by Figure 7.2. The ensemble average of this process at $t = t_1$ is given by

$$\langle x(t_1) \rangle_k = \lim_{N\to\infty} \sum_{1}^{N} \frac{x_k(t_1)}{N} \tag{7.36}$$

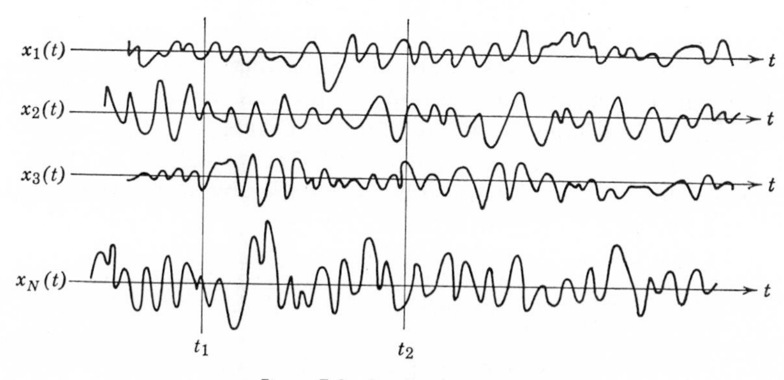

Figure 7.2 Random processes.

and that at $t = t_2$ by

$$\langle x(t_2)\rangle_k = \lim_{N\to\infty} \sum_{1}^{N} \frac{x_k(t_2)}{N}. \tag{7.37}$$

Other interesting functions are computed in a similar manner;

$$\langle x^2(t_1)\rangle_k = \lim_{N\to\infty} \sum_{1}^{N} \frac{x_k^{\,2}(t_1)}{N} \tag{7.38}$$

is called a mean square value and

$$\langle x(t_1)x(t_2\rangle_k = \lim_{N\to\infty} \sum_{k=1}^{N} \frac{x_k(t_1)x_k(t_2)}{N} \tag{7.39}$$

is called a correlation function. When the process is invariant to a time shift and all the statistical properties such as the correlation function

$$\langle x(t_1 + \tau)x(t_2 + \tau)\rangle_k = \langle x(t_1)x(t_2)\rangle_k \tag{7.40}$$

are the same for all τ, the process is said to be *stationary*. One may also compute the statistical properties of a single stationary process by similar methods

$$\langle x_k\rangle_t = \lim_{T\to\infty} \frac{1}{2T} \int_{-T}^{T} x_k(t)\, dt, \tag{7.41}$$

$$\langle x_k^{\,2}\rangle_t = \lim_{T\to\infty} \frac{1}{2T} \int_{-T}^{T} x_k^{\,2}(t)\, dt, \tag{7.42}$$

$$\langle x_k(t)x_k^*(t + \tau)\rangle_t = \lim_{T\to\infty} \frac{1}{2T} \int_{-T}^{T} x_k(t)x_k^*(t + \tau)\, dt, \tag{7.43}$$

where the asterisk indicates the complex conjugate. These are called the average, mean square, and autocorrelation functions respectively. If the process is stationary and the ensemble statistical properties are equal to the time averaged statistical properties then the process is termed *ergodic*. Although nonstationary processes are of engineering importance (such as during the launch phase of a large rocket) they are covered in other references [25,26] and will not be discussed in this chapter.

It is possible to use a function similar to (7.43) to establish a relationship between two signals. For instance it might be desirable to determine if the vibration at two points on a structure are related in some manner. Such a relation is called the *cross correlation* [24] and is given by

$$R_{xy}(\tau) = \langle x(t)y^*(t + \tau)\rangle_t$$

$$R_{xy}(\tau) = \lim_{T\to\infty} \frac{1}{2T} \int_{-T}^{T} x(t)y^*(t + \tau)\, dt, \tag{7.44}$$

and

$$R_{yx}(\tau) = \lim_{T\to\infty} \frac{1}{2T} \int_{-T}^{T} x^*(t)y(t-\tau)\,dt. \tag{7.45}$$

Both processes must be stationary for these equations to be valid. It is often convenient to normalize the correlation function by the root mean square value of each signal

$$\tilde{C}_{xy}(\tau) = \frac{R_{xy}(\tau)}{[\langle x^2\rangle\langle y^2\rangle]^{1/2}}. \tag{7.46}$$

This function is more useful in comparing two signals because the effect of amplitude is not present, and the function is restricted to lie in the range $-1 < \tilde{C}(\tau) < +1$.

A value of $+1.0$ would imply that the two signals are of any arbitrary shape but exactly alike, with just a time shift of τ between them. A correlation coefficient of -1.0 simply means that one signal is just the negative of the other.

7.2.2 *Power Spectral Density*

Any signal which extends from $-T$ to T (or repeats itself with a period $2T$) can be expressed in a Fourier series [24].

$$x(t) = \sum_{-\infty}^{\infty} a_n e^{i\omega_n t}, \tag{7.47}$$

where

$$a_n = \frac{1}{2T} \int_{-T}^{T} x(t)e^{-i\omega_n t}\,dt, \tag{7.48}$$

and

$$\omega_n = \frac{n\pi}{T}. \tag{7.49}$$

The coefficients a_n are one measure of the frequency content of the signal. Another measure which proves to be more useful is the energy content at each frequency. If $x(t)$ represents a current flowing through a unit resistance, then $x^2(t)$ is the instantaneous power dissipated. To find the average power dissipated we square (7.50) and average over time

$$\langle x(t)x^*(t)\rangle = \sum_{m-\infty}^{\infty} \sum_{n-\infty}^{\infty} \int_{-T}^{T} \frac{a_m a_n^* e^{i(\omega_m - \omega_n)t}}{2T}\,dt. \tag{7.50}$$

Parseval's formula [24] tells us that

$$\frac{1}{2T} \int_{-T}^{T} |x|^2\,dt = \sum_{-\infty}^{\infty} |a_n|^2. \tag{7.51}$$

Thus

$$\langle x(t)x^*(t)\rangle_t = \sum_{-\infty}^{\infty} |a_n|^2. \tag{7.52}$$

The average power which is dissipated at any frequency ω_n is just $|a_n|^2$.

As an example, consider an N wave as shown in Figure 7.3. This function is given mathematically by

$$x(t) = \frac{t}{T} \qquad -T < t < T, \tag{7.53}$$

and the coefficients are

$$a_n = \frac{1}{2F}\int_{-T}^{T} \frac{t}{T} e^{-in\pi t/T}\, dt \tag{7.54}$$

$$= -\frac{(-1)^{n-½}}{2n\pi} \qquad |n| > 0$$

$$= 0 \qquad n = 0.$$

Values of $4\pi^2\, |a_n|^2$ are shown in Figure 7.4.

It is possible to write (7.52) in a slightly different form as

$$\langle x(t)x^*(t)\rangle_t = \sum_{-\infty}^{\infty} \left[\frac{|a|_n{}^2}{\Delta\omega}\right] \Delta\omega. \tag{7.55}$$

In this form, the term in brackets represents the amount of power contained in a small bandwidth $\Delta\omega$. Because it is in terms of a power per unit bandwidth, it is appropriate to call the function a power spectral density $S(\omega_n)$. As the period T becomes large, $\Delta\omega$ decreases and the summation may be replaced by an integration

$$\langle x^2\rangle_t = \int_{-\infty}^{\infty} S(\omega)\, d\omega, \tag{7.56}$$

where

$$S(\omega) = \lim_{T\to\infty} \frac{|a_n|^2}{\Delta\omega}. \tag{7.57}$$

This rather heuristic development of the power spectral density may be supplemented by a more rigorous mathematical derivation [24]. Let a real

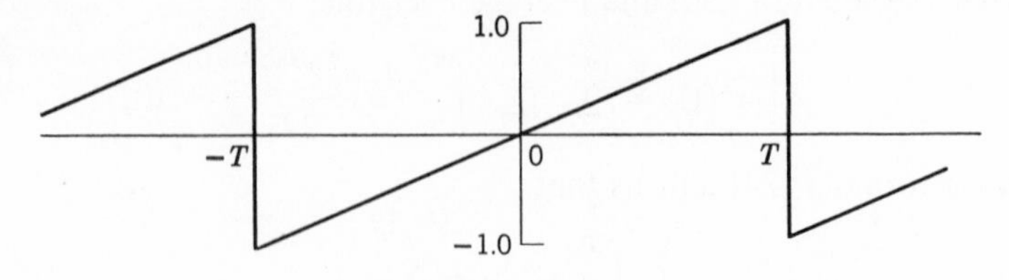

Figure 7.3 N wave.

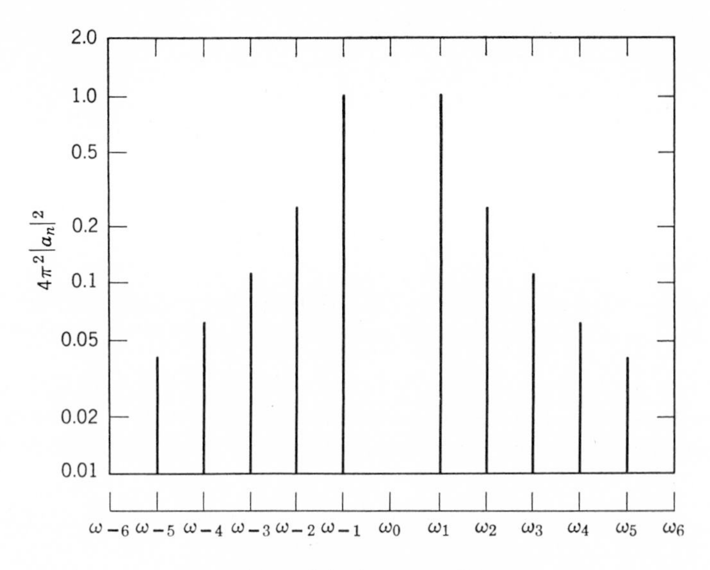

Figure 7.4 Spectrum of N wave.

function $x_T(t)$ be arbitrary in the interval $-T$ to T, and zero outside this interval. See Figure 7.5. The Fourier transform of this function is

$$X_T(\omega) = \int_{-\infty}^{\infty} x_T(t) e^{i\omega t}\, dt, \tag{7.58}$$

and the inverse transform is

$$x_T(t) = \frac{1}{2\pi} \int_{-\infty}^{\infty} X_T(\omega) e^{-i\omega t}\, dt. \tag{7.59}$$

The mean square value of $x_T(t)$ over the interval is

$$\langle x_T^2(t) \rangle = \frac{1}{2T} \int_{-T}^{T} x_T^2(t)\, dt. \tag{7.60}$$

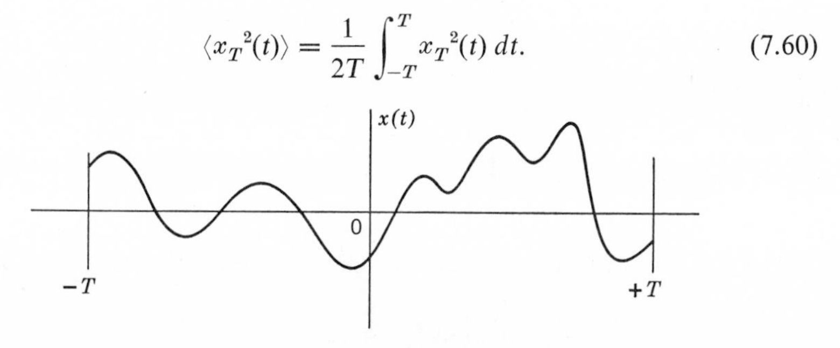

Figure 7.5 Truncated time function.

Parseval's theorem for integrals [24] states that

$$\int_{-T}^{T} x_T^2(t)\,dt = \frac{1}{2\pi}\int_{-\infty}^{\infty} |x_T(\omega)|^2\,d\omega. \tag{7.61}$$

Thus

$$\langle x_T^2(t)\rangle = \frac{1}{2\pi}\int_{-\infty}^{\infty} \frac{|X_T(\omega)|^2}{2T}\,d\omega. \tag{7.62}$$

As T becomes large we have

$$\langle x^2(t)\rangle = \lim_{T\to\infty}\frac{1}{2\pi}\int_{-\infty}^{\infty} \frac{|X_T(\omega)|^2}{2T}\,d\omega. \tag{7.63}$$

Because we are integrating over frequency, the integrand is a density function representing the contribution to the mean square-value per-unit bandwidth. Thus we again develop the power spectral density, but this time it is given by

$$S(\omega) = \lim_{T\to\infty}\frac{|X_T(\omega)|^2}{2T\times 2\pi}. \tag{7.64}$$

Let us now consider the relation between the correlation function $R(\tau)$ and the power spectral density function $S(\omega)$. The Fourier transform of the correlation function of the truncated signal $x_T(t)$ is

$$\int_{-\infty}^{\infty} R_T(\tau)e^{-i\omega\tau}\,d\tau = \int_{-\infty}^{\infty}\frac{1}{2T}\int_{-T}^{T} x_T(t)\,x_T(t+\tau)\,dt e^{-i\omega\tau}\,d\tau \tag{7.65}$$

$$= \frac{1}{2T}\int_{-\infty}^{\infty}\int_{-\infty}^{\infty} x_T(t)e^{i\omega t}x_T(t+\tau)e^{-i\omega(t+\tau)}\,dt\,d\tau$$

$$= \frac{1}{2T}\int_{-\infty}^{\infty} x_T(t)e^{i\omega t}\,dt\int_{-\infty}^{\infty} x_T(s)e^{-i\omega s}\,ds.$$

From (7.58) we get

$$\int_{-\infty}^{\infty} R_T(\tau)e^{-i\omega\tau}\,d\tau = \frac{1}{2T}X_T(\omega)X^*(\omega). \tag{7.66}$$

Taking the limit as $T\to\infty$ and using (7.64) we see that

$$\frac{1}{2\pi}\int_{-\infty}^{\infty} R(\tau)e^{-i\omega\tau}\,d\tau = S(\omega). \tag{7.67}$$

Thus the power spectral density of a function is given as the Fourier transform of the autocorrelation function. The inverse relation is

$$R(\tau) = \int_{-\infty}^{\infty} S(\omega)e^{i\omega\tau}\,d\omega. \tag{7.68}$$

Equations (7.67) and (7.68) are commonly referred to as the Wiener-Kinchin relations. From these equations it is seen that the power spectral density and the autocorrelation function contain the same information and that knowledge of one is equivalent to knowledge of the other.

It is important to note that the power spectral density as defined above is given in units of some quantity squared per radian per second, over both positive and negative frequencies. Physical processes are usually measured in different units, however, and it is most common to be in terms of some quantity squared per cycle per second, over only positive frequencies. The mathematical and experimental values are related by the equation

$$W(f) = 4\pi S(\omega). \tag{7.69}$$

In this notation the mean square value of a signal is given by

$$\langle x^2 \rangle_t = \int_0^{\infty} W(f)\, df. \tag{7.70}$$

In many instances it is necessary to know the relation between two signals. In a manner analogous to that used to derive the spectral density of one signal, we obtain the *cross-spectral density*

$$S_{xy}(\omega) = \int_{-\infty}^{\infty} R_{xy}(\tau) e^{-i\omega\tau}\, d\tau \tag{7.71}$$

and

$$S_{yx}(\omega) = \int_{-\infty}^{\infty} R_{yx}(\tau) e^{-i\omega\tau}\, d\tau. \tag{7.72}$$

Because the cross correlation function $R_{xy}(\tau)$ is not in general an even function of τ, the cross spectral densities will be complex quantities, and

$$S_{xy}(\omega) = S^*_{yx}(\omega). \tag{7.73}$$

7.2.3 *Probability Distributions*

The third property which we use to describe a random signal is its probability distribution. A random signal may have almost any kind of physically realizable probability distributions; however, it has been found by experience that many of the random physical processes occurring in nature have a normal or Gaussian probability distribution. In particular, the pressure in the sound field of a jet engine or on the surface under a turbulent boundary layer is approximately Gaussian. The statistical properties of a Gaussian signal and the modifications produced by filtering operations have been developed extensively by Rice [27].

The Gaussian probability distribution function is given by

$$p(x) = \frac{1}{\sqrt{2\pi\sigma^2}} \exp\left[-\frac{(x - \langle x \rangle)^2}{2\sigma^2} \right], \tag{7.74}$$

where

$$\sigma^2 = E[(x - \langle x \rangle)^2]$$

is the variance, and $E[\cdot]$ denotes an expected value. For a random signal with a zero mean the variance is just the mean square value and is given by

$$\sigma^2 = \int_{-\infty}^{\infty} S(\omega)\, d\omega. \tag{7.75}$$

Thus the probability distribution for Gaussian noise is only dependent upon the power spectral density.

7.2.4 *Typical Pressure Fields*

Although we are interested in the response of structures to general pressure fields, several pressure fields are commonly encountered and thus deserve special examination. These pressure fields and their cross power spectral densities are described in this section. In all cases of random pressures the pressure amplitude is assumed to be a random variable with a Gaussian distribution.

The first and most simple pressure field is a uniform pressure over the surface which oscillates at a fixed frequency and amplitude. This is typical of a sinusoidal sound wave falling on a surface whose dimensions are smaller than a quarter wave length. The cross-spectral density is composed of a frequency dependent part $S_p(\omega)$ and a spatially dependent part

$$S_p(\mathbf{x}, \mathbf{x}', \omega) = S_p(\omega)\tilde{C}_p(\mathbf{x}, \mathbf{x}', \omega), \tag{7.76}$$

where $\tilde{C}_p(\mathbf{x}, \mathbf{x}', \omega)$ is the normalized cross spectral density. For a single frequency wave

$$S_p(\omega) = \frac{A^2}{4}\,\delta(\omega - \omega_0) + \frac{A^2}{4}\,\delta(\omega + \omega_0), \tag{7.77}$$

where A is the amplitude of the sine wave and ω_0 is its frequency. As long as the dimension of the surface is less than one quarter wave length,

$$\tilde{C}_p(\mathbf{x}, \mathbf{x}', \omega) \simeq 1.0 \tag{7.78}$$

for all ω. Should the sound field consist of many waves, (7.76) remains true, but $S_p(\omega)$ could have many spikes or be a continuous function.

A second type of pressure field which is often used for analysis purposes is the completely random field. Here the pressure at any two points is completely uncorrelated, and the pressure at a single point has a flat power spectral density of constant value. Hence

$$S_p(\mathbf{x}, \mathbf{x}', \omega) = S_0\delta(|\mathbf{x} - \mathbf{x}'|). \tag{7.79}$$

Although physically impossible because of the infinite mean square value, it is useful in many analyses in which only a portion of the spectrum is of interest. This pressure field is sometimes referred to as "raindrops on the roof."

When a plane acoustic wave impinges upon a structure which is not small in comparison to a wave length, the instantaneous pressure is not uniform over the surface but varies in a sinusoidal manner. The pressure for the wave shown in Figure 7.6 is

$$p(x, t) = P_0 \cos\,[(k_0 \sin\theta)x - \omega t]. \tag{7.80}$$

This pressure field leads to a cross-spectral density of the form

$$S_p(x, x', \omega) = S_p(\omega) \cos\,[k_0(x - x') \sin\theta]. \tag{7.81}$$

If many independent waves of the same frequency are incident from all angles, the correlation function is obtained by integrating over all angles of incidence

$$\tilde{C}_p(x, x', \omega) = \int_0^{\pi/2} F(\theta)\tilde{C}_p(x, x', \omega, \theta)\, d\theta. \tag{7.82}$$

For a uniform distribution in angle $[F(\theta) = 1.0]$

$$\tilde{C}_p(x, x', \omega) = J_0[k_0(x - x')]. \tag{7.83}$$

We see that at large separation distances the correlation decreases to a small value, whereas the correlation for a single frequency wave oscillates forever.

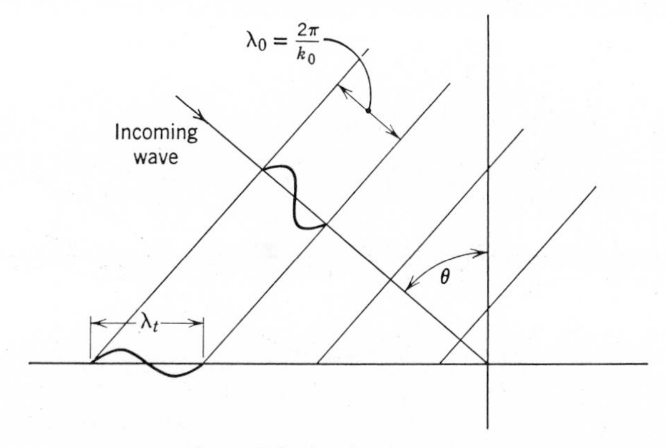

Figure 7.6 Incident acoustic wave.

By adding another dimension to the above situation we arrive at a three-dimensional wave field with uniform intensity in all directions. This is commonly called a *reverberent field.* This type of sound field is present in many large auditoriums and churches, and is used in some acoustic test facilities as a means of exposing a piece of equipment to an intense uniform acoustic field. The cross-spectral density is obtained in a similar fashion as above; however, in integrating over three dimensions the spatially dependent part becomes

$$\tilde{C}_p(\mathbf{x}, \mathbf{x}', \omega) = \frac{\sin(\omega/c)\,|\mathbf{x} - \mathbf{x}'|}{(\omega/c)\,|\mathbf{x} - \mathbf{x}'|}, \tag{7.84}$$

rather than a Bessel function. These two shapes are shown in Figure 7.7. We note that the two functions are similar in shape, and oscillate at about the same frequency.

The final pressure field of interest is that of convected boundary layer turbulence. This form of pressure, so important in aerospace and underwater vehicles, has not been sufficiently described mathematically to give an accurate prediction of its properties. Experimental results may be used to estimate its characteristics.

Several theories exist that may be used to predict the pressure field, but the predictions are not the same and there is insufficient experimental evidence to prove (or disprove) any theory [40]. Certain characteristics of boundary layer turbulence in air and fluids, over flat plates and in tubes or ducts, and at subsonic and supersonic speeds are common, however, and serve to describe the general process. The pressure field may be thought to consist of eddies of different sizes and vorticities which are swept along with the fluid

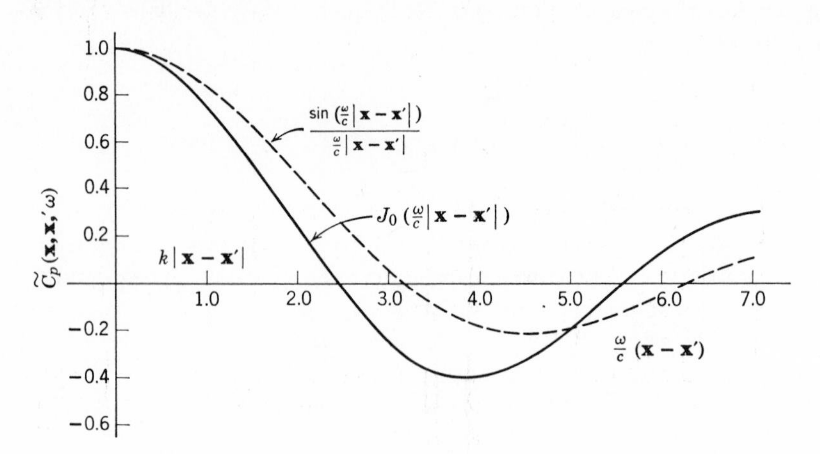

Figure 7.7 Correlation functions.

flow. It appears that the larger eddies are associated with the outer portion of the boundary layer and are carried along at a faster speed than the smaller eddies close to the surface. There is no definite stratification according to eddy size, but there is a strong tendency for the statistical averages to take this form. The eddies tend to lose coherence or be dissipated after traveling several eddy lengths, generally less than 5. This corresponds to a life time on the order of five times the boundary layer thickness. The convection velocity, or the apparent speed at which the eddies move past a point ranges from 0.9 for low frequencies to 0.5 at high frequencies. An average value of about 0.8 is sometimes used.

Because there is such a diversity in experimental results, if one wishes to apply the data to a problem, it is necessary to make a close correspondence between the specific problem and the experimental situation. Although no mathematical model fits all the experimental results simultaneously, it is possible to devise one which can be adjusted to comply with a specific situation. Such a model is

$$S_p(\mathbf{x}, \mathbf{x}', \omega) = S_p(\omega) \times \exp\left[-c_1\left|\frac{\omega}{U_c}(x_1 - x_1')\right| - c_2\left|\frac{\omega}{U_c}(x_2 - x_2')\right|\right] \exp\left[i\frac{\omega}{U_c}(x_1 - x_1')\right], \tag{7.85}$$

where c_1 and c_2 are the constants to be adjusted and U_c is the convection velocity. This function shows an exponentially decaying cosine form in the streamwise direction normal to the flow. Typical values for c_1 and c_2 are around 0.1 and 0.5, respectively. This indicates that the flow is more coherent in a streamwise direction than across it. Several typical curves of $S_p(\omega)$ are shown in Figure 7.8.

7.3 RESPONSE OF STRUCTURES TO RANDOM PRESSURES

The response of an elastic structure to a random pressure field may be obtained by two different methods; a Green's function approach and a normal mode approach. Powell has shown that they are in reality the same; however, the manipulations are different enough to consider them as two separate technqiues.

In this section a normal mode approach is used. Starting from a single-degree-of-freedom oscillator, the statistics of the response of a linear system to random forces are developed and examples worked out. Following this, the response of a distributed system is developed.

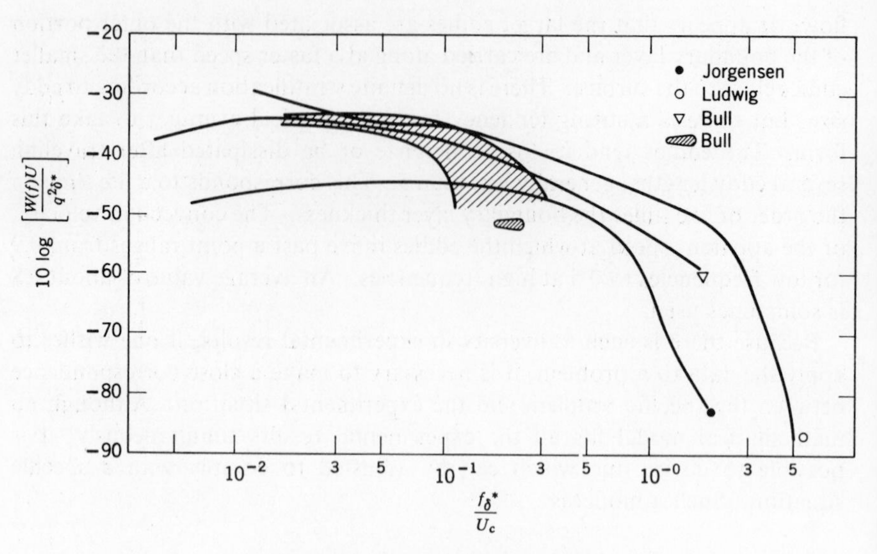

Figure 7.8 Boundary layer pressures.

7.3.1 Single-Degree-of-Freedom Oscillator

Consider the oscillator shown in Figure 7.9 to be acted on by a stationary random force $F(t)$ of power spectral density $S_F(\omega)$. The differential equation of motion of this system is

$$M\ddot{y} + R\dot{y} + Ky = F(t). \tag{7.86}$$

By the substitution

$$\omega_0^2 = K/M,$$
$$\zeta = R/2\sqrt{KM}, \tag{7.87}$$
$$g(t) = F(t)/M\omega_0^2,$$

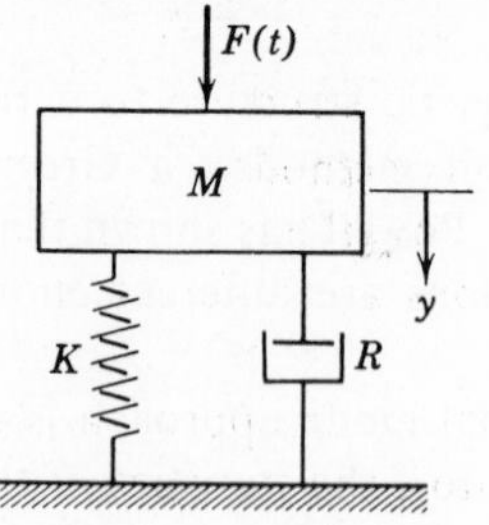

Figure 7.9

the equation may be put into a normalized form

$$\ddot{y} + 2\zeta\omega_0\dot{y} + \omega_0^2 y = g(t)\omega_0^2. \tag{7.88}$$

It is important to note that even if the system had been driven by base motion the differential equation of motion could be expressed in this form [28]. For a sinusoidal exciting function of the form

$$g(t) = Ae^{i\omega t}, \tag{7.89}$$

the response is given by

$$y(t) = B \exp\left[(-\zeta + i\sqrt{1-\zeta^2})\omega_0 t\right] + \frac{Ae^{i\omega t}}{1 - (\omega/\omega_0)^2 + i2\zeta(\omega/\omega_0)}. \tag{7.90}$$

The first term on the right-hand side is the transient term and is dependent upon the initial conditions. The second term is the steady-state response term. When only steady-state motion is of interest, (7.90) may be rewritten as

$$y(t) = H(\omega)A(\omega)e^{i\omega t} = \frac{H(\omega)F(\omega)}{M\omega_0^2}e^{i\omega t}, \tag{7.91}$$

where

$$H(\omega) = \left[1 - \left(\frac{\omega}{\omega_0}\right)^2 + i2\zeta\frac{\omega}{\omega_0}\right]^{-1}, \tag{7.92}$$

and $F(\omega)$ is the magnitude of the driving force at frequency ω. $H(\omega)$ is called the complex frequency response function and is represented by Figure 7.10.

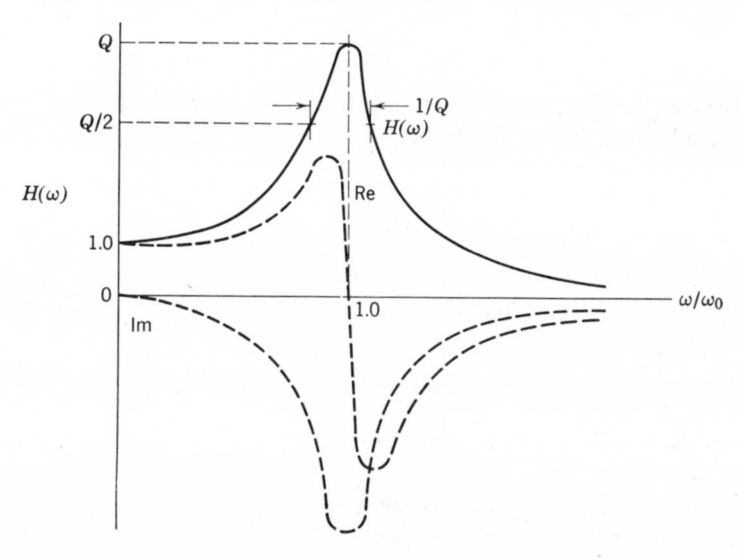

Figure 7.10 Response function.

Note that there is a sudden phase change in the response in the neighborhood of $\omega/\omega_0 = 1.0$. At $\omega/\omega_0 = 1.0$ the response lags the driving force by $\pi/2$.

Under conditions of small damping ($\zeta \ll 1$) the maximum of $|H(\omega)|$ occurs approximately at $\omega/\omega_0 = 1.0$. The value of $|H(\omega)|$ is approximately $1/2\zeta$ at this point. This peak value is commonly referred to as the Q of the system. In many situations it is convenient to define the points at which $|H(\omega)|^2$ is one half the peak value or $Q^2/2$. These points are approximately equidistant from $\omega/\omega_0 = 1.0$ and define a frequency band of width $\Delta(\omega/\omega_0) = 1/Q$, usually called the bandwidth of the system.

If the excitation is not a single frequency force but is composed of a sum of many single frequency components, the response is given by

$$y(t) = \sum_{j=1}^{N} H(\omega_j)\, A(\omega_j) e^{i\omega_j t}. \tag{7.93}$$

The autocorrelation of the response may be determined by the relation

$$\langle y(t) y^*(t+\tau)\rangle_t$$

$$= \lim_{T\to\infty} \frac{1}{2T} \left[\int_{-T}^{T} \sum_{j}^{N} \sum_{k}^{N} H(\omega_j)\, H^*(\omega_k)\, A(\omega_j)(A^*(\omega_k) e^{i\omega_j t} e^{-i\omega_k (t+\tau)} \right] dt \tag{7.94}$$

$$= \sum_{j=1}^{N} H(\omega_j)\, H^*(\omega_j)\, A(\omega_j)\, A^*(\omega_j) e^{-i\omega_j \tau}. \tag{7.95}$$

The mean square response is merely a special case of (7.95):

$$\langle y^2 \rangle_t = \sum_{1}^{N} |H(\omega_j)|^2\, |A(\omega_j)|^2. \tag{7.96}$$

An arbitrary forcing function $g(t)$ over the finite interval $-T < t < T$ can be expanded into a Fourier series [29]

$$g(t) = \sum_{-\infty}^{\infty} C_n e^{i\omega_n t}, \tag{7.97}$$

where

$$\omega_n = \frac{n\pi}{T} \tag{7.98}$$

$$C_n = \frac{1}{2T} \int_{-T}^{T} g(t) e^{-i\omega_n t}\, dt. \tag{7.99}$$

The response to this function is given by

$$y(t) = \sum_{-\infty}^{\infty} H(\omega_n) C_n e^{i\omega_n t}, \tag{7.100}$$

$$= \sum_{-\infty}^{\infty} H(\omega_n) \frac{1}{2T} \int_{-T}^{T} g(\tau) e^{-i\omega_n \tau}\, d\tau\, e^{i\omega_n t}. \tag{7.101}$$

The frequency increment from ω_n to ω_{n+1} is $\Delta\omega = \pi/T$, thus as $T \to \infty$ it is possible to replace the summation by an integration

$$y(t) = \int_{-\infty}^{\infty} H(\omega) e^{i\omega t} \int_{-\infty}^{\infty} \frac{g(\tau)}{2\pi} e^{-i\omega\tau}\, d\tau\, d\omega. \tag{7.102}$$

Equation 7.102 may be rewritten as

$$y(t) = \int_{-\infty}^{\infty} H(\omega) G(\omega) e^{i\omega t}\, d\omega, \tag{7.103}$$

where

$$G(\omega) = \frac{1}{2\pi} \int_{-\infty}^{\infty} g(\tau) e^{-i\omega\tau}\, d\tau. \tag{7.104}$$

The last relation is merely an expression for the Fourier transform of $g(t)$ [29]. The inverse transform is given by

$$g(t) = \int_{-\infty}^{\infty} G(\omega) e^{i\omega t}\, d\omega. \tag{7.105}$$

Applying (7.105) to (7.103) gives

$$Y(\omega) = H(\omega) G(\omega) = \frac{H(\omega) F(\omega)}{M\omega_0^2} \tag{7.106}$$

where $Y(\omega)$ is the Fourier transform of the response.

To further aid in understanding the response of the system it is convenient to introduce the *impulse response function* $h(t - \tau)$. This function represents the response of the system at a time t when it has been excited by a unit impulse at some earlier time τ. It may easily be done by letting $g(t) = \delta(t)$ in (7.104). This makes $G(\omega)$ take on the constant value $1/2\pi$. Inserting this value for $G(\omega)$ in (7.103) gives the response at time t to a unit impulse at $t = 0$.

$$y(t) = \frac{1}{2\pi} \int_{-\infty}^{\infty} H(\omega) e^{i\omega t}\, d\omega. \tag{7.107}$$

Thus

$$h(t) = \frac{1}{2\pi} \int_{-\infty}^{\infty} H(\omega) e^{i\omega t}\, d\omega \tag{7.108}$$

says that the unit impulse response function and the complex frequency response function are a Fourier transform pair.

The motion at any time t is given by†

$$y(t) = \int_{-\infty}^{t} h(t - \tau)\, g(\tau)\, d\tau, \tag{7.109}$$

† This fact may also be obtained in a different manner by the convolution integral formula [30]

$$\int_{-\infty}^{\infty} H(\omega)\, G(\omega) e^{-i\omega t}\, d\omega = \int_{-\infty}^{\infty} h(\tau)\, g(t - \tau)\, d\tau.$$

or

$$y(t) = \int_{-\infty}^{\infty} h(\tau)\, g(t - \tau)\, d\tau. \tag{7.110}$$

Because $h(\tau)$ is zero for $\tau < 0$, the upper limit may be replaced by ∞ with no change in validity.

The use of the unit impulse response function is shown in Figure 7.11. The response at a time t is made up of the sum of responses to a sequence of impulse of strength $g(t - \tau)$ occurring at a time $t - \tau$ earlier.

The autocorrelation function is defined by

$$R_y(\tau) = \langle y(t)\, y^*(t + \tau) \rangle_t$$

$$R_y(\tau) = \lim_{T \to \infty} \frac{1}{2T} \int_{-T}^{T} \int_{-\infty}^{\infty} \int_{-\infty}^{\infty} h(\tau_1)\, g(t - \tau_1)\, h^*(\tau_2)\, g^*(t + \tau - \tau_2)\, d\tau_1\, d\tau_2\, dt. \tag{7.111}$$

Interchanging the order of integration gives

$$R_y(\tau) = \iint_{-\infty}^{\infty} h(\tau_1) h^*(\tau_2) \left[\lim_{T \to \infty} \frac{1}{2T} \int_{-T}^{T} g(t - \tau_1) g^*(t + \tau - \tau_2)\, dt \right] d\tau_1\, d\tau_2. \tag{7.112}$$

The term in square brackets is merely the autocorrelation of the exciting force with a time lag of $\tau + \tau_1 - \tau_2$

$$R_g(\tau + \tau_1 - \tau_2) = \lim_{T \to \infty} \frac{1}{2T} \int_{-T}^{T} g(t - \tau_1) g^*(t - \tau - \tau_2)\, dt. \tag{7.113}$$

Thus the autocorrelation of the response is given in terms of the autocorrelation of the excitation

$$R_y(\tau) = \iint_{-\infty}^{\infty} h(\tau_1) h(\tau_2) R_g(\tau + \tau_1 - \tau_2)\, d\tau_1\, d\tau_2. \tag{7.114}$$

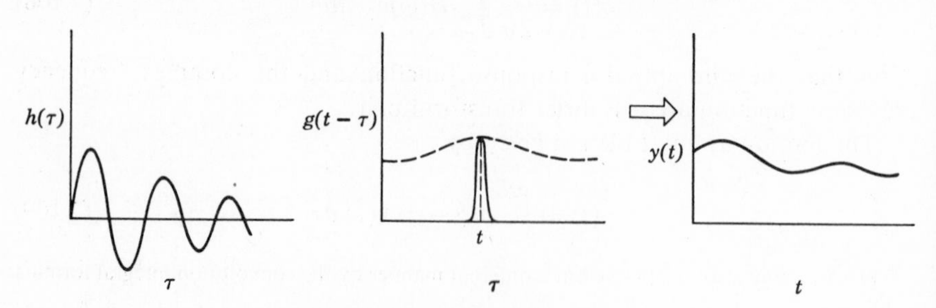

Figure 7.11 Impulse response function.

Direct evaluation of this integral is difficult. It has been done for the case of broad band random noise [31] which has an autocorrelation function of the form

$$R_g(\tau) = 2\pi S_g \delta(\tau), \tag{7.115}$$

where S_g is the power spectral density of the excitation acting on a single degree of freedom system. The result is

$$R_y(\tau) = \frac{\pi}{2} \frac{S_g}{\zeta \omega_n{}^3} e^{-\zeta\omega_n|\tau|}\left[\cos\left(\sqrt{1-\zeta^2}\,\omega_n\tau\right) + \frac{\zeta}{\sqrt{1-\zeta^2}} \sin\left(\sqrt{1-\zeta^2}\,\omega_n\tau\right)\right]. \tag{7.116}$$

This function is shown in Figure 7.12. This function is the typical correlation function for a narrow band random process. It is approximately an exponentially decaying cosine curve with a period of the natural period of the system.

The power spectral density of the response may be found in two different ways. By presenting both methods, the reader will be reminded that the mathematics is just a tool to explain the physical phenomena, and often many different tools will lead to the same result but with different interpretations of the intervening steps.

The Wiener-Kinchtine relations developed in the previous section state that

$$S(\omega) = \frac{1}{2\pi}\int_{-\infty}^{\infty} R(\tau) e^{-i\omega\tau}\, d\tau, \tag{7.117}$$

$$R(\tau) = \int_{-\infty}^{\infty} S(\omega) e^{i\omega\tau}\, d\omega. \tag{7.118}$$

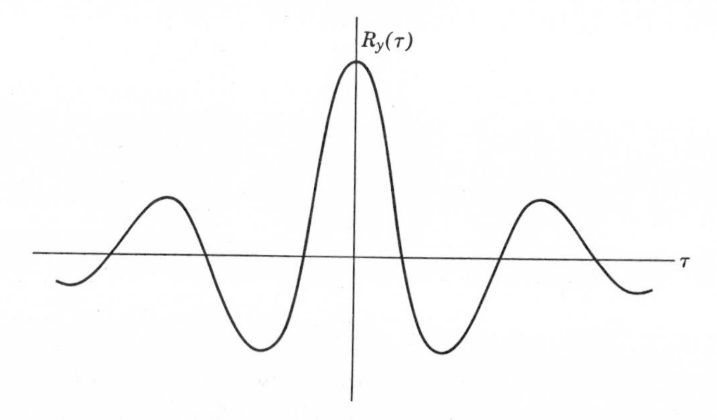

Figure 7.12 Autocorrelation function.

Thus the power spectral density of the response may be found by a Fourier transformation of the autocorrelation function of the response.

$$S_y(\omega) = \frac{1}{2\pi}\int_{-\infty}^{\infty} R_y(\tau)e^{-i\omega\tau}\,d\tau. \tag{7.119}$$

For the random excitation of a single-degree-of-freedom system which gives the autocorrelation of (7.116), the power spectral density is

$$S_y(\omega) = \frac{S_g}{[1-(\omega^2/\omega_0^{\,2})] + [2\zeta(\omega/\omega_0)]^2}. \tag{7.120}$$

This result may also be arrived at in a slightly different manner. Taking the Fourier transform of $R_y(\tau)$ and setting $\tau_3 = \tau + \tau_1 - \tau_2$ in (7.114) gives

$$S_y(\omega) = \int_{-\infty}^{\infty} h(\tau_1)e^{+i\omega\tau_1}\,d\tau_1 \int_{-\infty}^{\infty} h(\tau_2)e^{-i\omega\tau_2}\,d\tau_2 \int_{-\infty}^{\infty} \frac{R_g(\tau_3)}{2\pi}\,e^{-i\omega\tau_3}\,d\tau_3. \tag{7.121}$$

Each of the integrals may be identified as Fourier transforms,

$$\begin{aligned}\int_{-\infty}^{\infty} h(\tau_1)e^{+i\omega\tau_1}\,d\tau_1 &= H^*(\omega)\\ \int_{-\infty}^{\infty} h(\tau_2)e^{-i\omega\tau_2}\,d\tau_2 &= H(\omega)\\ \frac{1}{2\pi}\int_{-\infty}^{\infty} R_g(\tau_3)e^{-i\omega\tau_3}\,d\tau_3 &= S_g(\omega).\end{aligned} \tag{7.122}$$

The result is thus

$$S_y(\omega) = |H(\omega)|^2\, S_g(\omega). \tag{7.123}$$

This result is completely general, and is extremely useful in obtaining the spectral density of the response. It is far easier to compute this way than going through a complex integration of (7.114). The complex frequency response function for the oscillator has been found earlier to be

$$|H(\omega)|^2 = \frac{1}{[1-(\omega/\omega_0)^2]^2 + [2\zeta(\omega/\omega_0)]^2}. \tag{7.124}$$

Thus

$$S_y(\omega) = \frac{S_g(\omega)}{[1-(\omega/\omega_0)^2]^2 + [2\zeta(\omega/\omega_0)]^2}. \tag{7.125}$$

which is exactly the same as (7.120). These results show that either method is usable, and the choice should depend upon which is easiest to use in a given situation.

7.3.2 *Response of Continuous Systems to Random Pressure Fields*

The response of continuous and distributed systems to random pressure fields has been studied rather extensively and several different techniques have evolved. Miles' work on the topic considered only the first mode of a panel, thus it was still in reality a single-degree-of-freedom system. Powell [11] considered the vibrations of all the modes, and also took into account the spatial correlation of the pressure field in determining the modal response. Others [12,32] also used a normal mode approach, but with variations in the method of analysis. Lyon [33] took a unit impulse or Green's function [29] approach to the problem. This method determines the response at all points due to a load at one single point, then adds up the response due to the loading at all points. It may be shown that this method is essentially the same as the normal mode approach, because the Green's function method makes use of the eigenfunctions of the system; the intermediate steps having a very different appearance.

Corcos and Liepmann [34] and Ribner [35] examined the excitation of panels by a convected pressure field in an effort to determine the transmitted noise due to boundary layer turbulence. They considered the problem to be one of flexural waves in the panels excited by moving pressure waves. The solution by this infinite panel technique is applicable to finite size panels in the region where there are many wave lengths over the panel extent or if the modes are averaged over a frequency band [36].

In this section the general theory for the response of structure to random pressure fields is derived. Several specific cases are worked out and examples are presented to show how the method is applicable to a broad class of problems.

The differential equation of motion of a two-dimensional surface such as a panel is given by

$$m \frac{\partial^2 y}{\partial t^2} + b \frac{\partial y}{\partial t} + D \nabla^4 y = p(\mathbf{x}, t), \tag{7.126}$$

where m = surface mass density, including fluid loading,
b = surface damping, including fluid damping,
D = stiffness, $Et^3/[12(1 - \nu^2)]$.

The motion of the surface is taken to be made up of a series of the normal modes

$$y(\mathbf{x}, t) = \sum_{n=1}^{\infty} \phi_n(\mathbf{x})\, q_n(t). \tag{7.127}$$

where $\phi_n(x)$ is the mode shape and $q_n(t)$ is the generalized coordinate.

Substituting (7.127) into (7.126) gives

$$m \sum_{n=1}^{\infty} \phi_n(\mathbf{x})\, \ddot{q}_n(t) + b \sum_{n=1}^{\infty} \phi_n(\mathbf{x})\, \dot{q}_n(t) + D \sum_{n=1}^{\infty} \nabla^4 \phi_n(\mathbf{x})\, q_n(t) = p(\mathbf{x}, t) \quad (7.128)$$

Multiplying (7.128) by $\phi_m(x)$ and integrating over the surface gives a sequence of modal equations

$$M_n \ddot{q}_n + B_n \dot{q}_n + K_n q_n = p_n(t), \quad (7.129)$$

where

$$M_n = \int m \phi_n^{\,2}(\mathbf{x})\, dA,$$

$$B_n = \int b \phi_n^{\,2}(\mathbf{x})\, dA,$$

$$K_n = \int D \phi_n(\mathbf{x})\, \nabla^4 \phi_n(\mathbf{x})\, dA,$$

$$p_n(t) = \int p(\mathbf{x}, t) \phi_n(\mathbf{x})\, dA,$$

(the repeated subscript does not signify a summation). Note that K_n may also be obtained from the homogeneous form of (7.128), and is given by

$$K_n = M_n \omega_n^{\,2}$$

where ω_n is the natural frequency of the nth mode. Usually M_n, B_n, and K_n are termed the generalized mass, damping, and stiffness for the mode. It is assumed that the damping force is proportional to the mass density and there is no damping coupling between the modes.

The normalized modal equation is now

$$\ddot{q}_n + 2\zeta_n \omega_n \dot{q}_n + \omega_n^{\,2} q_n = \frac{p_n(t)}{M_n} = g_n(t) \omega_n^{\,2}, \quad (7.130)$$

which is identical in form to that of a simple oscillator.

Drawing on the relations developed in the previous section, the response may be immediately written down

$$q_n(t) = \int_{-\infty}^{\infty} h_n(\tau)\, g_n(t - \tau)\, d\tau \quad (7.131)$$

$$y(\mathbf{x}, t) = \sum_{1}^{\infty} \phi_n(\mathbf{x}) \int_{-\infty}^{\infty} h_n(\tau)\, g_n(t - \tau)\, d\tau. \quad (7.132)$$

The most general correlation function is given by

$$\langle y(\mathbf{x}, t)y^*(\mathbf{x}', t + \tau)\rangle_t$$
$$= \left\langle \sum_{m,n}^{\infty} \phi_m(\mathbf{x})\, \phi_n^*(\mathbf{x}') \iint_{-\infty}^{\infty} h_m(\tau_1)\, g_m(t - \tau_1)\, h_n^*(\tau_2)\, g_n^*(t + \tau - \tau_2)\, d\tau_1\, d\tau_2 \right\rangle_t . \tag{7.133}$$

The time average is only applicable to g_m and g_n, hence it may be put inside the integral. The time average of the product is

$$\langle g_m(t - \tau_1) g_n^*(t - \tau_2 + \tau)\rangle_t$$
$$= \lim_{T\to\infty} \left[\frac{1}{2T} \int_{-T}^{T} \iint_A p(\mathbf{x}, t - \tau_1)\, p(\mathbf{x}', t + \tau - \tau_2)\, \phi_m(\mathbf{x})\, \phi_n^*(\mathbf{x})\, dA\, dA'\, dt \right]. \tag{7.134}$$

This may be rewritten as

$$\langle g_m(t - \tau_1) g_n^*(t + \tau - \tau_2)\rangle_t = \iint_A \phi_m(\mathbf{x})\, \phi_n^*(\mathbf{x}')\, R_g(\mathbf{x}, \mathbf{x}', \tau + \tau_1 - \tau_2)\, dA\, dA', \tag{7.135}$$

where $R_g(\mathbf{x}, \mathbf{x}', \tau + \tau_1 - \tau_2)$ is the general cross-correlation. Equation (7.135) may be normalized by dividing by the mean square value of g and the area squared. The general correlation function of the response is now written in terms of the mode shapes, the unit impulse response function for each mode, and the correlation function of the pressure field.

$$\langle y(\mathbf{x}, t)y^*(\mathbf{x}', t + \tau)\rangle_t - \sum_{m,n}^{\infty} \phi_m(\mathbf{x})\, \phi_n^*(\mathbf{x}) \iint_{-\infty}^{\infty} h_m(\tau_1)\, h_n^*(\tau_2)$$
$$\times \iint_A \phi_m(\boldsymbol{\xi})\, \phi_n^*(\boldsymbol{\xi}')\, R_g(\boldsymbol{\xi}, \boldsymbol{\xi}', \tau + \tau_1 - \tau_2)\, dA\, dA'\, d\tau_1\, d\tau_2. \tag{7.136}$$

This equation is the most general form for the response. By setting $\tau = 0$ it gives the spatial correlation function, and setting $\mathbf{x} = \mathbf{x}'$ it gives the temporal correlation at a point. When the Fourier transform of (7.136) is taken, the power spectral density and cross power spectral density are found.

Except for some very simple cases, such as when the pressure correlation function is a constant and the pressure is uniform over the surface, or when it is a delta function and the pressure is completely uncorrelated (raindrops on the roof), the integrals in (7.136) cannot be evaluated analytically. By taking the Fourier transform of the correlation function one obtains the power spectral density, and this function is more amenable to computation.

Furthermore, it is one of the most frequently measured quantities in real vibration problems.

Following the procedure set forth in the section on the single-degree-of-freedom system, we set $\tau_3 = \tau + \tau_1 - \tau_2$ and compute

$$\int_{-\infty}^{\infty} h_m(\tau_1)e^{i\omega\tau_1}\,d\tau_1 \int_{-\infty}^{\infty} h_n^*(\tau_2)e^{i\omega\tau_2}\,d\tau_2 \int_{-\infty}^{\infty} \frac{R_g(\mathbf{x}, \mathbf{x}', \tau)}{2\pi} e^{i\omega\tau}\,d\tau = H_m(\omega)\, H_n^*(\omega)\, S_g(\mathbf{x}, \mathbf{x}', \omega). \quad (7.137)$$

This now transforms (7.136) into

$$S_y(\mathbf{x}, \mathbf{x}', \omega) = \sum_{m,n} \phi_m(\mathbf{x})\, \phi_n^*(\mathbf{x}')\, H_m(\omega)\, H_n^*(\omega) \times \iint_A \phi_m(\boldsymbol{\xi})\, \phi_n^*(\boldsymbol{\epsilon}')\, S_g(\boldsymbol{\xi}, \boldsymbol{\epsilon}', \omega)\, dA\, dA'. \quad (7.138)$$

The latter integral may be normalized by the power spectral density and the area squared. The resulting form is called the *joint acceptance* [11], and it is a measure of how well the pressure field couples with the modes

$$j_{mn}{}^2 = \frac{1}{S_g(\omega)A^2} \iint \phi_m(\boldsymbol{\xi})\, \phi_n^*(\boldsymbol{\xi}')\, S_g(\boldsymbol{\xi}, \boldsymbol{\xi}', \omega)\, dA\, dA', \quad (7.139)$$

where $S_g(\mathbf{x}, \mathbf{x}', \omega)/S_g(\omega)$ is the normalized cross spectral density. The cross power spectral density is now written as

$$S_y(\mathbf{x}, \mathbf{x}', \omega) = \sum_{m,n}^{\infty} \phi_m(\mathbf{x})\, \phi_n^*(\mathbf{x}')\, H_m(\omega)\, H_n^*(\omega) j_{mn}{}^2 A^2\, S_g(\omega). \quad (7.140)$$

The complex frequency response functions are related to the modal impedances by the relation†

$$H_n(\omega) = \frac{M_n \omega_n{}^2}{Z_n(\omega)\omega}\;. \quad (7.141)$$

Thus the cross spectral density of the response may be written as

$$S_y(\mathbf{x}, \mathbf{x}', \omega) = \sum_{m,n}^{\infty} \frac{\phi_m(\mathbf{x})\phi_n^*(\mathbf{x}')}{Z_m(\omega)Z_n^*(\omega)\omega^2} A^2\, S_p(\omega) j_{mn}{}^2. \quad (7.142)$$

Either (7.140) or (7.142) may be used, depending upon which is easier to compute. These two equations are the central result of this section. The remainder of this section will be devoted to particular cases and examples.

The mode shapes, impedances, and mean square pressure may be determined experimentally or analytically. These will be invariant under different

† Impedance being defined by the ratio of force to velocity.

types of pressure fields. The joint acceptance is a function of the specific pressure field and must be determined for each loading situation.

Consider first the case of a uniform pressure loading. The pressure correlation function is constant and the joint acceptance is

$$j_{mn}{}^2 = \frac{1}{A^2}\iint_A \phi_m(\mathbf{x})\, \phi_n^*(\mathbf{x}')\, dA\, dA'$$

$$= \frac{1}{A^2}\int_A \phi_m(\mathbf{x})\, dA \int_A \phi_n^*(\mathbf{x}')\, dA'. \tag{7.143}$$

Structures such as panels, beams, or cylinders generally have modes which are volume displacing (like $\sin \pi x_1/l_1 \sin \pi x_2/l_2$) or which have no net volume displacement ($\sin 2\pi x_1/l_1 \sin 2\pi x_2/l_2$). Only the volume displacing modes have a nonzero joint acceptance, and the joint acceptance is a decreasing function of mode number for this type of mode. The joint acceptance of a simply supported rectangular panel is of the form

$$j_{mn}{}^2 = \frac{16}{(n_1 n_2 \pi^2)^2}; \qquad n_1, n_2, \qquad \text{odd}, \tag{7.144}$$

where n_1 and n_2 are the number of half waves in the x_1 and x_2 directions, respectively.

If the pressure field is completely uncorrelated from point to point ("raindrops on the roof") the correlation function is

$$\tilde{C}_p(\mathbf{x}, \mathbf{x}', \omega) = \delta(\mathbf{x} - \mathbf{x}'), \tag{7.145}$$

and

$$j_{mn}{}^2 = \frac{1}{A^2}\int \phi_m(\mathbf{x})\, \phi_n^*(\mathbf{x})\, dA. \tag{7.146}$$

For normal mode this will, of course, be zero except when $m = n$. The simply supported rectangular panel with sinusoidal mode shapes gives a joint acceptance of

$$j_m{}^2 = \frac{1}{4A}. \tag{7.147}$$

If the modal damping is small, then the double summation may be approximated by a single summation of only those terms where $m = n$

$$S_y(\mathbf{x}, \mathbf{x}', \omega) = \sum_n \frac{\phi_n{}^2(x)}{|Z_n(\omega)|^2\omega^2} A^2\, S_p(\omega) j_n{}^2. \tag{7.148}$$

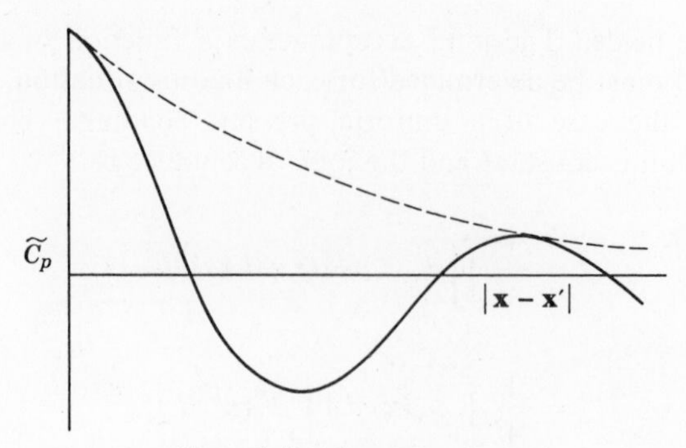

Figure 7.13 Correlation function.

The joint acceptance in this case is

$$j_n^2 = \frac{1}{A^2} \iint_{A'} \phi_n(\mathbf{x})\, \phi_n(\mathbf{x}')\, \tilde{C}_p(\mathbf{x}, \mathbf{x}', \omega)\, dA\, dA'. \tag{7.149}$$

This joint acceptance has been studied extensively for a variety of correlation functions. One of the most general conditions is a decaying progressive wave which propagates over a rectangular plate. The correlation function for this wave field is

$$\tilde{C}_p(\mathbf{x}, \mathbf{x}', \omega) = e^{-a_1|\mathbf{x}-\mathbf{x}'|} \cos\left[\frac{\omega}{U}(|\mathbf{x}-\mathbf{x}'|)\right]. \tag{7.150}$$

It is shown in Figure 7.13. The joint acceptance is factorable into the product of the joint acceptances of the longitudinal and lateral modes.

$$j_n^2 = j_n^2\,(\text{lat})\, j_n^2\,(\text{long}). \tag{7.151}$$

Powell [37] has shown the general conditions for the joint acceptance to be separable in this way to be that the modes of vibration be separable, and the correlation function be separable in the same coordinates. The joint acceptance for a wave propagating in the direction of one of the panel edges has been computed by Bozich [38] and for a more general case by White [39]. The result is tedious to evaluate, but is plotted in Figure 7.14. These figures demonstrate a very important feature of the coupling between the structural vibration and the pressure field. If the pressure field is strongly periodic there are very pronounced maxima and minima in the joint acceptance when the correlation period is about the same as the wave length in the structure. What this means is the pressure field is pushing the structure in its natural vibration in the best possible way. A sharply decaying pressure field is coherent over only a small area; thus it does not drive the mode very well.

The foregoing results are immediately applicable to boundary layer turbulence induced vibration. The pressure field may be resolved into a

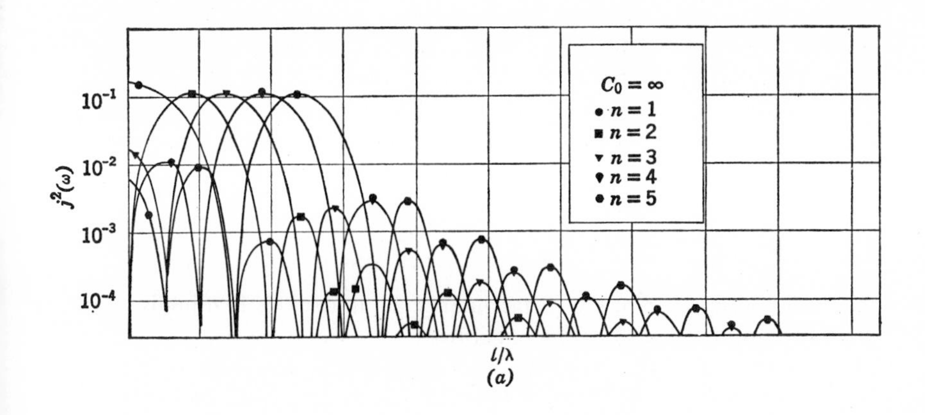

(a)

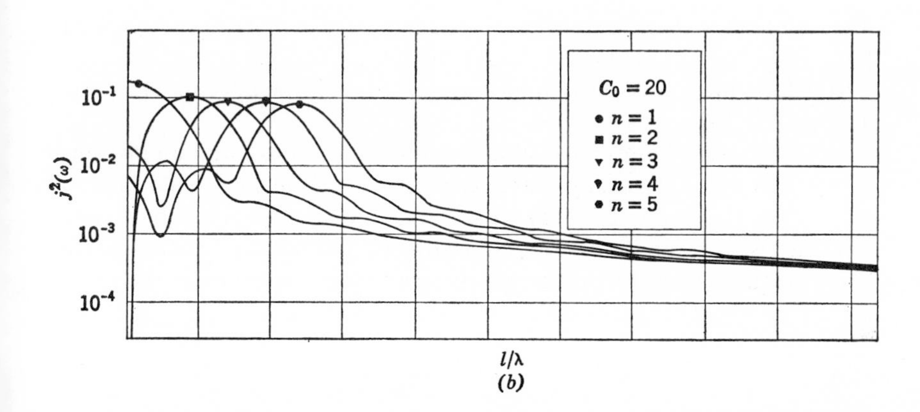

(b)

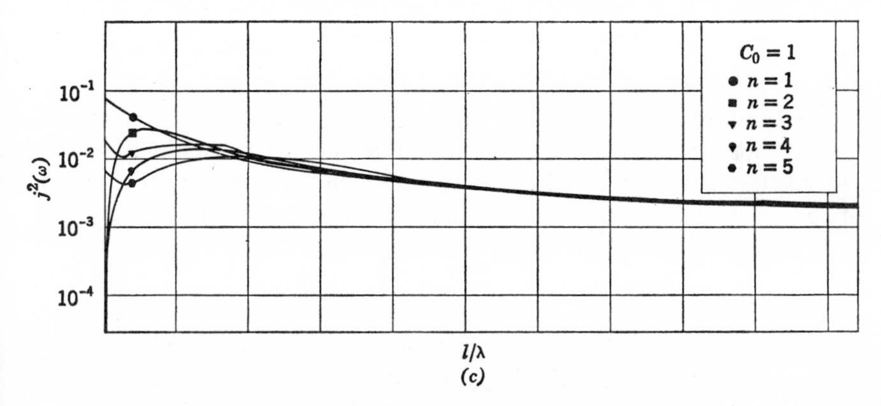

(c)

Figure 7.14 a, b, c.

convected series of decaying waves having a correlation function of a form similar to (7.150). This form has been used by White [40] to determine the noise radiated by boundary layer excited panels.

Often the sound field does not consist of an acoustic wave at some fixed angle of incidence but of many waves from many angles, thus creating a reverberent sound field. The correlation function for this pressure field is

$$\tilde{C}_p(\mathbf{x}, \mathbf{x}', \omega) = \frac{\sin(\omega/c)\,|\mathbf{x} - \mathbf{x}'|}{(\omega/c)\,|\mathbf{x} - \mathbf{x}'|}\,. \tag{7.152}$$

The joint acceptance of the rectangular panel under this loading is related to the radiation resistance of the panel by

$$j_n^{\,2} = \frac{\pi R^{\text{rad}}}{2\rho c k^2 A^2}\,. \tag{7.153}$$

Values of the radiation resistance for a large range of frequency and panel sizes have been computed by Maidanik [42]. The joint acceptance for a particular situation is easily found by substituting the appropriate value of R^{rad} into (7.153).

In many cases the response of a structure is measured with an instrument which indicates the mean-square response in a frequency band $\Delta\omega$. The majority of the response in that band is due to the modes which have natural frequencies there and are vibrating near their resonances. Each mode acts like a single-degree-of-freedom and its mean square response is [13]

$$\langle q_n^{\,2}\rangle_t = \frac{\pi S_p(\omega)}{2\zeta M_n \omega_n^{\,3}}\,.$$

The total response is

$$\langle y^2(\mathbf{x})\rangle_t \simeq \pi \sum_{n,\Delta\omega} \frac{\phi_n^{\,2}(\mathbf{x}) A^2\, S_p(\omega) j_n^{\,2}}{2\zeta_n M \omega_n^{\,3}}\,, \tag{7.154}$$

where the summation is taken only over those modes with natural frequencies in the band $\Delta\omega$. The modal mass, damping, and resonant frequency will not change appreciably over the frequency interval, hence they may be taken outside the summation. Also, if there are a large number (say 10 or more) of modes in the band, the summation may be approximated by

$$\langle y^2(\mathbf{x})\rangle_t \simeq \frac{\pi A^2 S_p(\omega)}{2\zeta_n M_n \omega_n^{\,3}} \langle \phi_n^{\,2}(\mathbf{x})\rangle_n \langle j_n^{\,2}\rangle_n n(\omega)\,\Delta\omega, \tag{7.155}$$

where $\langle \phi_n^{\,2}(\mathbf{x})\rangle_n$ = average value of $\phi_n^{\,2}(\mathbf{x})$ taken over many modes,
$\langle j_n^{\,2}\rangle_n$ = average value of $j_n^{\,2}$ taken over many modes,
$n(\omega)$ = modal density, modes/rad/sec.

Dividing by the bandwidth $\Delta\omega$ gives the power spectral density of the response

$$S_y(\mathbf{x}) \simeq \frac{\pi A^2 S_p(\omega)}{2\zeta M_n \omega_n^{\ 3}} \langle \phi_n^{\ 2}(\mathbf{x}) \rangle_n \langle j_n^{\ 2} \rangle_n n(\omega). \tag{7.156}$$

For the frequency range of applicability, and for plates and cylinders, $\langle \phi_n^2(\mathbf{x}) \rangle_n \approx \frac{1}{4}$. The modal density is determined by the size and shape of the structure. Values of $\langle j_n^2 \rangle$ have been determined by White for several types of pressure fields on a rectangular plate [39,40].

If the mode shapes and the pressure correlation function are complicated functions, making the joint acceptance difficult to compute, it is possible to obtain an upper bound for j_n^2. The bound for sinusoidal modes is given by

$$j_n^{\ 2} \leq \frac{1}{l^2} \int_0^l \phi_n^{\ 2}(x)\, dA \left[2 \int (l - x) \tilde{C}_p^{\ 2}(x)\, dx \right]^{1/2} \tag{7.157}$$

Other approximations have been computed by Mayer [43].

In this section we have derived the general form for the correlation function of the response of a continuous or multi-degree-of-freedom system to a distributed random pressure field (7.136). There is a functional dependence upon the mode shapes, the impulse response function, and the pressure correlation function. We have also derived the cross power spectral density of the response (7.138) or (7.142), which is a function of the mode shapes, complex frequency response functions, and cross power spectral density of the pressure. These two functions are a Fourier transform pair and contain the same information. The choice of usage is dependent upon the available data and the end to which the result will be put. Although we have derived the response characteristics in terms of displacement, the stress could just as well have been considered. For each displacement mode shape there will be a stress mode shape, which may be easily determined and substituted in the series.

7.3.3 *Statistics of the Response to Random Excitation*

Because the statistical characteristics of the response of a structure are of importance in determining its fatigue life, it is necessary to examine these properties. Not all theories of fatigue under random loading use the same statistical properties of the response; thus several different characteristics must be examined. In all cases it is assumed that the system is linear and time invariant and that the input is a stationary Gaussian process. This implies that the response is also a stationary Gaussian process [24].

The probability distribution of the instantaneous values of the response is Gaussian. In most cases the mean value is zero; hence the variance is simply the mean square response

$$\sigma_y^{\,2} = \langle y^2 \rangle_t = \int_{-\infty}^{\infty} S_y(\omega)\, d\omega. \tag{7.158}$$

The probability distribution is

$$p(y) = \frac{1}{\sqrt{2\pi\sigma_y}} \exp\left[-\frac{y^2}{2\sigma_y^{\,2}}\right]. \tag{7.159}$$

This is independent of the shape of the spectrum of the driving force and the response function $|H(\omega)|^2$, except as it influences the spectral density of the response $S_y(\omega)$. An infinite number of possible shapes for $S_x(\omega)$ and $|H(\omega)|^2$ give identically the same value for the integral in (7.158).

It is also of interest to determine the crossing rate, or the expected number of times per second that the response will exceed some fixed level. In order to do this, it is necessary to know the joint probability density function of the displacement and velocity of the response, $p(x, \dot{y})$. Consider a small portion of the signal as shown in Figure 7.15. Over the small interval Δt, $y(t)$ is essentially a straight line. The minimum slope which $y(t)$ must have in order to cross a with a positive slope during the time interval Δt is

$$\frac{\Delta y}{\Delta t} = \frac{a - y(t)}{\Delta t}. \tag{7.160}$$

Thus

$$\frac{\Delta y}{\Delta t} < \dot{y} < \infty. \tag{7.161}$$

The probability of a signal being in the region y to $y = \Delta y$ and simultaneously having a velocity in the region $\dot{y}$ to $\dot{y} + \Delta\dot{y}$ is $p(y, \dot{y})\, dy\, d\dot{y}$. By integrating overall possible combinations that satisfy (7.3.74) and (7.3.75) we get

$$\nu_a^{+}\, dt = \int_0^{\infty} d\dot{y} \int_{a - y\, dt}^{a} p(y, \dot{y})\, dy, \tag{7.162}$$

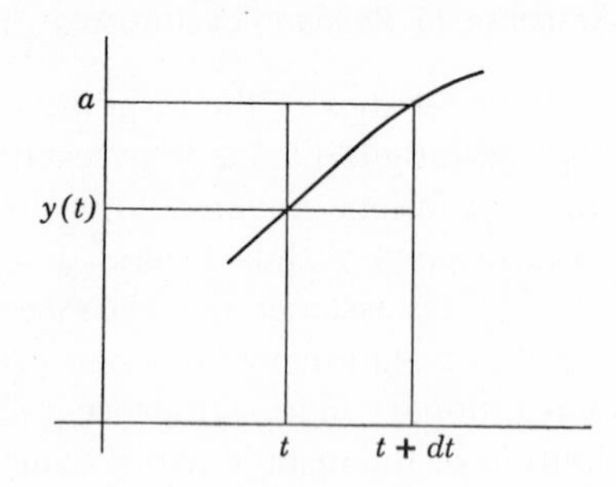

Figure 7.15

where ν_a^+ is the expected number of crossings of level a with a positive slope. Dividing by dt and letting $dt \to 0$ gives

$$\nu_a^+ = \int_0^\infty p(a, \dot{y})\, d\dot{y}. \tag{7.163}$$

This relation, attributed to Rice [27], applies to any stationary process and is not restricted to Gaussian distribution. Since many of the commonly encountered processes are Gaussian and the mathematical evaluation of (7.3.77) can then be easily made, we assume that the joint density function is of the form

$$p(y, \dot{y}) = \frac{1}{2\pi\sigma_y\sigma_{\dot{y}}} \exp\left[-\frac{1}{2}\left(\frac{y^2}{\sigma_{\dot{y}}^2} + \frac{\dot{y}^2}{\sigma_{\dot{y}}^2}\right)\right], \tag{7.164}$$

where σ_y^2 and $\sigma_{\dot{y}}^2$ are the variances of the displacement and velocity, and are given by

$$\sigma_y^2 = \int_{-\infty}^{\infty} S_y(\omega)\, d\omega = R(0),$$

$$\sigma_{\dot{y}}^2 = \int_{-\infty}^{\infty} \omega^2 S(\omega)\, d\omega = R''(0). \tag{7.165}$$

This equation is valid for a random Gaussian signal passed through any linear device. Hence

$$\nu_a^+ = \frac{1}{2\pi}\frac{\sigma_{\dot{y}}}{\sigma_y} \exp\left(\frac{-a^2}{2\sigma_y^2}\right). \tag{7.166}$$

By setting $a = 0$ we obtain the expected number of zero crossings per unit time

$$\nu_0^+ = \frac{\sigma_{\dot{y}}}{2\pi\sigma_y}, \tag{7.167}$$

when the process is narrowband, each crossing implies a complete cycle of oscillation and the expected frequency of oscillation is then given by (7.167).

The probability of a peak of a stationary random process occurring in the interval y to $y + \Delta y$, t to $t + \Delta t$ is [27]

$$\text{Prob (peak in } \Delta y\, \Delta t) = -dy\, dt \int_{-\infty}^{0} \ddot{y}\, p(y, \dot{y}, \ddot{y})_{\dot{y}=0}\, d\ddot{y}. \tag{7.168}$$

For a Gaussian distribution this is expressed as

$$\frac{dy\, dt\sqrt{-R''}}{2\sqrt{2\pi R}} \frac{y}{\sqrt{2R}} \exp\left(\frac{-y^2}{2R}\right)\left[1 + erf\frac{ky}{\sqrt{2R}} + \frac{1}{\sqrt{\pi}}\frac{\sqrt{2R}}{ky} \exp\left(\frac{-k^2y^2}{2R}\right)\right], \tag{7.169}$$

where

$$R = R(0),$$

$$R' = \left.\frac{dR}{d\tau}\right|_{t=0},$$

$$R'' = \left.\frac{d^2R}{d\tau^2}\right|_{\tau=0},$$

and

$$k^2 = \frac{R''^2}{(RR^{\text{IV}} - R''^2)}.$$

Using (7.169), the expected number of maxima per second of a Gaussian process is given by

$$n^+ = \frac{1}{2\pi}\left[\frac{\int_{-\infty}^{\infty} \omega^4 S(\omega)\, d\omega}{\int_{-\infty}^{\infty} \omega^2 S(\omega)\, d\omega}\right]^{1/2}. \tag{7.170}$$

We note that the integrals emphasize the high-frequency portion of the spectrum.

In the case of a narrowband Gaussian signal (7.169) may be simplified considerably. The narrowband random signal is approximately a sine wave at the center frequency of the band, which is randomly modulated. This is shown in Figure 7.16. The probability distribution of this envelope, and thus of the peaks, is

$$p(y) = \frac{y}{\sigma_y^{\,2}} \exp\left(-\frac{y^2}{2\sigma_y^{\,2}}\right). \tag{7.171}$$

This distribution is known as the Rayleigh distribution.

A signal such as shown in Figure 7.17 can be said to be made up of a sequence of half-cycles of an oscillating component about some mean value. The mean is the arithmetic average of the values of consecutive peaks and

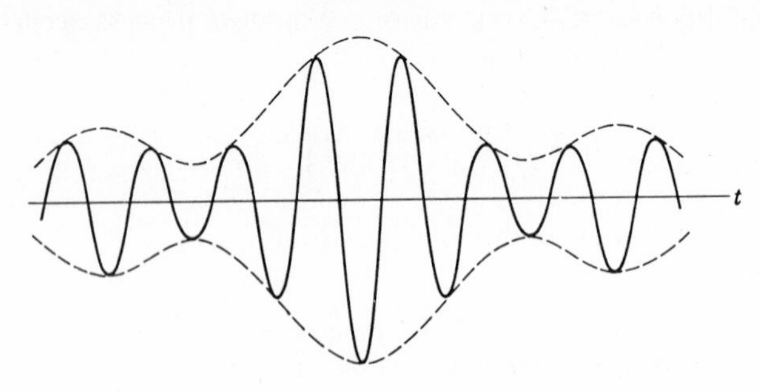

Figure 7.16 Narrow band random process.

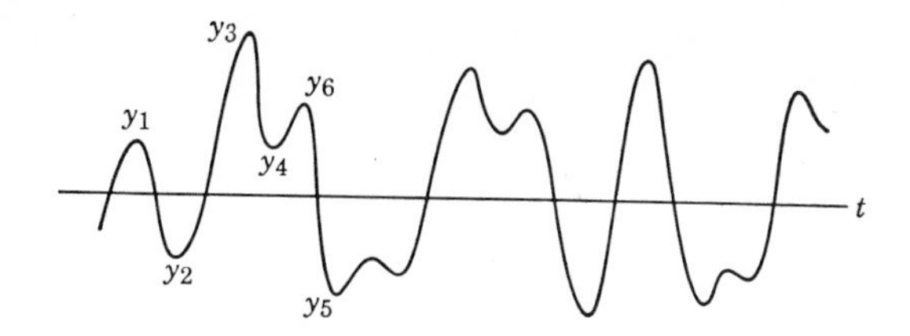

Figure 7.17 Maxima and minima.

troughs (or between consecutive maxima and minima); that is

$$y_{m,1} = \frac{y_1 + y_2}{2}$$
$$y_{m,2} = \frac{y_2 + y_3}{2}, \tag{7.172}$$
$$y_{m,3} = \frac{y_3 + y_4}{2}.$$

The oscillating portion is taken as half the difference

$$y_{0,1} = \frac{|y_1 - y_2|}{2},$$
$$y_{0,2} = \frac{|y_2 - y_3|}{2}, \tag{7.173}$$
$$y_{0,3} = \frac{|y_3 - y_4|}{2}.$$

An analytic form for the joint probability density of these variables has not been derived for the general case. When the process is a single narrowband random signal, the mean is approximately zero (or some constant) and the oscillating portion has a Rayleigh distribution. If there is a mixture of two narrowband processes, widely separated in frequency, the mean will have the characteristics of the low-frequency portion, and the oscillating part will have the nature of the high frequency section. In many cases the joint distribution is found by actually computing the mean and oscillating components of the signal and generating the distribution from a large sample of values.

7.4 CLOSURE

The material in this chapter has covered the common forms of acoustic fields which excite structures and cause sonic fatigue. We have presented methods for determining the response of structures to such random pressure fields and the statistics of the response. These results must then be integrated with the theories of fatigue presented in other chapters in order to estimate the fatigue characteristics of a given situation.

REFERENCES

[1] M. J. Lighthill, "On Sound Generated Aerodynamically," *Proc. Roy. Soc.* (*London*), **A211,** 564 (1952).

[2] M. J. Lighthill, "On Sound Generated Aerodynamically, Part II, Turbulence as a Source of Sound," *Proc. Roy. Soc.* (*London*), **A222,** 1 (1954).

[3] Alan Powell, "On the Mechanism of Choked Jet Noise," *Proc. Phys. Soc. London*, **B66,** 1039 (1953).

[4] I. Proudman, "The Generation of Noise by Isotropic Turbulence," *Proc. Roy. Soc.* (*London*) **A214,** 119 (1952).

[5] Alan Powell, "A Schlieren Study of Small Scale Air Jets and Some Noise Measurements on Two Inch Diameter Jets," *Aero. Res. Council*, Paper 14726 (1951).

[6] Alan Powell, "Survey of Experiments on Jet Noise," *Aircraft Engrg.* **26,** 2 (1954).

[7] L. W. Lassiter, and H. H. Hubbard, "Experimental Studies of Noise from Subsonic Jets in Still Air," NACA TN 2757, August 1952.

[8] J. N. Cole *et al.* "Noise Radiation from Fourteen Types of Rockets in the 1000 to 13,000 Pounds Thrust Range," WADC TR 57-354, 1957.

[9] Alan Powell, "On the Generation of Noise by Turbulent Jets," *Am. Soc. Mech. Engr.*, Aviation Conference Paper 59-AV-53, Los Angeles, March 1959. (Also see "Theory and Experiment in Aerodynamic Noise, with a Critique of Research on Jet Flows and Their Relationship to Sound," Second Symposium on Naval Hydrodynamics, 1958, ACR-38, Office of Naval Research, Department of the Navy, Washington, D.C.)

[10] J. W. Miles, "On Structural Fatigue Under Random Loading," *J. Aero. Sci.* **21,** 753–762 (1954).

[11] Alan Powell, "On the Fatigue Failure of Structures Due to Vibrations Excited by Random Pressure Fields," Aeronautical Research Paper No. 17925, 1955. Also *Jr. Acoust. Soc. Am.*, **30,** 1130 (1958).

[12] W. T. Thomson, and M. N. Barton, "Response of Mechanical Systems to Random Excitation," *J. Appl. Mech.*, **24,** 248, (1957).

[13] S. H. Crandall, Editor, *Random Vibration*, Vol. I, Technology Press, Cambridge, Mass., 1958.

[14] S. H. Crandall, Editor, *Random Vibration*, Vol. II, M.I.T. Press, Cambridge, Mass., 1963.

[15] R. W. Hess, R. W. Fralich, and H. H. Hubbard, "Studies of Structural Failure Due to Acoustic Loading," NASA Tech. Note No. 4050, July 1957.

[16] R. W. Hess, R. W. Herr, and W. H. Mayes, "A Study of the Acoustic Fatigue Characteristics of Some Flat and Curved Aluminum Panels Exposed to Random and Discrete Noise," NASA Tech. Note No. D-1, August 1959.

[17] G. S. Johnson, "Skin Cracking Due to Jet Noise," McDonnell Aircraft Company Report 4885, August 1957.

[18] H. C. Schjelderup, and J. C. McClymonds, "Structural Fatigue by Jet Engine Noise on RB/66 Aircraft," Douglas Aircraft Company, Report LB-25187.

[19] W. J. Trapp, and D. M. Forney, Jr., "WADC-University of Minnesota Conference on Acoustical Fatigue," WADC Tech. Report 59-676, March 1961.

[20] U.S. Air Force Conference on Acoustical Fatigue, Wright-Patterson AFB, Ohio, April 1964.

[21] L. E. Kinsler, and A. R. Frey, *Fundamentals of Acoustics*, Wiley, New York, 1959.

[22] J. W. S. Rayleigh, *Theory of Sound, Vol. II*, Dover, New York, 1945.

[23] L. L. Beranek, *Acoustics*, McGraw-Hill, New York, 1954.

[24] J. S. Bendat, *Principles and Applications of Random Noise Theory*, Wiley, New York, 1958.

[25] J. S. Bendat and G. P. Thrall, "Spectra of Nonstationary Random Processes," AFFDL TR-64-198, Wright-Patterson AFB, Ohio, August 1964.

[26] J. H. Laning, Jr. and R. H. Battin, *Random Processes in Automatic Control*, McGraw-Hill, New York, 1956.

[27] Nelson Wax, Editor *Selected Papers on Noise and Stochastic Processes*, Dover, New York, 1954.

[28] A. G. Piersol, "The Measurement and Interpretation of Ordinary Power Spectra for Vibration Problems," NASA Contractor Report NASA CR-90, September 1964.

[29] P. M. Morse, *Vibration and Sound*, McGraw-Hill, New York, 1948.

[30] I. N. Sneddon, *Elements of Partial Differential Equations*, McGraw-Hill, New York, 1957.

[31] S. H. Crandall and W. D. Mark, *Random Vibration in Mechanical Systems*, Academic, New York, 1963.

[32] A. C. Eringen, "Response of Beams and Plates to Random Loads," *J. Appl. Mech.*, **24**, 46–52, 1957.

[33] R. H. Lyon, "Response of Strings to Random Noise Fields," *J. Acoust. Soc. Am.*, **28**, 391 (1956).

[34] G. R. Corcos and H. W. Liepmann, "On the Contribution of Turbulent Boundary Layers to Noise Inside a Fuselage," NASA TM-1420, 1958.

[35] H. S. Ribner, "Boundary Layer Induced Noise in the Interior of Aircraft," University of Toronto UTIA Report No. 37, 1956.

[36] Alan Powell, "On the Approximation to the Infinite Solution by the Method of Normal Modes for Random Vibration," *J. Acoust. Soc. Am.*, **30**, 1136 (1958).

[37] Alan Powell, "On the Estimation of the Generalized Force Due to Random Pressures and on Necessary Modes," *J. Acoust. Soc. Am.*, **36**, 783 (1964).

[38] D. Bozich, "Spatial Correlations in Acoustic-Structural Coupling," *J. Acoust. Soc. Am.*, **36**, 52 (1964).

[39] P. H. White and Alan Powell, "Transmission of Random Sound and Vibration Through a Rectangular Double Wall," *J. Acoust. Soc. Am.*, **40**, 1966.

[40] P. H. White, "The Transduction of Boundary Layer Noise by a Rectangular Panel," *J. Acoust. Soc. Am.*, **40**, 1966.

[41] P. W. Smith, Jr., "The Response and Radiation of Structural Modes Excited by Sound," *J. Acoust. Soc. Am.*, **34**, 640 (1962).

[42] G. Maidanik, "Response of Ribbed Panels to Reverberant Sound Fields," *J. Acoust. Soc. Am.*, **34**, 809 (1962).

[43] F. S. Mayer, "Calculation of the Generalized Load for Random Pressure Fields with Applications," Douglas Aircraft Co. Engr. Paper 1739, 1963.

[44] Donald W. Jorgensen, "Measurements of Fluctuating Pressures on a Wall Adjacent to a Turbulent Boundary Layer," David Taylor Model Basin Report 1744, July 1963.

[45] G. R. Ludwig, "An Experimental Investigation of the Sound Generated by Thin Steel Panels Excited by Turbulent Flow (Boundary Layer Noise)," University of Toronto, Institute of Aerophysics, UTIA Report No. 87, November 1962.

[46] M. K. Bull, "Properties of the Fluctuating Wall-Pressure Field of a Turbulent Boundary Layer," NATO A.G.A.R.D. Report 455, April 1963.

[47] M. K. Bull and J. L. Willis, "Some Results of Experimental Investigations of the Surface Pressure Field Due to a Turbulent Boundary Layer," Wright-Patterson Air Force Base, Ohio Report No. ASD-TDR-62-425, August 1962.

8

Fatigue Analysis and Measurement of Random Loading

WILLIAM H. RASER, JR.

8.1 INTRODUCTION TO THE DAMAGE EVALUATION PROBLEM

To the extent that fatigue damage can be considered to be a measurable quantity, its measurement is constantly being sought by aeronautical, mechanical and civil engineers interested in obtaining safe and reliable structures. There exists no clearly-defined physical quantity which measures exactly the fraction of useful life consumed as a result of vibration on a structure. Theoretically, however, it is possible to calculate an approximate, but useful *damage-to-life fraction* (N/N_L) from measurable quantities and events provided the conditions permitting an application of this fraction are understood to be limited. For example, they are limited to only one point on each of two probability distributions shown in Figure 8.1. For a given specimen and a given loading, fatigue damage can be expressed at any particular time at a certain confidence level for a particular theory of fatigue failure. Systems which continuously evaluate fatigue damage are becoming feasible and are being demanded by increasing requirements for reliable helicopters, VTOL aircraft, supersonic transports, and even certain types of farm machinery.

The distributions shown in Figure 8.1 represent normal distributions of variations of strength and loading, each in terms of its standard deviation identified as σ. Actually, no such orderly distributions are likely to exist. Random loading is sometimes called acoustic loading in order to emphasize

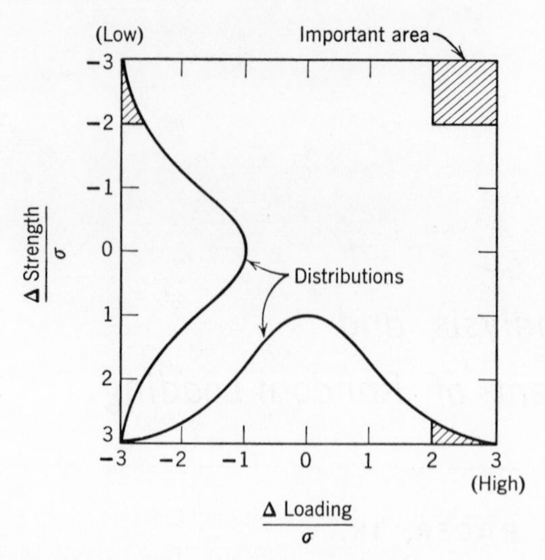

Figure 8.1 Two distributions that should be considered before an attempt is made to relate fatigue damage to sevice life.

the fact that, in general, no degree of randomness should be implied. Usually, the only characteristic implied by the term, random loading, is that the load history is not a pure sine wave. Whatever the distributions, the results should be related to something like the worst specimen and the worst loading. In the worst specimen, scratches or material tolerances may be such as to produce abnormally low fatigue limit (endurance limit); that is, the probability of occurrence of a fatigue limit could be as small as that of a 3σ point.

Figure 8.2 shows a measured distribution of helicopter rotor hub bending loads and Figure 8.3 shows the distribution of fatigue limits for 18 helicopter hub specimens. Both are from a rather extensive fatigue study [1] which appears to produce straight lines on special probability-scaled graph paper which, for the case of a normal distribution, would produce a perfectly straight line. However, if we extrapolate to the "3σ" locations, we find ourselves at the circles on these lines and, as often happens at 3σ points, wanting a little more data.

Figure 8.4 shows how the test data used to obtain Figure 8.2 can be obtained. It shows a number of damage indicating systems, each being a path going from testing to complete safe-life substantiation. It illustrates over a dozen different combinations of data flow lines and represents over a dozen different alternative damage indicating systems.

The simplest one and the first one of these to become widely used is the combination of a strain gage, a recording oscillograph, and hand calculations.

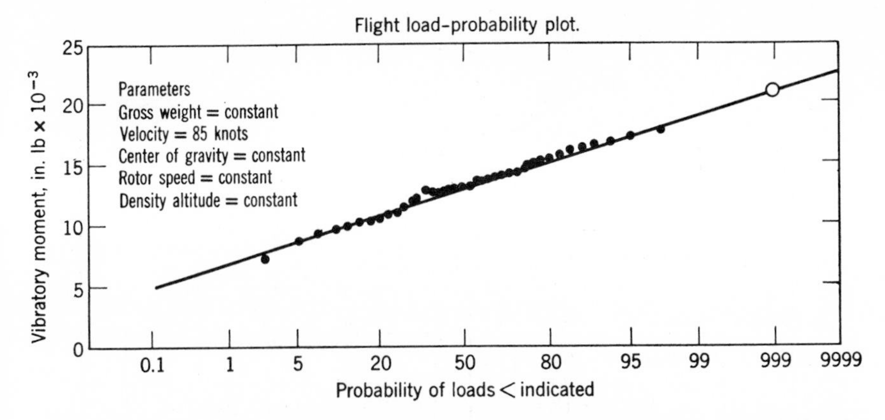

Figure 8.2 Some measured helicopter rotor hub loadings (courtesy of Kaman Aircraft Corp.)

Fatigue evaluation by this method is prohibitively tedious when large amounts of data are involved. When very large amounts are involved, it is sometimes desirable to record the information on magnetic tape rather than on paper.

Instead of the strain gages and material-simulating gages shown in Figure 8.4, the transducer may be an accelerometer, a microphone, or even a resistance-measuring or other crack-detecting system in the original structure. Certain generalizations have become evident as research has progressed on these transducers and on other components. Three of the more obvious of these generalizations relate to accuracy and are shown in Table 8.1. According to the first of these, fatigue data should be in the form of a stress record rather than in the form of certain statistical quantities sometimes measured with electronic intruments and used in *spectrum-type laboratory tests* or acoustic tests. Such can include root-mean-square (RMS) stress, decibels of noise, and power spectral density.

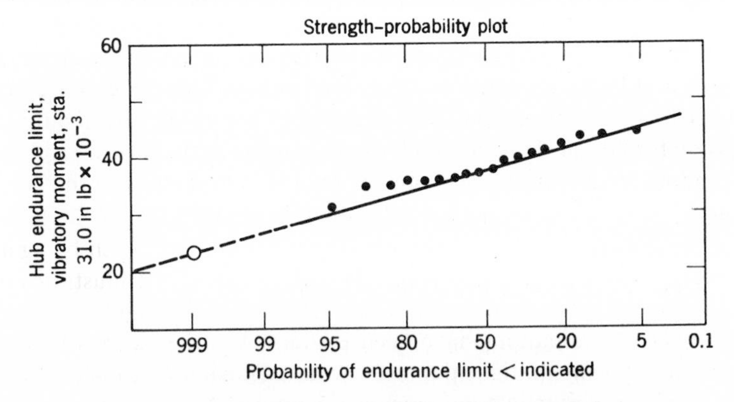

Figure 8.3 Measured fatigue strengths of some rotor hubs.

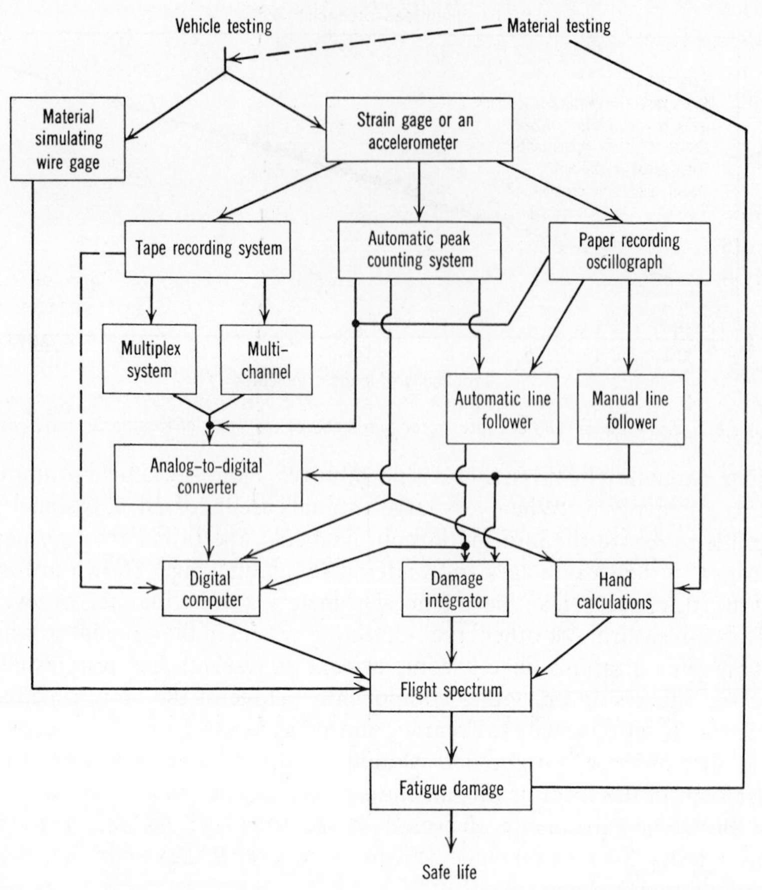

Figure 8.4 Alternative fatigue damage indicating systems for a life substantiation program.

For spectrum-type tests test variables such as certain frequency effects, stress intervals, and loading sequences affect the results. These effects are investigated briefly in a later section. The basic tool for such an investigation is an expression for fatigue damage which can be applied to the case of random and near random stress variations. This expression is developed in the next two sections.

8.1.1 Stress Concentration and Fatigue Theory

A number of evaluations have been made to test the various hypotheses and theories by which fatigue damage can be calculated. In one of the more extensive of these investigations, one which treated 266 specimens (including full scale P-51 and C-46 airplane wing tests) all tested with 78 different loading patterns, Crichlow [2] kept records of the tested number of cycles to produce failure and that predicted by each of ten different fatigue theories. The ratio of the two numbers is called C_N.

$$C_N = N_{\text{test}}/N_{\text{predicted}}. \tag{8.1}$$

For each theory evaluated, a certain number of the specimens were observed to survive an arbitrary value of C_N. As C_N was increased, fewer survived. Figure 8.5 shows the percentage surviving of the total number of specimens tested as a function of C_N for each of several theories used. Several theories produced such similar results that they can be grouped together. Broken lines show the upper and lower boundaries of this group. This group includes two theories by Lundberg [3], one by Henry [4], one by Corten and Dolan [5], and one by Shanley [6]. Theories by Miner [7], Foster and Wells [8,9], Crichlow [2], and the refinement of Miner's theory using stress concentration factors [2] are marked on the figure. Perfect agreement would produce a vertical line at C_N equal to unity. The figure shows Miner's theory to be approximately as good as any other theory evaluated for the theories and tests employed. The theories and tests employed were quite numerous and varied. However, this variation was not without limitations.

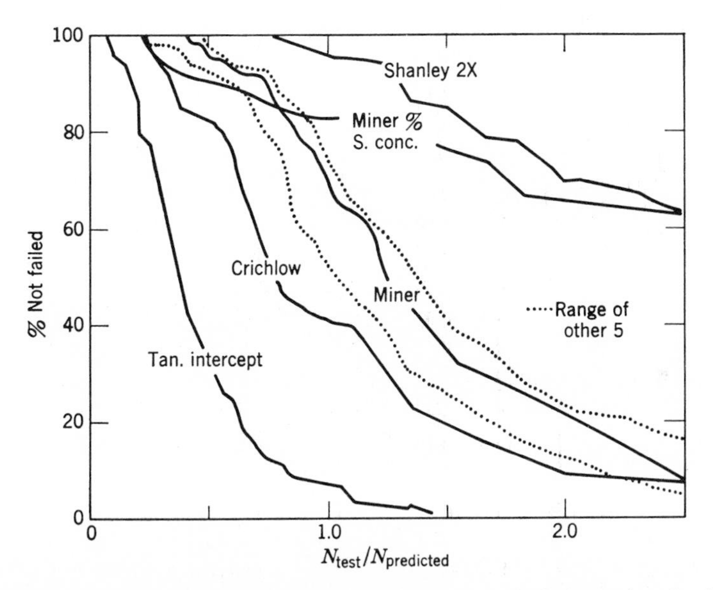

Figure 8.5 Percent of specimens surviving throughout tests in which the number of cycles is related to the predicted life by various theories.

For example, no theories were evaluated which utilize the inclusion of residual stresses from high-load plastic yielding or of the time rate of change of stress. Some recent theories which include one or more of these features are believed to be capable of showing greater accuracy for the higher stress concentration type of specimens used in the study. Early versions of such theories have been described by Smith [10] and by Valluri [11]. It remains to be seen whether the additional accuracy of these refinements justifies the additional complexity. Published evidence supporting the refinements has so far been limited to relatively simple types of loading.

Figure 8.5 also shows that, in structure where stress concentration is significant, the simple process of refining Miner's theory by introducing a calculated stress concentration factor is not very accurate. See second line from the top. This applies not only to the case where an unrecorded stress concentration is introduced as a stress multiplying factor but also to the case where it is used to select an appropriate *S-N* curve from a graded set for notched specimens of the material. When the strain gage is so located that only the point of maximum stress concentration is gaged, there is reason to believe that Miner's theory gives good results when operating within the elastic region of the material but such location is often difficult to arrange.

Figure 8.6 shows a family of *S-N* curves for smooth axially-loaded 7075-76 alloy specimens and shows a cross hatched region beyond the yield stress. It is in this area that the stress concentration difficulties arise. The nature of the difficulty in this area is best seen by looking at a life prediction problem involving stress concentration [12].

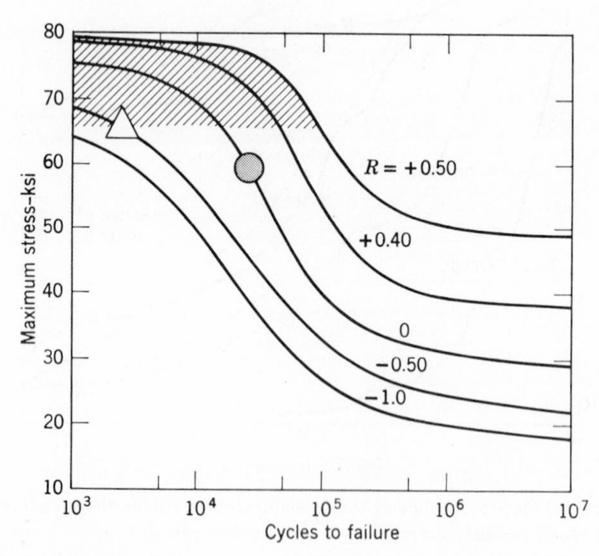

Figure 8.6 Some S-N curves for 7075-T6 aluminum and the yield stress area.

EXAMPLE 8-1

Compare the stress reductions required to effect a 10:1 life increase in the high and middle portions of Figure 8.6 if a moderate stress concentration is present.

This family of curves tends to have a lower slope, m_1 in the cross-hatched region and a higher slope magnitude, m_2 between 10^4 and 10^5 cycles. Offhand, it might appear that the required stress reduction in the cross-hatched region would be only m_1/m_2 times as much as in the m_2 region. This would be true if the stress were proportional to the load. But a stress concentration factor of less than 10 can be assumed to put a small amount of material in plastic deformation in one case and not in the other.

Fatigue failures tend to occur at the location of the stress concentration. A starting point for this discussion could be the encircled point in Figure 8.6; it represents 25,000 cycles of repeated loading causing 59,000 psi stress. In this figure, R represents the stress ratio defined as

$$R = \frac{\text{minimum stress in any cycle}}{\text{maximum stress in that cycle}}. \tag{8.2}$$

For cycles wholly below the shaded area, that is, for maximum stresses below 66,000 psi yield stress, repeated loading means that the minimum stress and R are zero. From this starting point, the two regions of interest can be investigated by considering factors which first multiply and, second, divide the number of cycles by 10.

In the first case, R remains zero. Reading along the same curve, the stress is seen to drop to 35,000 psi, a drop of 41%. This point is marked with an "X."

In the second case, if we continued with the zero-R curve, the 1/10 factor would produce a point in the shaded area corresponding to 2500 cycles; but at the place of which the stress concentration occurs local yielding limits the maximum stress to 66,000 psi. Instead of being zero, R must be read from the intersection of 66,000 psi maximum stress and 2500 cycles. This intersection is indicated by a triangle in the figure and represents an R value of -0.5. The yielding has occurred during the time of high tensile stress and introduces a residual stress in the compressive direction equal to -0.5 times 66,000 or $-33{,}000$ psi. This residual stress has not affected the stress concentration factor and has not disturbed the ratio between stress range and load range. For this point, the total cyclic stress range is 66,000 plus 33,000 or 99,000 psi. Compared to this value, the original stress range of 59,000 psi represents a reduction of 41%.

Therefore, although the S–N curves have different slopes in the high and middle portions of Figure 8.6, the effect of a moderate stress concentration can cause the load reduction required to effect a certain life increase in the two portions to be the same.

The above example illustrates that for materials having a plastic range, the effect of a stress concentration can act like a change in the S-N family. Therefore, it is not surprising that the curve in Figure 8.5 representing Miner's theory with stress concentration shows relatively poor agreement with test results.

A different type of evaluation of stress theory by Kaechele [13] and an earlier evaluation by Honaker [14] also tend to agree with Crichlow in supporting use of Miner's theory. Furthermore, the simplicity of Miner's theory is an important factor in its adoption. The Federal Aviation Administration also accepts its use [15]. But its accuracy should always be questioned when appreciable stress concentration factors are present which cause the material to enter the plastic range.

8.1.2 A Random Load Fatigue Damage Expression

Most of the commonly used theories of fatigue failure can be expressed as equations and solved for N, the number of cycles expected to cause failure. According to most of these theories, the relative damage per cycle is $1/N$. Therefore, any integral of fatigue damage can be thought of as a progressive summation of $1/N$'s. Each $1/N$ is a function of either the stress or the strain values occurring within a cycle. Of the two, stress is more commonly used. However, each has its advantages.

The main advantage of evaluating fatigue damage in terms of stress rather than strain is that most available fatigue data for various materials is in terms of stress. Other advantages are of minor importance; for example, Freudenthal [16] has demonstrated some interaction between fatigue life and creep life; to the extent that creep phenomena interfere with strain and not stress measurements, stress appears to be the sounder choice.

The main advantage of evaluating fatigue damage in terms of strain is the improved linearity and regularity of the fatigue data which results. Not only do strain-versus -N curves tend to be straighter than S-N curves, as can be demonstrated by a procedure similar to Examples 8.1; they also develop less variation between different materials. Depending upon what equipment is used to process the load data, this improved linearity and regularity can increase the feasibility and simplicity of certain methods of fatigue analysis. For example, the fatigue-life measuring strain gage developed by Harting [17] appears to be dependent upon the existence of some simple universal relationship between N and strain, ϵ such as

$$\epsilon_N^{0.32} = 0.15, \tag{8.3}$$

for at least an important part of the range of load amplitudes. This range includes the range of "short term" or "low cycle" fatigue from 10^3 to 10^4 cycles. Several investigations [18–20] show this representation to be valid for 19 different materials tested, including five steels, five nickel-base, six aluminum base, and three copper base alloys. The prospects of a corresponding universal stress-N curve seem dim indeed.

In fact, the prospect of finding (8.3) to be applicable over a wide range of life values also appear dim. One reason for this is that the mechanism of short-term fatigue may not be the same as that causing long term fatigue. In terms of stress versus N, evidence of this difference sometimes appears in the form of a discontinuity at the "knee" of a number of experimental S-N curves. See Figure 8.7. A common explanation for this discontinuity is that two separate phenomena exist simultaneously, each with its own occurrence probability as shown in Figure 8.8. To support this explanation using test data, Cicci [21] has made some linear regression analyses of specimen lives as

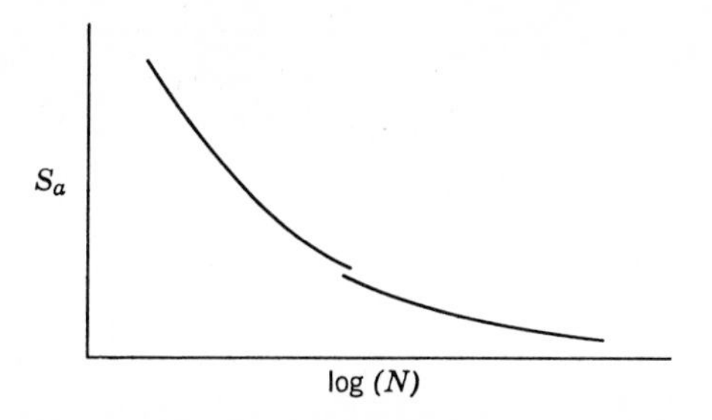

Figure 8.7 A representation of the "knee" of an S-N curve by means of a discontinuity.

they occur (single distribution) and as they can be divided into short-term (STF) and long-term fatigue (LTF), that is, two distributions. The results, shown in Figure 8.9, lead to high correlation coefficients (0.99) for the two-distribution representations but only a low correlation coefficient (0.86) for the single distribution. This two-separate-phenomena theory tends to suggest that a wide-range universal strain-N curve could not exist.

Actually, for damage indicating systems to be useful, it is not necessary for such a curve to have a wide range. The requirements for accuracy may lie well within a logarithmic decade on a life scale as long as the required accuracy does not justify consideration of cycles of smaller amplitude and as long as the few, if any, cycles with amplitudes larger than this can be observed or corrected separately. On the other hand, however inconsequential this limitation may sometimes be in practice, it is still a limitation.

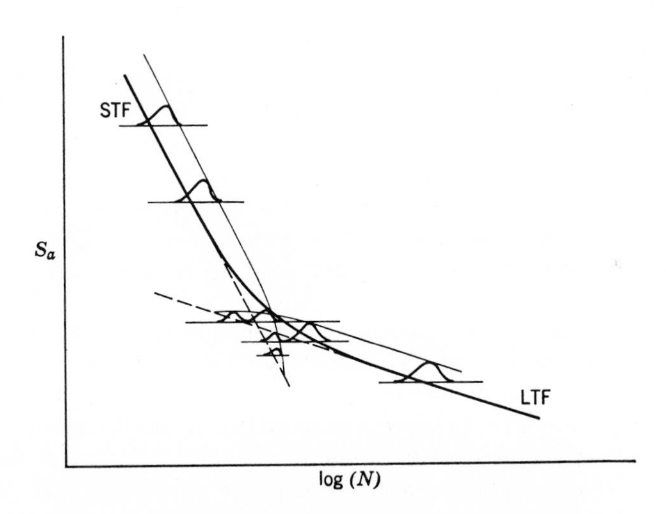

Figure 8.8 The two-distribution presentation of the S-N curve.

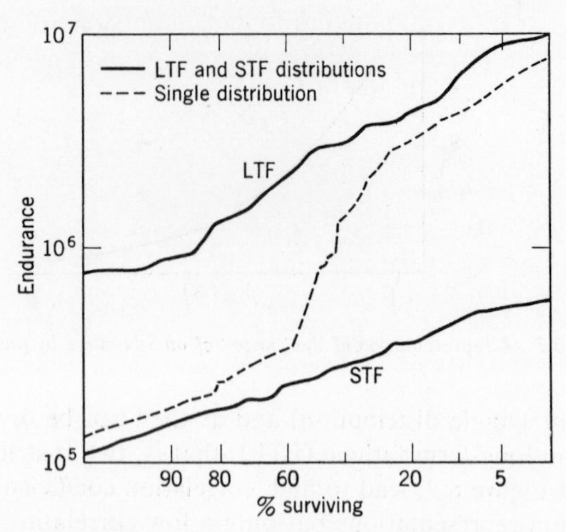

Figure 8.9 Log-normal probability plotting of 115-ksi amplitude endurances with single and split distributions.

Therefore the remainder of the section will be concerned with fatigue data in terms of stress rather than of strain.

The purpose, then, of the remainder of this section is to obtain expressions of the form

$$\frac{1}{N} = f(\text{maximum stress, minimum stress}) \tag{8.4}$$

for any cycle considering several commonly-used materials. Such expressions form an important part of the operation of some of the equipment to be described later.

The following symbols are used:

a, b, c	constants among the material fatigue characteristics;
A, B, C	other constants;
d, Δ	differential and finite difference, respectively;
D	fatigue damage per cycle, usually given by $1/N$;
$\mathscr{D}$	fatigue damage of entire record or test;
$f(\ )$	function of;
K_T	theoretical stress-concentration factor;
KSI	stress unit of 1000 pounds per inch² (1000 psi);
n	number of similar cycles endured;
N	life of fatigue specimen in cycles;
$p(\ )$	probability distribution;
P	probability of probable damage;
ΔP	change in probability due to neglected cycles;

R	ratio of minimum to maximum stress during one cycle;
S	stress as a function of time, t;
S_f	fatigue limit (endurance limit);
S_m, S_a	cycle mean and alternating stresses, respectively;
S_{max}, S_{ma}	primary and secondary maximum stresses, respectively;
S_{min}, S_{mi}	primary and secondary minimum stresses, respectively;
S_{mean}	mean stress considering entire stress record;
S_u	ultimate tensile strength;
u	bivalued factor: unity when $dS/dt > 0$ and 0 otherwise;
v	bivalued factor: 1 when within limits and 0 otherwise;
σ	standard deviation.

A variety of forms have been used for curvefitting S-N curves ranging from a simple two-straight-line approach to some rather elaborate expressions. Two examples of the latter are one used by Lariviere [22],

$$\frac{S_{max}}{S_f} = 1 + a \log (1 + bN^{-c}), \tag{8.5}$$

and one used by Stussi and Fuller [23],

$$\frac{S_{max} - S_f}{S_u - S_{max}} = bN^{-c}. \tag{8.6}$$

For most materials,

$$\tfrac{1}{4} < c < 1 \tag{8.7}$$

If c is taken to be unity, the latter becomes

$$\frac{S_{max} - S_f}{S_u - S_{max}} = \frac{b}{N}. \tag{8.8}$$

In the cases in which $b/N \ll 1$, further approximation is possible.

$$\frac{S_{max}}{S_f} = 1 + \frac{bS_u}{S_f N} = 1 + \frac{\text{constant}}{N}. \tag{8.9}$$

However, these equations apply to a single S-N curve, such as when the stress ratio, R is -1.

As in the case of Figure 8.6, most materials change S_{max} by equal amounts as the R is changed by a fixed amount, i.e., the R curves are "parallel." This suggests that as R varies from -1, a quantity proportional to $R + 1$ should be added to S_{max}. However, S-N curves are not completely parallel but tend to converge upon a point off the diagram to the left. It can be shown that convergence upon unity N can be introduced if the $R + 1$ term is added to

$f(N)$ instead of to $S_{\max}$. Let the addition be such that (8.6) and (8.8) become

$$\frac{S_{\max} - S_f}{S_u - S_{\max}} = \frac{b_1}{N^c} + b_2 \frac{S_{\max} + S_{\min}}{2S_u}. \tag{8.10}$$

$$\frac{S_{\max} - S_f}{S_u - S_{\max}} = \frac{b_3}{N} + b_2 \frac{S_{\max} + S_{\min}}{2S_u}. \tag{8.11}$$

For example, for 4130 normalized steel at room temperature, the expression

$$\frac{S_{\max} - 34{,}000}{123{,}000 - S_{\max}} = 3410N^{-0.87} + \frac{2.38}{2}\frac{S_{\max} + S_{\min}}{123{,}000}. \tag{8.12}$$

produces fairly good agreement with the data of [9] as indicated in Table 8.1. In this table, the calculated stress is the maximum stress obtained from (8.12). Errors for 10,000 and 20,000 psi mean stress are generally less than those for 0 and 30,000 psi using this equation. The average of the last column listings in this table is an error of 3.4%, but, since there is almost an equal distribution here of positive and negative errors, the application of (8.12) to a random variation of stress should prove quite satisfactory.

TABLE 8.1

ACCURACY OF EQUATION 8.10 FOR 4130 STEEL

Mean Stress	Cycles	Max. Stress	Calc. Stress	Rel. Error, $S_{\max}/S_{\max}$
30,000 psi	10^4	90 KSI	91 KSI	1%
	5×10^4	80	77	3
	10^5	76	73	3
	5×10^5	66	65	2
	10^6	64	64	0
	10^7	61	62	2
0	10^4	72	81	12
	5×10^4	55	53	4
	10^5	49	46	6
	5×10^5	40	38	5
	10^6	38	37	3
	10^7	35	35	0

From the data of [22] and [23], the fatigue properties of 2024-$T3$ aluminum and PH 15-7 Mo (1050) stainless were curvefitted similarly. For these three materials, the fatigue properties required by (8.11) are summarized in Table 8.2. In the event of an apparent conflict of certain values, (8.11) should be considered to be the defining equation for all parameters. For example, in

TABLE 8.2

TYPICAL SETS OF CONSTANTS FOR EQUATION 8.11

Material	S_f	S_u	b_3	b_2	S_f/S_u
4130 Steel	34,000 psi	123,000 psi	$1.5(10^4)$	1.68	0.276
Ph15-7Mo Stainless	79,000	200,000	63,240	0.80	0.395
2024-T3 Alum.	28,100	72,000	29,300	2.20	0.390

order to satisfy (8.11), S_u may have to be given a value somewhat greater than the actual ultimate tensile strength of the material.

Equation 8.11 and Table 8.3 together produce less accurate representations than if (8.10) were used, particularly at extreme values. Figure 8.10 illustrates a comparison of actual 4130 Steel *S-N* data with that suggested by (8.11) and Table 8.3.

8.1.3 Primary and Secondary Cycles

The instrumentation engineer involved in a fatigue measurement problem frequently finds that, while he is taking only a small fraction of the data which was originally wanted, it turns out to be more data than is ever used. Numerous simplifying assumptions are often needed to speed up the analyzing of data. Using the approximate *S-N* expressions obtained above, the effect of some of these assumptions will now be discussed. Mainly, these assumptions have to do with primary and secondary cycles. The following four conventions serve to establish and define primary and secondary cycles:

1. A mean value of stress, S_{mean} for the entire record is approximated. Although a time average and a peak point average generally yield different average values, the two may be assumed to be equal.

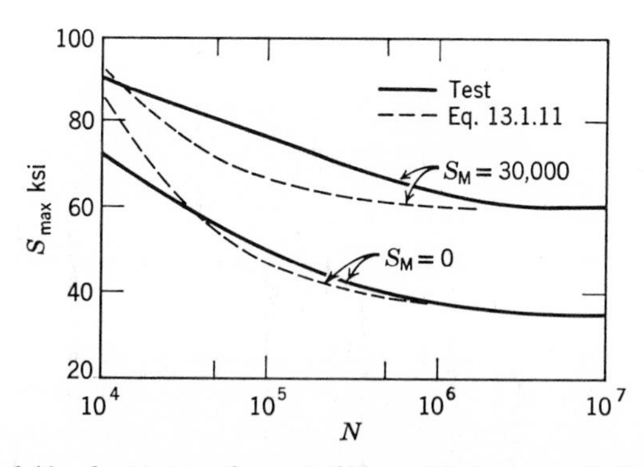

Figure 8.10 Comparison of actual 4130 steel S-N curves with (8.11) curves.

2. The maximum stress occurring between a positive S_{mean} crossing (positive in slope) and a negative (negative slope) S_{mean} crossing is called S_{max}. All other maximum stress values are designated as S_{ma}.

3. The minimum stress occurring between a negative S_{mean} crossing and a corresponding positive crossing is designated as S_{min}. All other minimum stress values are S_{mi} points. See Figure 8.11

4. The combination of an S_{min} and the next S_{max} form a primary cycle. The combination of an S_{mi} and the next S_{ma} form a secondary cycle.

Accurate evaluation of fatigue damage usually requires that all primary cycles within a given period be considered but it rarely requires that all secondary cycles be considered. The effect of neglecting secondary stresses in random data will be investigated in terms of the following proposition:

PROPOSITION I *If peaks (maximums) and troughs (minimums) are normally distributed about a mean stress, the damage per cycle caused by all cycles which do not cross the mean stress line can be neglected.*

This proposition can be evaluated either by examining a large number of cycles and comparing neglected damage with the total damage or by considering one cycle and comparing integrals of the product of the two following terms: (1) the probability of the cycle having certain peaks and (2) the damage of a cycle having these peaks. The latter approach is more direct. It represents a double integration because both maximum and minimum stress values are variables and because all combinations must be considered.

Either of (8.10) or (8.11) can be expressed in the form of damage per cycle,

$$\frac{1}{N} = f([S_{max} \text{ or } S_{ma}], \quad [S_{min} \text{ or } S_{mi}]). \tag{8.13}$$

For simplicity, let the above be expressed as

$$\frac{1}{N} = D(S_{max}, S_{min}), \tag{8.14}$$

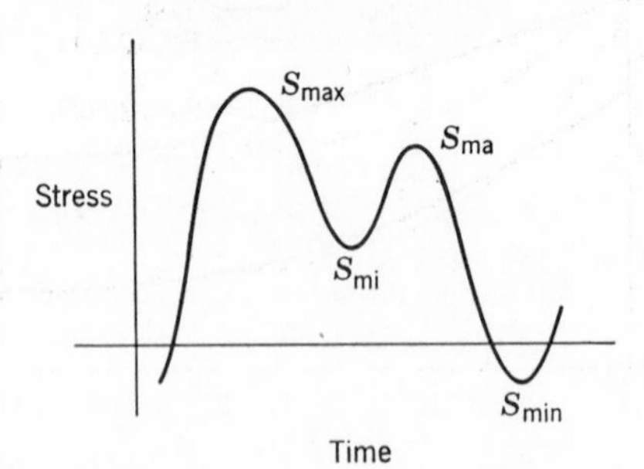

Figure 8.11 A part of a stress record.

after material properties are substituted. Both S_{max} and S_{min} (or S_{ma} and S_{mi}) are values represented by the statistical properties, S_{mean} and σ clearly, S_{max} is the higher of the two values. Let $p(S_{max}, S_{min})$ represent the probability density of any combination of S_{max} and S_{min} in accordance with the well known normal distributions and let P represent the total probable damage of the cycle. Then

$$P = \int_{-\infty}^{\infty}\int_{-\infty}^{S_{max}} P(S_{max}, S_{min})D(S_{max}, S_{min})\, dS_{min}\, dS_{max}, \tag{8.15}$$

$$\frac{dP}{dS_{max}} = \int_{-\infty}^{S_{max}} P_{S_{max}}(S_{min})D_{S_{max}}(S_{min})\, dS_{min}, \tag{8.16}$$

where S_{max}, as a subscript, is used to indicate a nonvariance, i.e., in this case, P is no longer a function of it. Likewise, a change in total probability, ΔP can be evaluated to represent the effect of not including the cycles neglected. It is the same integral but with different (closer) limits. In both cases the limits of integration must exclude all no-damage conditions including those mathematically equivalent to negative damage.

EXAMPLE 8-2

Find the relative error introduced by neglecting all secondary cycles in a random-peak test of 2024-$T3$ where the standard deviation of stress peaks is $S_u/4$ and the mean stress is zero.

When S_{max} is σ, the standard deviation, it is slightly less than the fatigue limit given in Table 8.3. Since, for this value, only very low (and unlikely) values of S_{min} can produce damage, σ can be considered a value of S_{max} where integration should be started, approximately.

When S_{max} is 2σ, $D_{S_{max}}(S_{min})$ is given by Equations (8.11) and (8.14) and can be substituted into Equation (8.16). The probability distributions are assumed to be normal distributions.

$$\frac{dP}{dS_{max}} = \int_{-\infty}^{(S_u/2)=36000} \frac{\exp[-(4S_{min}/S_u)/2}{\sqrt{2\pi}} \left[\frac{\frac{1}{2}-\frac{S_f}{S_u}}{b_3(1-\frac{1}{2})} - \frac{b_2}{b_3}\frac{\frac{1}{2}+\frac{S_{min}}{S_u}}{2}\right] dS_{min}$$
$$= 0.00570S_u/b_3\ldots, \tag{8.17}$$

where the constants are obtained from Table 8.2.

By the time S_{max} reaches 3σ, the smallness of the S_{max} probability returns this integral to nearly zero. The small losses obtained by neglecting everything beyond the 1σ–3σ range can be offset to some extent by using steep rises in the solid curve drawn through the 1-, 2-, and 3-σ points in Figure 8.12.

Now consider the change due to the abridgment. The change affects cycles above and below zero stress but not cycles crossing zero; affected cycles above zero will be considered to represent half of the effect.

$$\frac{\Delta P}{2} = \int_{0}^{\infty}\int_{0}^{S_{max}} P(S_{max}, S_{min})D(S_{max}, S_{min})\, dS_{min}\, dS_{max}. \tag{8.18}$$

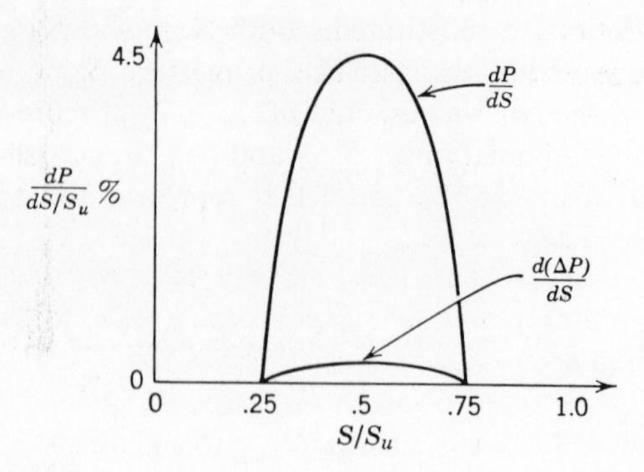

Figure 8.12 First-integration probability distribution for Example 8.2.

Again let the damage per cycle go to zero at $S_u/4$ and the probability density go to zero at $3S_u/4$. When $S_{\max}$ is 2σ or $S_u/2$,

$$\frac{1}{2}\frac{d\Delta P}{dS_{\max}} = \int_0^{(S_u/2)=36000} (2\pi)^{-1/2} \exp\left[-\frac{1}{2}\left(\frac{4S_{\min}}{S_u}\right)^2\right] \cdot \left(\frac{6.5 - 276}{0.5b_3} - \frac{0.5 + S_{\min}/S_u}{2_3{}^b/b_2}\right) dS_{\min}$$

$$= 0.000099 S_u/b_3 \ldots, \tag{8.19}$$

where, as before, the 36 KSI limit eliminates negative damage. Numerous published mathematical tables give areas under the normal curve. The second integrations can be performed graphically using Figure 8.12 which is a plot of both first integrations vs. stress using three points (the integrations at $0.5S_u$ and the two assumed zeros). The results of the integrations are

$$P = 0.0140 \qquad \Delta P = 0.00073.$$

Therefore, the relative error, $\Delta P/P$ is 5.2 percent.

There are many sources of large errors in any attempt to predict fatigue life. By comparison, this five percent error is small and Proposition I appears to be justified in most cases. The above problem also illustrates that ΔP is not particularly sensitive to small changes in the lower limit of integration in (8.19). Therefore in applying either this proposition or the one which follows extreme accuracy in determining the mean stress is not required. This is fortunate because there are several different ways of defining mean stress, some of which are more conveniently evaluated than others. See preceding Convention 1.

In Figure 8.12, the mean stress was given as zero. If it had been greater than zero, the errors due to neglected "positive" and "negative" secondary cycles would not have been equal, i.e., the secondary cycles above the mean would have proved to be more significant than those below. This suggests an alternative proposition which is more accurate than the first:

PROPOSITION II *If peaks and troughs are normally distributed about a positive mean stress, the damage per cycle caused by all cycles which neither cross nor exceed the mean stress line can be neglected.*

A compromise between Proposition I and II sometimes takes the following form:

PROPOSITION III *If peaks and troughs are normally distributed about a positive mean stress, the only time a secondary cycle needs to be considered is when it lies above the mean stress, when it occurs just after* S_{max} *has been established, and when it reaches the lowest* S_{mi} *established since the last mean stress crossing.*

Usually, for random data, one of these propositions can be considered valid. None of the three implies that upper and lower half cycles may be detached from each other. That is, if upper and lower primary half cycles are analyzed statistically as separate groups, the information obtained from these groups will not be sufficient to indicate fatigue damage with any accuracy. In a study by Naumann [24], a comparison was made of random tests having zero mean stress and a certain peak distribution with tests in which all of the negative half cycles were rearranged. This half-cycle rearranging consisted of "restraining" or assigning to each negative half-cycle the same magnitude as the preceding positive half-cycle. Since the original positive half-cycle and negative half-cycle peak distributions were equal, no change was made in the negative half-cycle group characteristics. The results, as shown in Figure 8.13, indicate that rearranging the large half-cycles to match large half-cycles and small to small does decrease fatigue life but only by some 20 or 30%. In other words, there are times (e.g., when mean stresses are zero) when peak statistics by themselves can be somewhat meaningful. Figure 8.14 illustrates several more or less useful approximations, some of which are generally valid, and all of which are usually more valid than this half-cycle rearrangement described above. In each case, a heavy base line is used to indicate a mean stress line.

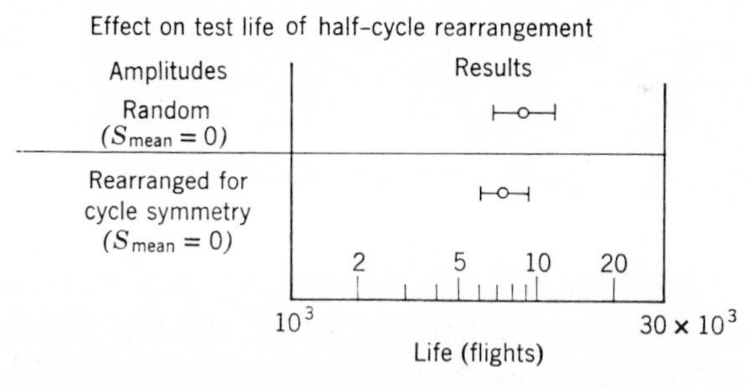

Figure 8.13 Effect on test life of half-cycle rearrangement.

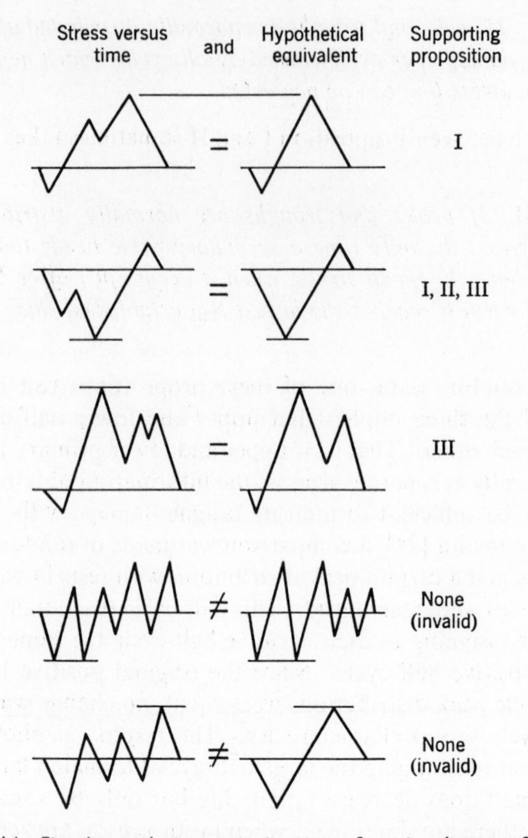

Figure 8.14 Five data-reducing approximations of varying degrees of validity.

Consider the two inequalities shown as cases 4 and 5 in Figure 8.14, the latter of which is widely used in periodic-vibration fatigue analysis such as for helicopter blades. Although neither is strictly valid, together, they can be used to bracket the damage. Their occasional usefulness comes from the fact that they can employ simpler peak analyzing techniques than the others; they can be valid when there is a small spread of damage so bracketed. Often one of these, the last one shown, is used by itself in spite of its nonconservative nature; for example, damage per rotor revolution of a helicopter blade is often approximated by ignoring this inequality.

8.1.4 The Binary Function Damage Integral

As was shown by Figure 8.4, fatigue damage indicating systems can have a variety of forms. Some of these forms can be better understood in terms of an explicit expression for cumulative damage. One such explicit expression is a line integral whose differential element is the change of stress, dS. The general form of such an expression for total damage can be either a time-limited or a stress-limited line integral,

$$\mathscr{D} = \int_{\text{time 1}}^{\text{time 2}} \frac{dD}{dS}\, dS = \sum \int_{S_1}^{S_2} f_1(S)\, dS. \tag{8.20}$$

The latter form is more useful when S-N data has been reduced to expressions like (8.10) or (8.11) since, for each cycle, (8.14) applies. If the function, $f_1(S)$ is multiplied by certain factors having values of either zero or unity, the integrand performs its own limit-setting operation and the integral becomes continuous. Let such an integrand be $f_2(S)$ and let a given stress-time record be considered to return to its origin along a line of zero stress. Such a record becomes a closed curve and the corresponding integral once around the curve is called a closed-line integral.

$$\mathscr{D} = \oint f_2(S)\, dS \tag{8.21}$$

For any one cycle,

$$D = \int_{S_1}^{S_2} f_1(S)\, dS = \int_{S_{\min}}^{S_{\max}} f_2(S)\, dS. \tag{8.22}$$

This simply means that $f_2(S)$ is zero between certain stress levels such as between $S_{\min}$ and S_1. The factors represented by f_2/f_1 are binary functions of $S_{\max}$, $S_{\min}$, S_{ma}, S_{mi}, etc. Equation 8.21 is called a *binary function damage integral*.

For example, consider (8.22). The first step in converting it into a binary function damage integral is to solve for $1/N$ and equate the result to (8.22). Next, arbitrary values or functions are assigned to S_1, S_2, and $f_1(S)$ until there are as many equations as there are unknowns:

$$S_2 = S_{\max} \tag{8.23}$$

$$f_1(S) = \frac{dD}{dS}, \qquad R = -1. \tag{8.24}$$

The results are then substituted into (8.22).

$$D = \frac{S_u - S_f}{b_3} \int_{\frac{b_2 S_m + 2S_f}{2 + b_2 (S_m/S_u)}}^{S_{\max}} \frac{dS}{(S_u - S)^2}. \tag{8.25}$$

When integrated, this produces the original equation (8.11),

$$D = \frac{S_{\max} - S_f}{b_3(S_u - S_{\max})} - \frac{b_2 S_m}{2 b_3 S_u}. \tag{8.26}$$

Any convenient binary function which, when introduced into the integrand, effectively establishes these limits will convert this to the closed form, (8.21). An index which serves to register whether any part of the record is part of a primary cycle is the variable, v. Anytime v is unity, the stress forms part of a primary cycle; any time it is zero, the stress is part of a secondary cycle.

8.1.6 Peak Counting Techniques

Figure 8.15 shows the three most commonly used cycle counting methods among a considerably greater number which have been proposed and used. Some of the others are modifications of the ones shown. For example, the first one is frequently modified so that positive and negative half cycles are counted and tabulated separately. The inaccuracy of this has been discussed previously. The second method pairs off succeeding positive and negative amplitudes so that each cycle mean can be calculated.

The third method without refinement is not very accurate unless only one frequency is present. If a low and a high frequency appear with equal amplitudes, the high-frequency low-range (amplitude) activity will mask the low-frequency high-range component; the result will be indistinguishable from a high-frequency high-amplitude loading which is very damaging. See Case 3 of Figure 8.14.

However, an important refinement to the third method of Figure 8.15 is the use of zones and the counting of passages from one zone to some other, particularly to a nonadjacent zone. This method, which is listed as method 3a in the figure, is equivalent to method 2 if a large number of zones and zone separations are used and properly recorded.

8.2 INDIRECT OR STATISTICAL METHODS OF RANDOM LOAD ANALYSIS

Most of the fundamental concepts of fatigue indicate that some type of cycle counting technique is needed to evaluate the fatigue damage caused by random loading. Direct techniques of this type, such as peak counts and range counts are simple and accurate but often quite cumbersome to apply. Other, less direct methods are sometimes sought in order to estimate the same information less laboriously. This section considers two indirect methods in which sampling or alternative measurements are used and converted into a useful form by statistical means.

In actual practice the quantity of data available may be quite large. This has led to many attempts to utilize laboratory equipment already available for rapid measurement of statistical parameters of random data, particularly

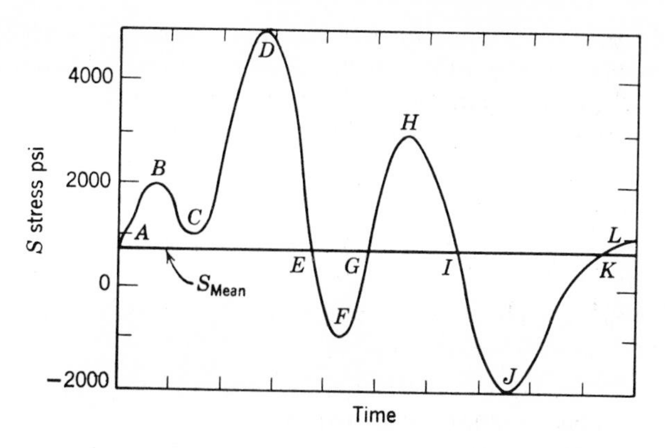

	Counting method	Quantities measured or counted	Equivalent waveform
1	Mean Crossing Peak Count	n d, f, h, j, etc.	D, d, f, F, h, H, J, j
2	Paired Range Count	n $S_{M_{FH}}$ and FH $S_{M_{JL}}$ and JL	H, $S_{M_{FH}}$, FH, F, L, JL, $S_{M_{JL}}$, J
3	Level Crossing Count	$n_{4000} = 1$ $n_{2000} = 3$ $n_0 = 2$ etc.	4000, 2000, 0
3a	Zone Passing Count	Same as method 3 except each crossing is a count only when preceded by a crossing at each of several variously separated levels and is categorized according to separation.	

Note. n is a number of positive-slope designated-level crossings (if without subscript, the designated level is S_{mean}).

Figure 8.15 Application of some cycle counting methods to a typical stress record (lower table illustrates each method in terms of an equivalent record).

frequency-oriented characteristics. Basically, these have been attempts to prove that power spectral density or other frequency parameters have a valid measureable correlation with cumulative fatigue damage. While these attempts have not been very successful, the efforts have generated widespread interest in the concept that they should be used even if relatively low standards for accuracy have to be accepted. The result is what can be called the autocorrelation-power spectral approach to random fatigue analysis.

The second statistical approach to be considered is somewhat less indirect. It consists of counting only the relatively small number of large-amplitude cycles and of using this information to estimate the effect of the others. When the cycles counted are the cycles which exceed a prescribed range, it is called the statistical range sampling method. Both methods will be discussed in terms of established statistical parameters. A more detailed derivation and discussion of the methods is given by Bendat [25], Rice [26], and Crandall [27].

8.2.1 Autocorrelation or Power Spectrum Method

Consider a random variable to be stationary, to have a Gaussian distribution, and to have a standard deviation σ. Consider part of the history of this variable, $y(t)$, shown in Figure 8.16. It is assumed that $y(t)$ has a mean value of zero in this particular distribution. First, consider a function $\phi(\tau)$ of the time history defined by

$$\phi(\tau) = \lim_{T\to\infty} \frac{1}{T} \int_{-T/2}^{T/2} y(t)y(t+\tau)\,dt, \tag{8.27}$$

where, as shown in Figure 8.16, $y(t)$ and $y(t+\tau)$ are values at times t and $t+\tau$, respectively. Notice that for $\tau = 0$, $\phi(\tau)$ reduces to:

$$\phi(0) = \lim_{T\to\infty} \frac{1}{T} \int_{-T/2}^{T/2} [y(t)]^2\,dt. \tag{8.28}$$

The above equation is equivalent to $-y^2$, the mean square value (and to σ^2 in any zero-mean-valued time history). The function, $\phi(\tau)$ is called the autocorrelation function, [28].

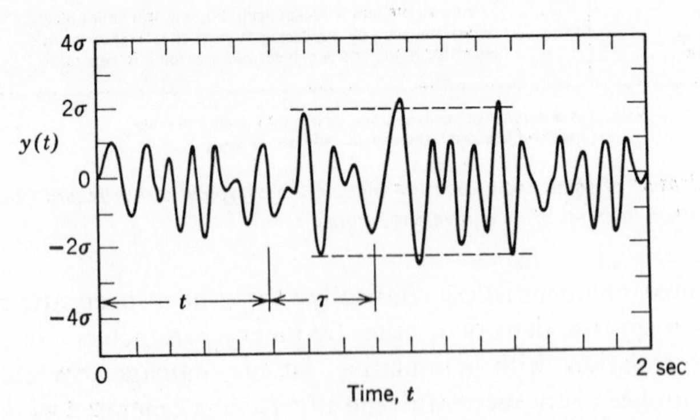

Figure 8.16 A given random variation of y with time.

The Fourier transform of the autocorrelation function as defined in (8.27), when multiplied by 2 is

$$S(f) = 2\int_{-\infty}^{\infty} \phi(\tau)e^{-2\pi i f \tau}\, d\tau \tag{8.29}$$

and is called the power spectral density. Here f is frequency. Just as the autocorrelation function has considerable mathematical importance, the power spectral density has considerable physical significance. Some quantities, like electrical current through a resistance and velocity across a dashpot, have the characteristic of being proportional to the square root of power. If $y(t)$ represents any of these, $y^2(t)$ is proportional to power and $S(f)$ is the part of that power at frequency, f. If $y(t)$ represents a quantity not possessing this characteristic, such as stress, strain, acceleration, etc., the ordinate of a plot of $S(f)$ against frequency is just the mean square value per unit of frequency and $S(f)$ can better be called mean square spectral density. Its units are the square of the quantity represented by y divided by the unit of frequency.

Power spectral density can be measured in a number of ways. An approximate method, which becomes exact if the power spectral density is uniform with respect to frequency (like "white noise"), involves a one-degree-of-freedom system such as a damped spring-mass combination. If this spring-mass combination has an undamped natural frequency, f_n, damping ratio, ζ, and a mean square displacement response, $\bar{x}^2$, then at this frequency, the power spectral density of the excitation must be

$$S(f_n) = \frac{4\zeta\bar{x}^2}{\pi f_n} \tag{8.30}$$

The electrical equivalent of this with f_n as an independent variable forms a convenient laboratory technique for measuring the power spectrum. Figure 8.17 shows a spectrum obtained experimentally from the stress record, of which Figure 8.16 is a part.

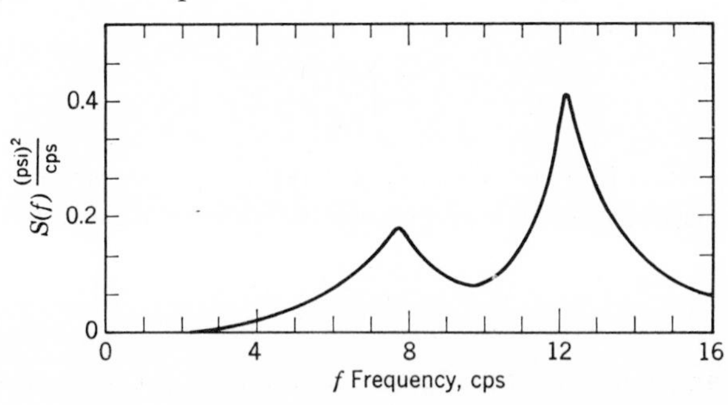

Figure 8.17 The power spectrum of y(t).

By a process of inverse transformation, integration, and substitution of zero for τ, the complete ($-\infty$ to ∞) integral of (8.29) can be put into the form of (8.28). Therefore, the area under a curve of $S(f)$ is the mean square value of $y(t)$. This is important because the power spectral density function is rather easily measured by means of (8.30). Figure 8.17 is the power spectrum of the random-time history in Figure 8.16. In the latter, a root mean square unit is the given unit of stress. In contrast, the power spectrum of a sinusoidal time history would result in an infinite peak occurring at the sinusoidal frequency.

For a stationary and Gaussian random-time history, several parameters which describe the nature of the time history can be computed in terms of the moments of the power spectrum. The nth-order moment of the power spectrum about the zero frequency axis is defined by the following expression:

$$M_n = \int_{-\infty}^{\infty} f^n \, S(f) \, df. \tag{8.31}$$

Rice [26] has shown that the expected average number of crossings per unit time, N_α, with positive or negative slope, of a level α is related to the zeroth and second moments according to the following equation:

$$N_\alpha = \left(\frac{M_2}{M_0}\right)^{1/2} \exp\left(-\frac{\alpha^2}{2M_0}\right). \tag{8.32}$$

An average frequency of the time history results when $\alpha = 0$.

$$N_0 = \left(\frac{M_2}{M_0}\right)^{1/2}. \tag{8.33}$$

This is simply the number of zero or mean crossings, with positive or negative slope, of the time history per unit time.

Bendat [25] has shown that the expected average number of maxima or minima per unit time, Q, is a function of the second and fourth moments of the power spectrum; that is,

$$Q = \left(\frac{M_4}{M_2}\right)^{1/2}. \tag{8.34}$$

In addition, Beer [29] has determined that the average rise or fall, $\bar{h}$, of the time history between maxima and successive minima is expressed as

$$\bar{h} = M_2\left(\frac{2\pi}{M_4}\right)^{1/2}. \tag{8.35}$$

Also, the mean value of the time history, if not known, can be measured from the time history.

The above quantities, when computed from the power spectrum, might be applied to the problem of evaluating fatigue in one of the three following ways:

1. The information can be converted into a loading spectrum.
2. The random variable time history can be assumed to be equivalent to a sine wave having an amplitude equal to some constant times the RMS value of $y(t)$ and having a frequency equal to N_0.
3. The random variable time history can be assumed to be equivalent to a sine wave having an amplitude equal to some constant times $\bar{h}$ and having a frequency equal to Q.

The first of these could be the most accurate but requires some additional information. While the other two have been used many times, they produce results which are applicable only to preliminary or order of magnitude structural evaluations.

PROBLEM 8-3

Using the autocorrelation power spectral density method, find an "equivalent" sine wave of the following:

$$f(\omega t) = \sin \omega t + \sin 2\omega t + \sin 3\omega t.$$

If $f(\omega t)$ is squared and integrated between zero and 2π, the resulting area is 3π. The root mean square value is the square root of $3\pi/2\pi$ or 1.2248. This is also the area under the power spectral density curve. This curve has spikes of equal areas at 1, 2, and 3 units of frequency. Therefore, the area of each spike is 0.4083. The moments which result are as follows:

$$M_0 = 0.4083$$

$$M_2 = 0.4083(1^2/3 + 2^2/3 + 3^2/3) = 1.9054$$

$$M_4 = 0.4083(1^4/3 + 2^4/3 + 3^4/3) = 13.438.$$

One solution to this problem is a sine wave having an amplitude equal to 1.2248 and a frequency, N_0 equal to 2.16. A second solution uses $\bar{h}$ or 1.303 and a frequency, Q equal to 2.645. See (8.31) and (8.33) to (8.35).

Some rather extensive comparisons between the results of this method and the actually counted characteristics have been made by Smith [30]. Five of his comparisons used experimental stress measurements consisting of over 1000 cycles each. Of these five, three were found to have single-peaked spectrums and two were found to have two-peaked spectrums. The few cycles investigated in Problem 8.3 represent a three-peaked spectrum.

Figure 8.18 plots the accuracy in terms of the actual rms value (rms of all peak amplitudes) of the rms value obtained from the power spectrum for all six cases mentioned above; for example, for the three-peak spectrum problem the fundamental cycle has maximum and minimum values as follows:

$$2.4999, \quad -0.2412, \quad 0.6354, \quad -0.6354, \quad 0.2412, \quad -2.4999.$$

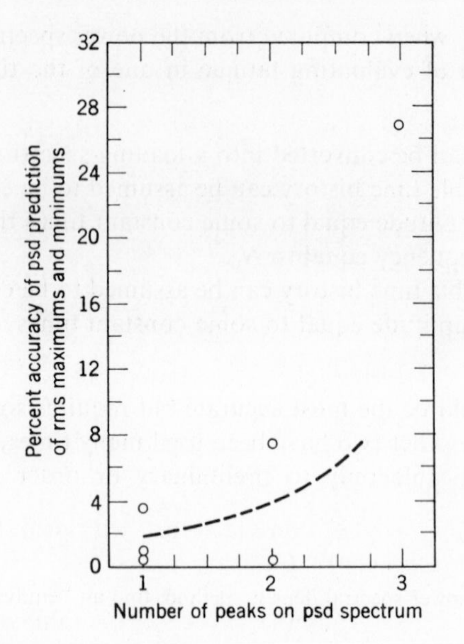

Figure 8.18 A possible relationship between the shape of the power spectrum and its value in estimating rms amplitude.

The rms value of these six amplitudes is 1.553, which is 26.9% greater than the rms value found from the power spectrum. There is nothing to indicate that if the complexity of the power spectrum increases to many peaks, the accuracy improves. In fact, the only time that the power spectrum appears likely to be of value in fatigue analysis is when it shows a single peak, suggesting that the stress history resembles a sine wave. Perhaps such histories are less likely to contain unusually large amplitude cycles.

8.2.2 A Statistical Range-Sampling Method

The statistical range-sampling method of random-time history analysis combines a limited amount of peak counting with a determination of additional parameters to describe the nature of the random process. This method is simple in application and is demonstrated and discussed by considering some details of the random-time history in Figure 8.16.

The analysis technique is to first observe from a sample length of record the greatest maximum and minimum and to measure their range, Y. Cycles are then counted which have ranges which equal or exceed one or more convenient percentages of Y.

For a stationary, Gaussian, and narrow band process, the probability distribution of the cycle ranges (rises or falls) is a Rayleigh distribution. The

Rayleigh distribution function, $P(X)$, of a time history variable X, is expressed as

$$P(X) = \frac{X}{\sigma^2} e^{-(X^2/2\sigma^2)} \qquad \text{for } X \geq 0, \tag{8.36}$$

where σ^2 is the mean square value of the time history and, initially, is unknown. However, a value of σ is obtained for each percentage of Y used in counting. If all of the σ values agree, the distribution must be a Rayleigh distribution.

Assume that three measurements are made on the original stress history as follows:

1. The largest amplitude on the record Y.
2. The number of cycles counted on the record with peak-to-peak amplitude equal to or greater than $0.9Y$. Call this N_1.
3. The number of cycles counted on the record with peak-to-peak amplitude equal to or greater than $0.8Y$. Call this N_2.

Two quantities are desired as follows: $Z_1 =$ the rms value of peak-to-peak range obtained from measurements 1 and 2, and $Z_2 =$ the rms value of peak-to-peak range obtained from measurements 1 and 3.

Item 1 can be interpreted to mean that only one cycle was measured with a range equal to Y. Then, using 2, N_1 is not only the ratio of the number of cycles at $0.9Y$ to those at Y but also the ratio of the Rayleigh-distribution probabilities of the two. This ratio is formed by substituting from (8.36) twice.

$$N_1 = \frac{P(0.9Y)}{P(Y)} = 0.9e^{0.095(Y/Z_1)^2} \tag{8.37}$$

$$0.095\left(\frac{Y}{Z_1}\right)^2 = \log e^{1.1111N_1}. \tag{8.38}$$

Likewise, using item 3 instead of 2,

$$0.18\left(\frac{Y}{Z_2}\right)^2 = \log e^{1.25N_2}. \tag{8.39}$$

Table 8.3 presents values of the ratios of Z_1 and Z_2 to Y obtained in this manner. Actually, a slightly different calculation for these ratios was employed in which, for the numerator of (8.37), a definite integral between limits was used in place of a single point on the probability density curve. However, the difference turns out to be negligible for most N values.

TABLE 8.3

RATIO OF PEAK-COUNTED RMS AMPLITUDE TO GREATEST OBSERVED AMPLITUDE AS INDICATED BY A COUNT OF *N* CYCLES OF *X* (OR GREATER) AMPLITUDE PER CYCLE OF GREATEST (*Y*) AMPLITUDE

N (Cycles counted)	Z_1/Y (Ratio when X/Y is 90%)	Z_2/Y (Ratio when X/Y is 80%)
1.0	0.6524	0.8981
2.0	0.3219	0.4432
3.0	0.2680	0.3690
4.0	0.2429	0.3344
5.0	0.2276	0.3134
6.0	0.2171	0.2988
7.0	0.2092	0.2880
8.0	0.2031	0.2795
9.0	0.1981	0.2727
10.0	0.1939	0.2669
11.0	0.1903	0.2620
12.0	0.1872	0.2578
13.0	0.1845	0.2540
14.0	0.1821	0.2507
15.0	0.1800	0.2478
16.0	0.1780	0.2451
17.0	0.1763	0.2426
18.0	0.1746	0.2404
19.0	0.1731	0.2383
20.0	0.1717	0.2364
21.0	0.1705	0.2347
22.0	0.1693	0.2330
23.0	0.1681	0.2315
24.0	0.1671	0.2300
25.0	0.1661	0.2286
26.0	0.1651	0.2273
27.0	0.1643	0.2261
28.0	0.1634	0.2250
29.0	0.1626	0.2239
30.0	0.1619	0.2228
31.0	0.1611	0.2218
32.0	0.1604	0.2208
33.0	0.1598	0.2199
34.0	0.1591	0.2191
35.0	0.1585	0.2182
36.0	0.1579	0.2174
37.0	0.1574	0.2166
38.0	0.1568	0.2159
39.0	0.1563	0.2152
40.0	0.1558	0.2145
41.0	0.1553	0.2138
42.0	0.1548	0.2131
43.0	0.1544	0.2125
44.0	0.1539	0.2119
45.0	0.1535	0.2113
46.0	0.1531	0.2107
47.0	0.1527	0.2102
48.0	0.1523	0.2096
49.0	0.1519	0.2091
50.0	0.1515	0.2086

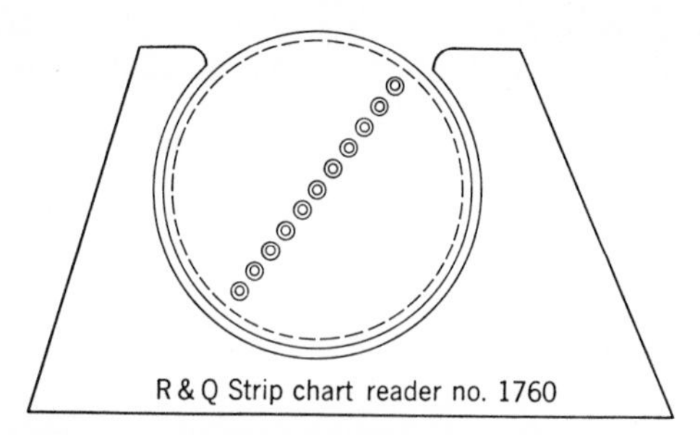

Figure 8.19 A plastic line-drawing guide for adding a grid.

Incidentally, note that the values of Z/Y when N is unity are not themselves unity, contrary to what might be suspected if the reversing of direction of the Rayleigh distribution is not considered.

In applying this method, it is convenient to use some kind of device for dividing the maximum stress range, Y into 10 or 20 equal zones. At least, it is desirable to draw lines corresponding to 5, 10, 90, and 95% of the height of Y. Two dotted lines bracketing the central 90% of Y (corresponding to 5% and 95% of the range of the largest cycle) are shown in Figure 8.16. A convenient plastic line-drawing guide for this purpose is shown in Figure 8.19. It consists of a plastic circular guide [31] with 11 equally-spaced holes mounted in a straight-edged frame. The circular part is first rotated to a position where the vertical projection of the distance between the two outermost holes is just equal to either Y or $\frac{1}{2}Y$. Then, as with a lettering guide, a pencil can be placed in each hole to produce the needed lines. With these lines, one can count N_1, N_2, etc., by counting full excursions; for example, in Figure 8.16, while two cycles run between the two (90% range) dotted lines, there would be three complete cycles if the dotted lines were narrowed to the 80% range. In other words, N_1 is 2 cycles and N_2 is 3 cycles. According to the table, the root mean square value is, therefore, either 34.5 or 35.6% of the maximum peak-to-peak (double) amplitude. Using the latter and converting (halving) to obtain single amplitude, the rms value is 0.356 times $4.6\sigma/2$ or about 82% of the "1σ" value or 82% of the value found by power spectral density methods for a long record, of which Figure 8.16 is a relatively small and possibly unrepresentative part. Also, the near equality of Z_1/Y and Z_2/Y suggest that, at least for the cycles of larger amplitude, the peak distribution approximates a Rayleigh distribution very closely.

These counts can be used to obtain a distribution in either of two ways. The counted cycles could be tabulated and a calculated distribution used to extend the tabulation down to small amplitude levels. First, all but one count can be discarded and the distribution obtained from just the rms value and this one count. Second, this one count could be at the 100% amplitude level.

Although believed to be the less accurate of the two, the latter approach is now illustrated because it is somewhat simpler and because it comes closer to what can be done in those cases in which the power spectral density approach has been chosen.

8.2.3 A Typical Damage Calculation

This section illustrates a more or less typical damage calculation. The last two sections presented two different statistical or partly statistical approaches for the reduction of stress or strain data into a distribution. In both cases, a root-mean-square value is found, the maximum-damage cycle is either known or can be measured, and for at least the second of the two approaches, the distribution of peak stresses can be compared with a Rayleigh distribution. Usually, if this comparison is negative, means can be devised to approximate the actual distribution after the differences have been studied. However, in this discussion, a Rayleigh distribution will be assumed as exemplified by Figure 8.20. This distribution can be approximated from Figure 8.16 using (8.36) and (8.38). For this problem, let the largest cycle in Figure 8.16 correspond to a maximum strain at some point of

$$(\epsilon_T)_{\max} = 0.01. \tag{8.40}$$

For simplicity, the material properties are assumed to be expressed in terms

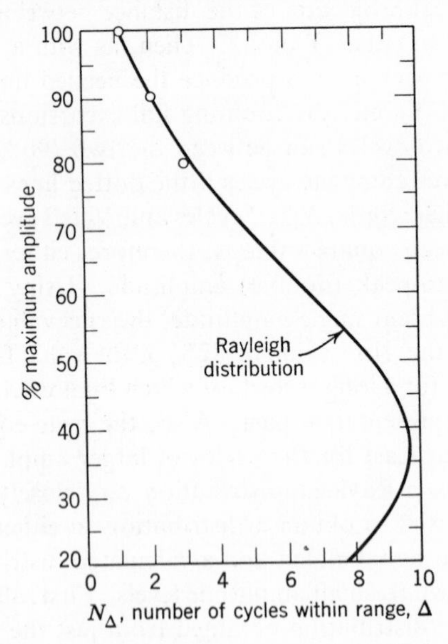

Figure 8.20 A theoretical distribution and some actual counts of amplitudes, using Figure 8.16.

of strain rather than stress as this results in a more linear relationship. The theory used will be Miner's hypothesis. The material properties are as given by Harting [17] for completely reversed cycles considering a variety of materials,

$$\epsilon_T N^{0.32} = 0.15. \tag{8.41}$$

Consider some arbitrary strain amplitude increment, Δ. The first step, then, is to obtain a Rayleigh distribution with the P or probability scale expressed in number of cycles per Δ unit of amplitude, with the amplitude scale expressed in percentage of the amplitude of the largest cycle, and with both so arranged that just one cycle is expected within $100\% \pm \Delta/2$ of this largest-cycle amplitude. If, as has been established for Figure 8.16, the rms value is 35.6% of maximum amplitude, then 100% amplitude corresponds to 2.809 times the rms value, σ.

If we wished to plot amplitude points at 10% intervals, $\Delta/2$ would be 5% and one cycle would correspond to

$$P\left(0.95 < \frac{X}{Y} < 1.05\right) = \int_{.95(2.809\sigma)}^{1.05(2.809\sigma)} \frac{X}{\sigma^2}\left[\exp -\left(\frac{X^2}{2\sigma^2}\right)\right] dX = 0.01547, \tag{8.42}$$

as a result of integrating (8.36). If, instead of the above limits of integration, 0 and ∞ were used, the integral would be unity. The reciprocal of 0.01547 would be the Rayleigh prediction of the allowable number of largest cycles in Figure 8.16, and illustrates one way to establish a scale factor for the number of cycles. However, it is simpler not to perform this integration at each point since the expression for the distribution can be used instead. At 100% amplitude (8.36) yields

$$\begin{aligned} P(100\% \times 2.809\sigma) &= \frac{2.809}{\sigma^2}\, e^{-\frac{1}{2}(2.809)^2}. \qquad (8.43) \\ &= 0.0543/\sigma. \end{aligned}$$

Therefore, the number of cycles, n at each point (i.e., within each of the ten increments) is approximately $P(X)$ divided by the result of (8.43). Let A represent the percentage of full amplitude, i.e., percentage of 2.809σ. Using (8.36), each number of cycles is,

$$n = \frac{\sigma}{0.0543} P(X) = 51.69e^{-3.9452A^2} \ldots. \tag{8.44}$$

This relationship is plotted in Figure 8.20. The three counts observed in Figure 8.16 are indicated by the small circles.

The final step in applying Miner's criteria,

$$\sum_i \frac{n_i}{N_i} < c, \qquad c \approx 1.0 \tag{8.45}$$

is a routine summation of damage at the ten arbitrarily-chosen values of the relative cycle amplitude, A. The calculations are performed in Table 8.4 and result in a damage summation or cycle ratio equal to 0.21 % of the expected life of the material. It is interesting to observe that the total damage is more than ten times that of the largest cycle. If the variable had been a quantity proportional to stress, the *S-N* curve would probably be steeper in the low life region and, for a loading pattern which produces the same distribution of amplitudes, the total damage for such a short history would probably be less than ten times that of the largest cycle. For example, for the approximate Rayleigh distribution of stresses considered by Mains [32], the total damage was slightly less than twice the damage of the cycles in the largest 10 % of the stress range.

TABLE 8.4

CUMULATIVE DAMAGE CALCULATIONS USING MINER'S THEORY

A — Load per maximum cycle load	n[a] (8.44) Number of cycles within $A \pm 0.05$ load range	ϵ_T (8.40) Strain amplitude (in./in.)	N (8.41) Number of cycles to failure	n/N — Damage cycle ratio
1.0	1.00	0.010	4,790	0.000209
0.9	1.904	0.009	6,700	0.000329
0.8	3.325	0.008	9,500	0.000352
0.7	5.25	0.007	14,200	0.000369
0.6	7.51	0.006	24,000	0.000312
0.5	9.63	0.005	38,000	0.000250
0.4	11.01	0.004	64,000	0.000172
0.3	10.88	0.003	115,000	0.000087
0.2	8.80	0.002	230,000	0.000038
0.1	4.97	0.001	490,000	0.000010
0	0	0	∞	0.000000

$\sum n/N$ or $\sum n_i/N_i = 0.002128$

[a] Plotted in Figure 8.20.

8.3 DAMAGE MEASURING EQUIPMENT AND SYSTEMS

The combination of a strain gage system, a data-type tape recording system, and a compatible digital computer provides sufficient capability for implementing what has been discussed in the preceding sections of this chapter.

However, for a number of reasons discussed later, this combination has not been very widely used for the purpose of evaluating fatigue.

The measured or counted quantity should be stress or strain for maximum accuracy. Sometimes, for convenience, economy, or other reasons, the measured quantity is an acceleration of a part of the vehicle or other structure. The combination of an accelerometer and a recording system have been used. Variations of a peak counting accelerometer system have also appeared on the market for this purpose both in United States and in Europe.

The usual form of the counting accelerometer system is a battery powered unit weighing several pounds. Acceleration along one axis causes a progressive actuation of electrical contacts until the peak acceleration is reached. Selection of *G* value for each contact point is based on a "weighting" taken from the *S*/*N* curve of the aircraft. Each acceleration-level contact produces a corresponding driving pulse to an electromechanical events counter. The number of events recorded on the counter gives some indication of the fatigue state of the aircraft structure.

8.3.1 *Material-Simulating or Fatigue Life Gages*

Quite different systems from the one above are being developed [33] by The Boeing Company and by Republic Aviation. These systems use special gages. They can be of wires drawn from the same material as used in the structure being evaluated, but with careful attention directed toward either eliminating or compensating for the changes in the fatigue properties of the material introduced by the wire-drawing process. Compensating for this has proved more feasible than eliminating it; the compensation consists of etching, distorting or prestressing the wires to introduce fatigue damage and cause the wires to simulate the *S*-*N* data of the structure.

Figure 8.21 shows a number of these material-simulating wires placed on a fatigue specimen. Figure 8.22 shows a typical result of an attempt to obtain an *S*-*N* curve from the wire gages which match a given *S*-*N* curve of 7075-*T*6 aluminum extrusions. The dotted lines represent the gages and show the scatter of 89 test points. This scatter is somewhat more than had been expected but could be subject to some improvement with refinements in the compensating techniques. Figure 8.23 shows that if a gage consists of ten of these wires, the wires tend to fail as predicted; that is, the total number of failures increases as cycles are added more or less in accordance with a calculated probable number of failures based on *S*-*N* data of the wires. But they do this only up to the time when the specimen on which they are mounted begins to fail. Just before this happens, the remaining wires fail provided the wires are on or near the resulting crack. If the wires were continually monitored for electrical continuity, this could provide an in-flight failure warning system. However, this is not the main purpose of these material-simulating wire gages.

Figure 8.21 Material-simulating wires form a fatigue gage.

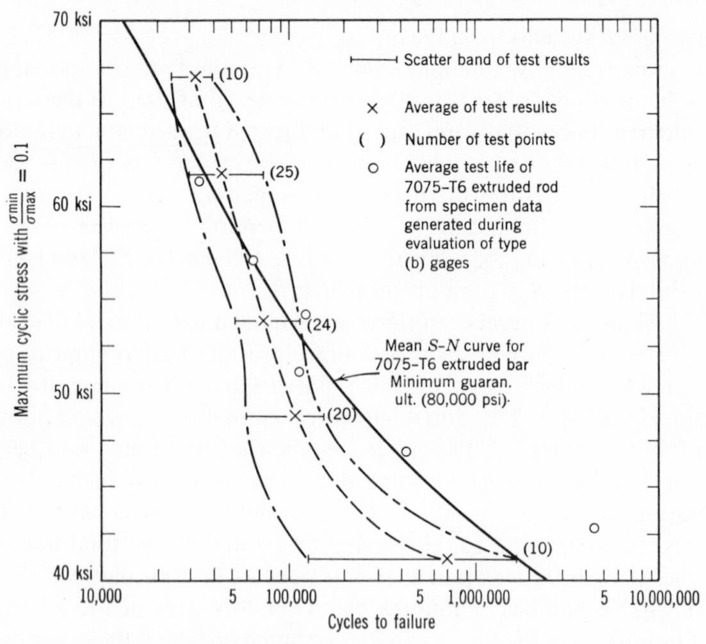

Figure 8.22 Comparison of some material-simulating wire gage test ranges with actual 7056-T6 S-N curve. Note: Ordinate equals product of elastic strain in carrier specimen and Young's modulus (10.3×10^6 psi) of 7075-T6.

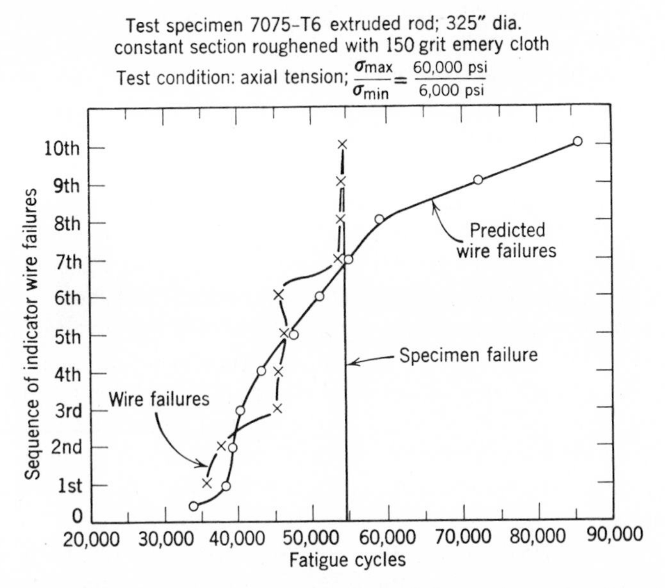

Figure 8.23 Sequence of actual and predicted wire failures using cycles of constant stress.

Their main purpose is to provide a flight-by-flight or week-to-week check of cumulative damage. For this purpose, no airborne electrical connections are required. Instead, a portable electrical continuity tester is wheeled into place after a flight and the number of broken wires is counted at each gage location. This system is adaptable to the simultaneous use of many gages but not to the shifting of gage locations.

The fatigue-life gages being developed at the Boeing Company [17] are different from the above material simulating gages in two respects. First, they are designed to operate with a variety of different materials. Second, resistance variation rather than just electrical continuity is measured. The variety-of-material compatibility feature is accomplished by taking advantage of the previously mentioned nearly universal strain relationship given in (8.3) and by designing the gages to have what could be called a high indicated strain disturbance as a function of fatigue. With other strain gages, fatigue, like temperature, causes an unwanted resistance change. With the fatigue gage, the fatigue effect is large enough to be measured. In other words, a measurement of resistance is made to become a measurement of fatigue damage.

8.3.2 *Line Following Integrators*

Although an increasing fraction of all strain gage data obtained from flight test is recorded on tape rather than on oscillograph paper, a certain amount of this taped data is transcribed back onto paper for inspection, checking, and visual interpretation. In this section all data are assumed to be either recorded or transcribed by an oscillograph. The needed information generally requires some form of automatic line following equipment; however, for short records manual line following is feasible.

Manual line following is commonly used to cause wheel rotation in an area-measuring planimeter. Like a planimeter, the fatigue integrator has a tracing point and an integrating wheel with markings. A number equal to or proportional to the calculated fatigue damage is read on the wheel markings to correspond to a traced section of oscillogram. Figure 8.24 shows a simplified schematic of the mechanism used to interconnect the tracing point and the wheel to integrate damage according to (8.3). Manual displacement of the tracing point by an amount ϵ causes rotation of both a cam and a disk proportional to a higher power of ϵ and of the marked wheel to a still higher power of ϵ. Although it is not shown in the figure, there is a sprag clutch or over-running device on the integrating wheel to permit only one-directional wheel rotation.

Part of a more versatile data-analyzing damage indicator [34], which mechanizes (8.25) is shown in Figure 8.25. This part is a computer for the integral,

$$D_{\text{part}} = \frac{S_u - S_f}{b_3} \int_{S_{\min}}^{S_{\max}} \frac{dS}{(S_u - S)^2}. \tag{8.46}$$

If every cycle considered were a completely reversed cycle, all mean stresses would be zero. Then

$$\left. \frac{b_2 S_m + 2S_f}{2 + b_2 S_m / S_u} \right|_{S_m = 0} = S_f, \tag{8.47}$$

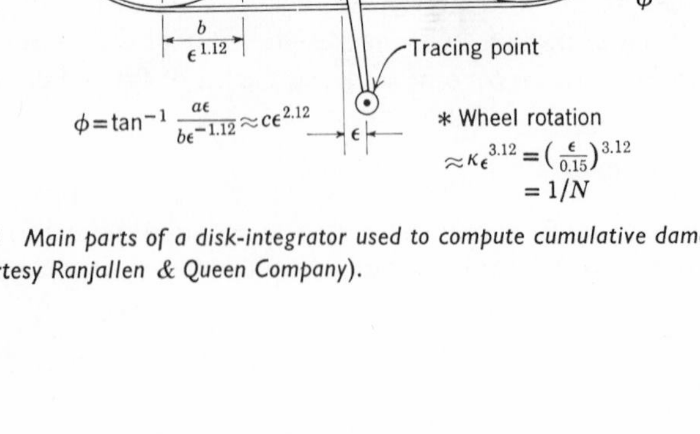

ϕ Disk $r\epsilon$ * $a\epsilon$ Cam ϕ $\frac{b}{\epsilon^{1.12}}$ Tracing point

$\phi = \tan^{-1} \frac{a\epsilon}{b\epsilon^{-1.12}} \approx c\epsilon^{2.12}$ ϵ

* Wheel rotation $\approx \kappa_\epsilon^{3.12} = \left(\frac{\epsilon}{0.15}\right)^{3.12} = 1/N$

Figure 8.24 Main parts of a disk-integrator used to compute cumulative damage from recorded peaks (courtesy Ranjallen & Queen Company).

and, for large values of S_u,

$$\int_{S_f}^{S_{\max}} \frac{dS}{(S_u - S)^2} = \int_0^{S_{\max}} \frac{dS}{(S_u - S)^2} \approx \frac{1}{2} \int_{S_{\min}}^{S_{\max}} \frac{dS}{(S_u - S)^2}. \tag{8.48}$$

Therefore for this case (8.46) is equal to (8.25). The part shown in Figure 8.25 can be called a reversed-stress-cycle damage integrator. This part reads the damage on a wheel and disc integrator, the operation of which is stopped by a computer-operated solenoid-driven clutch. This clutch, during each cycle, prevents integration whenever stress rises from $S_{\min}$ to the lower integration limit of (8.25). The computer for this clutch, with its binary function output, is small and is sometimes mounted directly on the instrument. The computer input is the output of a potentiometer which reads S at every instant. A second binary function is used, namely, a sprag clutch which prevents dS from going negative. Thus the damage from $S_{\max}$ to $S_{\min}$ is always zero. If v and u represent these two binary functions,

$$\mathscr{D} = \frac{S_u - S_f}{b_3} \oint \frac{uv\, dS}{(S_u - S)^2}, \tag{8.49}$$

Figure 8.25 A damage-computing integrator which constitutes the complete analyzing system in the case of reversed-stress cycles (courtesy Los Angeles Scientific Instrument Company).

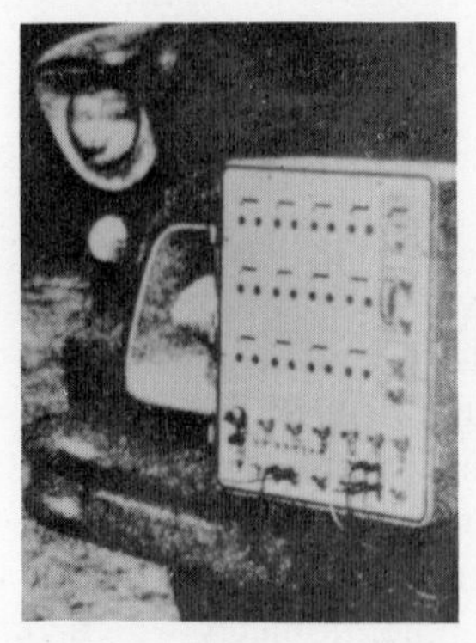

Figure 8.26 A Hughes peak analyzer measuring fatigue loads in a truck (courtesy Hughes Aircraft Corporation).

where the constants are obtained from the *S-N* curves or from the oscillogram scale factor and are adjusted directly into the instrument.

The manual operation of this instrument resembles that of the instrument represented by Figure 8.24. Instruments somewhat like these have been developed which have a photoelectric line follower and a small electrical servo to provide automatic line following. One of these is the United Gas Corporation's Electroscanner.

The design problems of the line-integrating systems and of the material simulating gages are rather similar. In both cases the main design task is to curvefit the *S-N* data using forms which fit the particular analog used. For the wire gage analog, the agreement which can be expected is typified by Figure 8.23 or better; for the linkage and computer analog, it is typified by Figure 8.10 or better. The two appear to have roughly equal orders of accuracy, although the line integrating system has less scatter. The integrator-computer system is more expensive but easier to adapt to new materials, to new locations, and, of course, to stationary or desk-top analysis of oscillograms.

8.3.3 Direct Strain Gage Output Analyzers

Sometimes neither an oscillogram nor a magnetic tape recorder are employed. For example, the Hughes peak distribution analyzer is a portable electronic instrument which can analyze a random voltage from either a strain gage or tape and instantly present certain data in numerical form. See Figure 8.26. The output is divided into equal increments for readout on 12 counters and may be adjusted to indicate any one of the following:

1. The numbers of peaks and valleys of selected magnitudes that occurred during a recorded time interval. This mode can be used to bracket fatigue damage in accordance with Case 4 (Figure 8.14).

2. The time spent in each of 12 selected increments during a recorded time interval. This can be used to determine S_{mean}.

3. The numbers of times the output traveled through each of 10 selected peak-to-valley ranges during a recorded time interval. For completely reversed cycles this could be used in method 2 of Figure 8.15.

4. The number of times the output traveled through the ranges represented by selected equivalent ranges about selected means during a recorded time interval. This, the most useful of all, performs the paired range count of Figure 8.15.

8.3.4 *Magnetic Tape Instrumentation Systems*

Tape recording systems have the potentiality for recording vast quantities of strain gage data. Were it not for the high cost of this equipment, the combination of tape recorder and digital computer would become a more or less standard method of measuring and analyzing fatigue damage. This section will attempt to summarize the technical problems which arise and which cause tape instrumentation systems to be quite complex and expensive.

There are four basic methods of recording information on tape. These are described briefly as follows:

1. Direct analog. This is the simplest method and represents the type of recording commonly used in homes and studios. A plastic tape with a coating of magnetic material is passed along the poles of an electromagnetic tape head so that, at any instant, a small section of the tape can be magnetized. At the instant represented in Figure 8.27, the poles of the recording head are directing a magnetic flux to pass between points (1) and (2) in the tape, some of which remains after the tape section leaves the head; for example, if, at some previous time, a pulse of current had flowed through the coils of the electromagnetic head, there would be a section of tape between points (3) and (4) which would resemble a permanent magnet. When this section passes an electromagnetic head similar to the recording head, an electrical signal can be produced which has certain characteristics in common with the original signal.

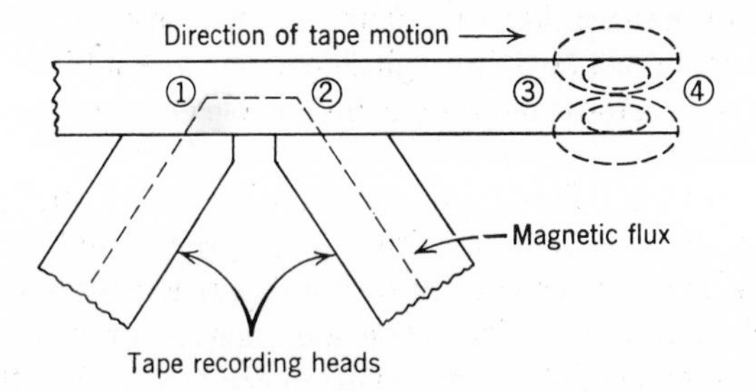

Figure 8.27 Essential elements of a tape recording.

It will not be identical to the original signal; for example, if the pulse is of long duration, it will simply put a direct current into the recording head, whereas the playback signal will be zero during most of this pulse duration. In other words, this type of system is incapable of recording direct current.

In the output of a strain gage the mean and amplitude stress values correspond to direct and alternating current, respectively. Both are important. Therefore this type of magnetic tape system can not generally be used in fatigue work.

2. Pulse-duration modulation analog. This is basically a time-shared or data-sampling type of system using pulses. In this system, the input-information amplitude content is converted into the width of a pulse and the slope of the leading edge of the pulse contains the frequency content. At a later time the input information is sampled again. Meanwhile, other signals can be sampled. As many as 86 signals have been multiplexed on one track [35], each of which can have a direct-current value. For fatigue work, in which capturing the peak values is of utmost importance, the sampling rate must be high. This type of tape system has a limited high-frequency response. Therefore in its present form this system is not very satisfactory for this kind of data.

3. Frequency modulation analog. In this type of recording one or more carrier frequencies are chosen, each with sideband or frequency-deviation limits. The input signal voltage or current at any instant of time is converted into the frequency by which the magnetic flux to the tape is changed. At the playback head a voltage is produced at the same frequency. Circuitry is used to convert this frequency into the corresponding original amplitude, a process known as demodulation. In principle, a high degree of precision is obtainable through the FM process.

In practice, this precision is not always attainable. In fatigue work we might wish to limit the error in large amplitudes to 2.5%, but FM recording is inherently sensitive to tape speed variations. A 1.0% error in either the recording tape speed or the tape speed at playback will violate this requirement. For laboratory-type equipment this requirement can be met by using separate carrier frequency reference tracks in a tape speed servo loop, by using a high quality tape transport system, and by some additional circuitry. When the equipment must be lightweight, portable, and subject to a vibratory environment, however, these requirements can be met only by some extremely high-priced equipment.

4. Digital recording. This type of recording offers the possibility of producing tape that is compatible with a high-speed digital computer. The speed, versatility, and convenience of such an arrangement in fatigue evaluation is extremely attractive. The strain gage data, however, must first be digitized, for only binary bits of information are recorded in the tape. The

type of versatile, reliable, airborne digitizing system required still appears to be unavailable, but it is being developed. Its use can be expected to have a considerable impact on the evaluation of fatigue and the testing of structures.

REFERENCES

[1] J. J. Schauble and P. F. Maloney, "An Approach to Helicopter Structural Reliability and Fatigue Life," *J. Am. Helicopter Soc.*, **9,** 37–38 (1964).

[2] W. J. Crichlow, A. J. McCulloch, L. Young, and M. A. Melcon, Report No. ASD-TR-61-434, "An engineering Evaluation of Methods for the Prediction of Fatigue Life in Airframe Structures," March, 1962, Aeronautical Systems Division, W-PAFB Ohio.

[3] F. Lundberg, "Fatigue Life of Airplane Structures," *J. Aero. Sci.*, **22,** 349 (1955).

[4] D. L. Henry, "A Theory of Fatigue-Damage Accumulation in Steel," *ASME Trans.*, **77,** 913 (1955).

[5] H. T. Corten and T. J. Dolan, "Cumulative Fatigue Damage," Session 3, Paper 2 of the International Conference on Fatigue of Metals at the IME, London, September 10–14, 1956, and at the ASME, N.Y., November 28–30, 1956.

[6] F. R. Shanley, "Fatigue Analysis of Aircraft Structures," Rand Corp. Paper No. RM1127, 31, July, 1953, AD 210794.

[7] M. A. Miner, "Cumulative Damage in Fatigue," *J. Appl. Mech.*, **12,** A-159 (1945).

[8] A. J. McCulloch, M. A. Melcon, W. J. Crichlow, H. W. Foster, and J. Rebman, Investigation of the Representation of Aircraft Service Loading in Fatigue Tests, WADD Contract No. AF 33(616)-6575, Project No. 1367, Task 14025, ASD TR 61-435, September, 1961.

[9] Aircraft Industries Associations, ARTC/W-76 Aircraft Structural Fatigue Panel, *Aircraft Fatigue Handbook*, Vol. III, January, 1957.

[10] C. R. Smith, "Prediction of Fatigue Failures in Aluminum Alloy Structures," *SESA Proc.* **12,** 21 (1955).

[11] S. R. Valluri, "A Unified Engineering Theory of High Stress Level Fatigue," *Aerospace Engrg.* (1961).

[12] C. R. Smith, "A System for Estimating Cumulative Fatigue Damage by Using the Miner Rule Corrected for Residual Stress," Society of Automotive Engineers, Preprint No. 353B for 1961 National Aeronautical Meeting.

[13] L. Kaechele, "Review and Analysis of Cumulative-Fatigue-Damage Theories," Rand Corp. Memo RM-3650-PR, August, 1963.

[14] J. P. Honaker, "An Evaluation of Procedures for Calculating Allowable Fatigue Stresses in Aircraft Design," ASTIA Document AD 69464, 1955.

[15] Federal Aviation Agency, Appendix A to Part 6 of the Civil Air Regulations (14 CFR Part 6).

[16] A. M. Freudenthal and R. A. Heller, "On Stress Interaction in Fatigue and a Cumulative Damage Rule, Part 1—2024 Aluminum and SAE 4340 Steel Alloys," WADC Tech. Report 58-69-Pt. 1, June, 1958. AD 155687.

[17] D. F. Harting, "The *S-N* Fatigue Gage, Second International Symposium on Experimental Mechanics, *SESA Proc.*, Figs. 5, 6, and 7 (1965).

[18] J. Morrow, "Fatigue Properties of Metals," Manual for SAE, ISTC Div. 4, April and October, 1964

[19] S. S. Manson, "Fatigue: A Complex Subject—Some Simple Approximations," *Experim. Mech.*, (1965).

[20] M. R. Gross, "Low Cycle Fatigue of Materials for Submarine Construction," U.S.N.E.E.S., ASTIA No. 013146, 1963.

[21] F. Cicci, "An Investigation of the Statistical Distribution of Constant Amplitude Fatigue Endurances for a Maraging Steel," University of Toronto Technical Note UTIAS No. 73, July, 1964.

[22] J. S. Lariviere, Method of Calculation to Determine Helicopter Blade Life, *J. Am. Helicopter Soc.*, **6,** 24–25 (1961).

[23] J. R. Fuller, "Research on Techniques of Establishing Random Type Fatigue Curves for Broad Band Sonic Loading," Air Force ASD-TDR-62-501, Oct., 1962. OTS (Department of Commerce, Washington, D.C.) No. AD 290799.

[24] E. C. Naumann, "Evaluation of the Influence of Load Randomization and of Ground-Air-Ground Cycles on Fatigue Life," NASA TN D-1584, October, 1964.

[25] J. S. Bendat, *Principles and Applications of Random Noise Theory*, Wiley, New York, 1958.

[26] S. O. Rice, "Mathematical Analysis of Random Noise," Bell System Tech. J. (1944).

[27] S. H. Crandall and W. D. Mark, *Random Vibration In Mechanical Systems*, Academic, New York, 1963.

[28] W, Barrois and E. L. Ripley, *Fatigue of Aircraft Structures*, Pergamon, New York, 1963, pp. 115–149.

[29] F. P. Beer, P. C. Paris and L. Y. Bahar, *An Approach to the Study of Crack Growth Under Random Loading*, Institute of Research, Lehigh University, Bethlehem, Pa., 1961.

[30] W. J. Trapp and D. M. Forney, Jr., Proceedings of 2nd International Conference on *Acoustical Fatigue In Aerospace Structures*, Syracuse University Press, Syracuse, N.Y., 1965, pp. 331 to 359.

[31] Strip Chart Reader, New Products Section, *Experim. Mech.*, p. 10A (1966).

[32] S. H. Crandall, *Random Vibration*, Wiley, New York, 1958.

[33] J. M. Kossar, "A Feasibility Study on the Development of a Pre-Crack Fatigue Damage Indicator," Wright-Patterson AFB, Ohio, Aiv Force Tech. Rep. ASD-TR-61-719 (Commerce Dept. OTS 2772105). April, 1962.

[34] *Mathematical Engineering Instruments*, Los Angeles Scientific Instrument Company, Los Angeles, California, 1965.

[35] B. B. Bycer, *Digital Magnetic Tape Recording: Principles and Computer Applications*, Hayden, New York, 1965.

9

Detail Design and Manufacturing Considerations

MITCHELL H. WEISMAN

The fatigue strength of a material may be defined as the maximum stress that can be applied repeatedly to the material without causing failure in less than a certain finite number of cycles. The endurance strength is that maximum stress that can be applied repeatedly to a material for an indefinitely large number of cycles without causing failure. The relationship of the magnitude of repeatedly applied stress to the number of cycles to failure is conventionally presented in the graphical form known as the *S-N* curve.

The preceding chapters have described *S-N* curves, the theories by which they are constructed, and equations that have been developed to express them mathematically. This chapter discusses the factors that affect the shape of the *S-N* curves, the relationship between these factors as they are influenced by design and manufacturing conditions, and the effects of such conditions on the fatigue properties of materials, components, and structures.

9.1 *S-N* CURVES AND COMPARISON OF TYPES OF LOADING

The *S-N* curve is a convenient representation of the relationship of the levels of stress and the number of repetitions of such stresses that will cause a material to fail. Empirical results have shown that fatigue life is an exponential function of the level of repeated stress. In recent years a number of investigations have shown that fatigue life is more likely a function of the level of repeated applied total strain rather than of applied stress. The total strain includes both the elastic and the plastic components of strain. Empirical relations developed for strain-cycle fatigue show that cyclic life is related to the applied strain range by a power law; that is, it is also an exponential function [34]. Therefore *S-N* curves are usually presented in semilogarithmic or double-logarithmic form. I recommend the double-logarithmic or log-log form of graphical presentation because it gives a more linear plot and permits more accurate extrapolation or interpolation of the data.

This discussion considers fatigue properties in terms of repeated applied stress since this is the form most familiar to design engineers. The fatigue properties of a material are determined by a number of factors. Therefore the shape of the *S-N* curve, which is a representation of the fatigue life, will vary according to the conditions represented. Factors influencing the shape of the *S-N* curve are the following:

1. Type of load:
   a. Tension.
   b. Compression.
   c. Torsion.
   d. Combined loads.
2. Relationship of maximum to minimum loads.
3. Manner of load application:
   a. Axial.
   b. Flexural.
   c. Torsional.
4. Rate of load application.
5. Frequency of repetition of loads.
6. Temperature of the material under load.
7. Environment:
   a. Corrosive.
   b. Abrasive.
   c. Inert.
8. Material condition:
   a. Prior heat treatment.
   b. Prior cold work.
9. Design of part or structure.
10. Fabrication techniques.

9.1.1 S-N Curves

The shape of the fatigue curve is influenced by the type of loading conditions to which the material is subjected. The loads applied may vary from zero to a maximum, or may be both tension and compression loads which may or may not be equal. Loads may be applied in tension, bending, torsion, or combinations of these.

A unidirectional axial tension fatigue curve is shown in Figure 9.1.

A significant feature that may be noted in this curve is that lifetimes of several hundred to several thousand cycles may be attained at loads approximating the ultimate strength of the material. This results from the fact that the ultimate strength of a material is a statistical number representing the results of several test pieces. The particular pieces used for the fatigue tests may be slightly stronger than those used to determine the tensile strength of that lot of material.

A comparison of unidirectional axial load and flexural bending fatigue tests is shown in Figure 9.2.

This graph shows that at lifetimes greater than approximately 20,000 cycles the two curves blend into each other. At shorter lifetimes the flexural curve indicates applied stresses exceeding the ultimate tension strength of the material by as much as 150%. These apparent high flexural stresses result from the elastic formulas which are used in the calculations and are actually much lower due to plastic deformation of the surface layers of the material when the applied load exceeds the yield strength.

Fatigue test results for similar types of materials group into well-defined bands when compared as a percentage of some property such as the ultimate tension strength. Rotating beam test data for a number of low-alloy steels, plotted as a percentage of the ultimate tension strength, are shown in Figure 9.3.

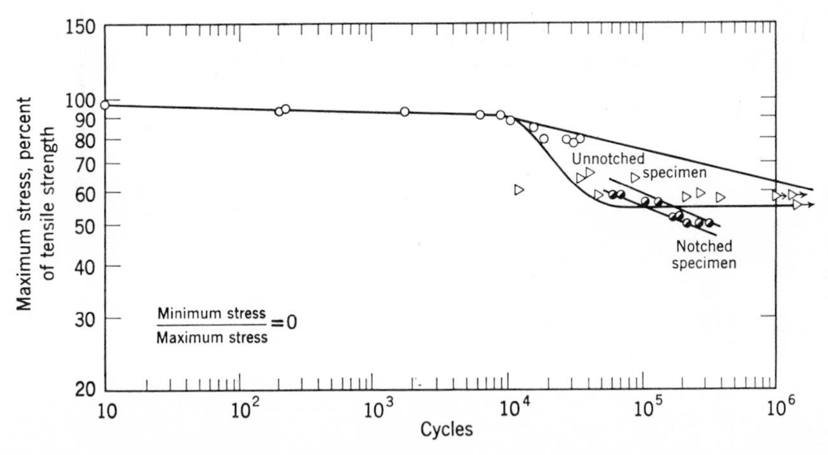

Figure 9.1 Relationship of fatigue strength to life from a compilation of axial loading test results (stress range from zero to maximum tension, R = 0) [1].

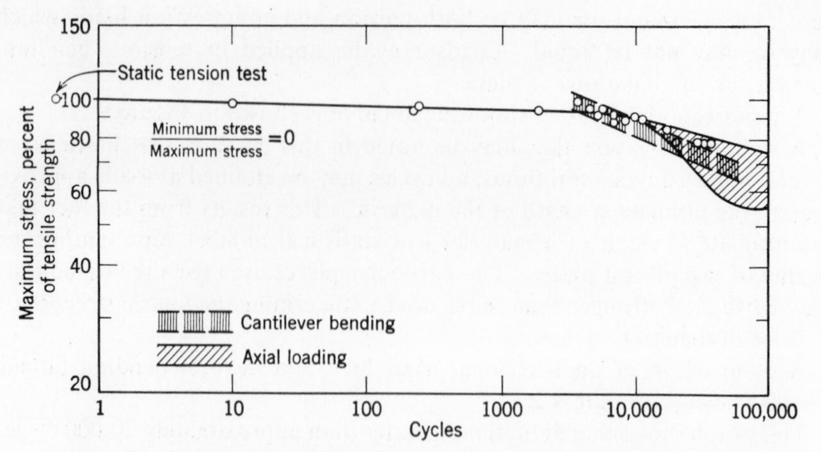

Figure 9.2 Comparison of unidirectional axial loading and flexure fatigue tests [1].

The upper *S-N* curve band represents unnotched specimen tests, the lower *S-N* curve band represents tests of notched specimens (K_t from 2.5 to 4.0). A similar grouping of axial loading fatigue test data, also for low-alloy steels, is shown in Figure 9.4.

It will be noted that the axial loading fatigue curves are similar in shape to the rotating beam fatigue curves, but the lower limits of the axial loading bands are displaced to lower stress values for a given fatigue life than are the lower limits of the rotating beam bands. The reason for this is that rotating

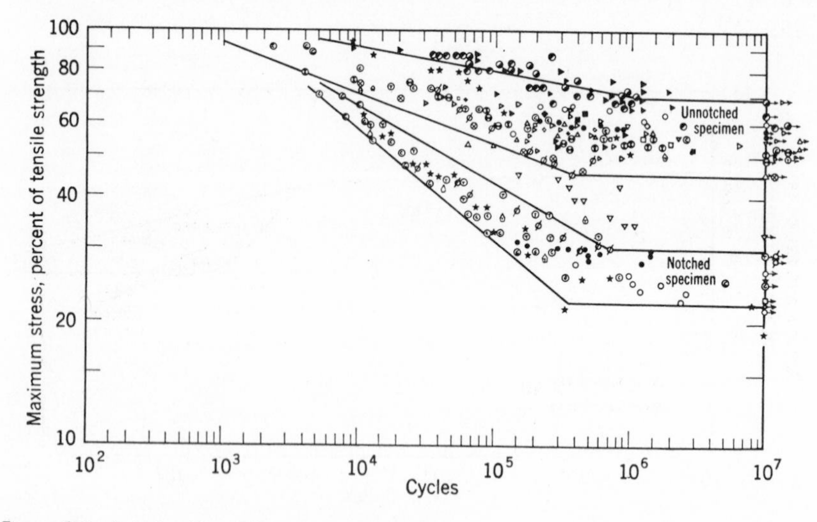

Figure 9.3 Relationship of fatigue strength to life from a compilation of rotating beam test results. [1].

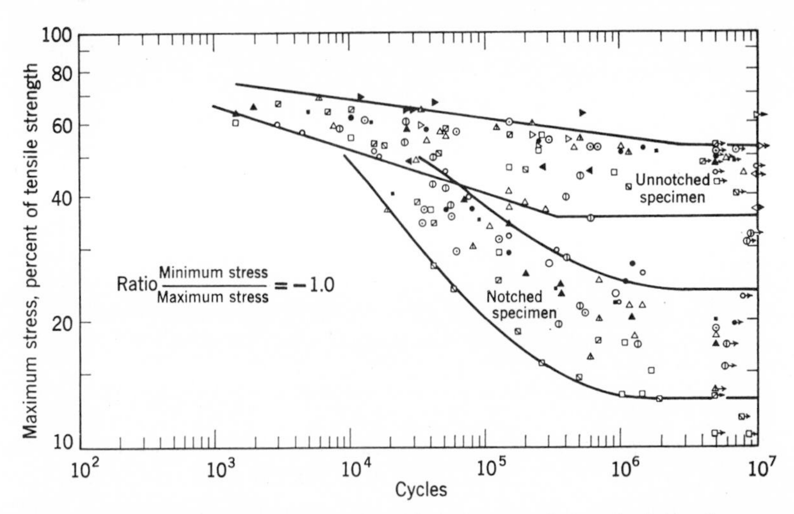

Figure 9.4 Relationship of fatigue strength to life from a compilation of axial loading test results (reversed stress, R = 1.0) [1].

beam tests subject only a narrow thin band of material to the maximum stress, whereas axial loading tests subject the entire cross section of the specimen to the maximum applied stress. Thus, although axial loading specimens should theoretically last as long as rotating beam specimens in fatigue tests at stresses less than the yield strength on the material, on a practical basis the axial loaded specimens have a much greater volume of material subjected to the full test stress and, therefore, a greater statistical chance for a material flaw or other defect to initiate fatigue failure. The upper limit of the rotating beam and axial loading fatigue test *S-N* bands will be approximately the same when plotted as a percentage of the ultimate tensile strength for families of comparable materials. The lower limits of the *S-N* bands for axial loading fatigue tests will be approximately 80% of the comparable stress value at a given lifetime of the lower limits of the *S-N* bands for rotating beam fatigue tests.

Another feature shown by the *S-N* curves in Figures 9.3 and 9.4 is that the notched bands appear to blend into the unnotched bands at short lifetimes. This is not necessarily the case, although it has been conventional to represent the curves that way. Even as the unnotched tensile strength of a material is the limiting factor in fatigue tests of unnotched specimens (a tensile test is a one-half-cycle fatigue test if one wants to look at it that way), the notched tensile strength is the limiting factor in fatigue tests of notched specimens. Therefore for materials with a notch tensile ratio greater than 1.0 the notched *S-N* curve will cross the unnotched *S-N* curve somewhere between 10^3 and 10^4 cycles and will lie above it for the short lifetimes, extending back to the

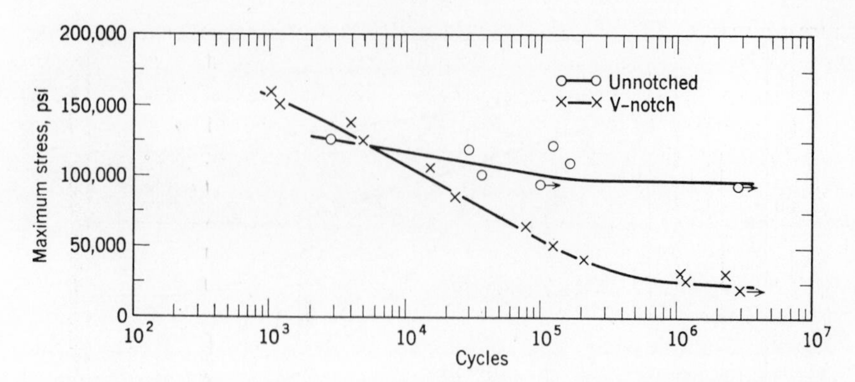

Figure 9.5 Material with notched to unnotched tensile strength ratio greater than 1.0 [2].

notch tensile strength. For materials with a notch tensile ratio less than 1.0 the notched *S-N* curve will always be below the unnotched *S-N* curve, even at very short lifetimes. These two different conditions are shown in Figures 9.5 and 9.6 for a material whose notched and unnotched *S-N* curves cross and one whose do not cross.

It is frequently necessary to furnish fatigue strength values for design use for materials for which only limited test data are available. Care must be exercised in such cases to avoid unsafe extrapolations. Predictions or extrapolations based on comparison with supposedly similar materials can be misleading if the two materials differ greatly in some condition, such as heat treatment. This is illustrated by the family of points in Figure 9.3 which lie on or above the upper limit of the unnotched *S-N* curve band. These particular points represent tests of annealed steel. The other points plotted in this

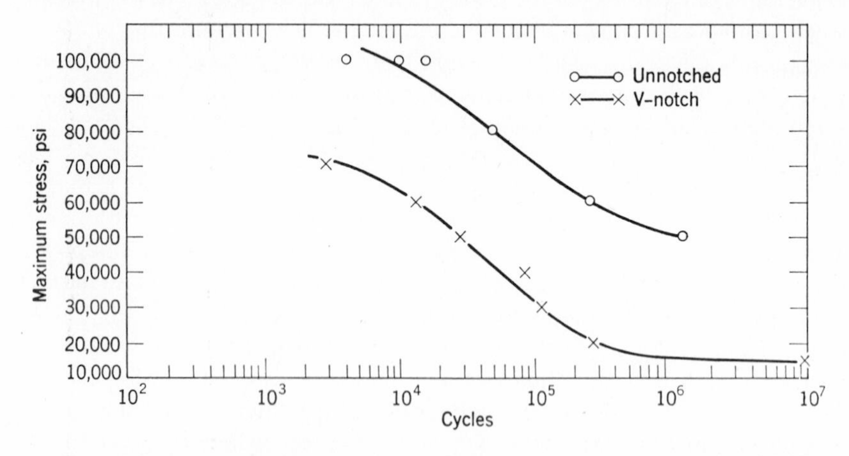

Figure 9.6 Material with notched to unnotched tensile strength ratio less than 1.0 [2].

figure represent tests of either normalized steels or steels which have been hardened by quenching and tempering. The annealed steel is subject to strain hardening during the repeated loads of a fatigue test. Therefore, the test specimens will become stronger and will have a longer life at a given stress value than will the normalized or quenched and tempered steel specimens which do not work harden. These results are similar to those of a recent study of two corrosion resistant steels, annealed 304 and AM350, which were shown to be subject to metallurgical transformation during cyclic stress testing [34]. Thus, at each stress (or strain) level, a somewhat different material is being tested.

Another example where fatigue data must be extrapolated with caution and understanding is the prediction of notch fatigue data where only unnotched data are available. One method of extrapolating a notched *S-N* curve is to divide the unnotched fatigue strength at various lifetimes by the notch stress concentration factor (K_t). Figure 9.7 shows the actual unnotched and notched *S-N* curves, and also the notched *S-N* curve predicted by the use of the stress concentration factor for a precipitation hardening corrosion resistant steel. The predicted notched *S-N* curve corresponds closely to the actual notched *S-N* curve for lifetimes in excess of approximately 50,000 cycles and is conservative for shorter lifetimes. However, a similarly developed notched *S-N* curve for a heat and corrosion resistant steel, shown in Figure 9.8, is conservative to lifetimes of approximately 500,000 cycles but then becomes too high, lying above the actual notched *S-N* curve. Use of such information could result in failure of notched components made from this material.

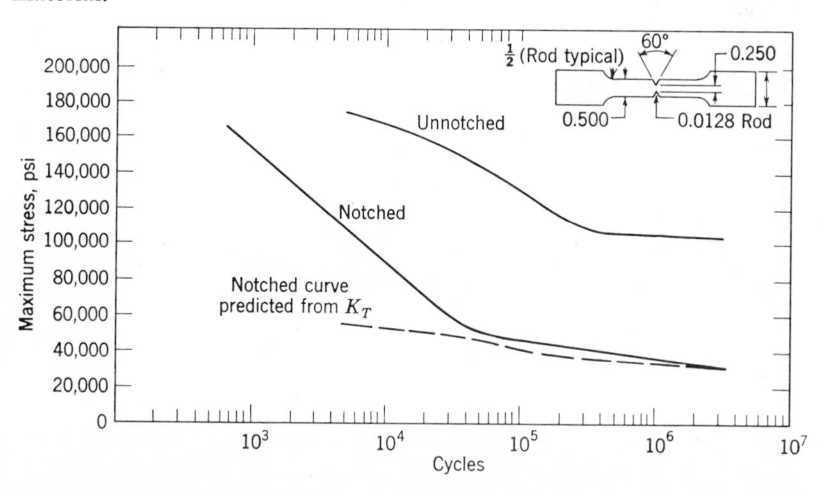

Figure 9.7 Comparison of experimental and K_T predicted notched fatigue curves for AM350 steel, condition SCT825, $K_t = 3.2$ [2].

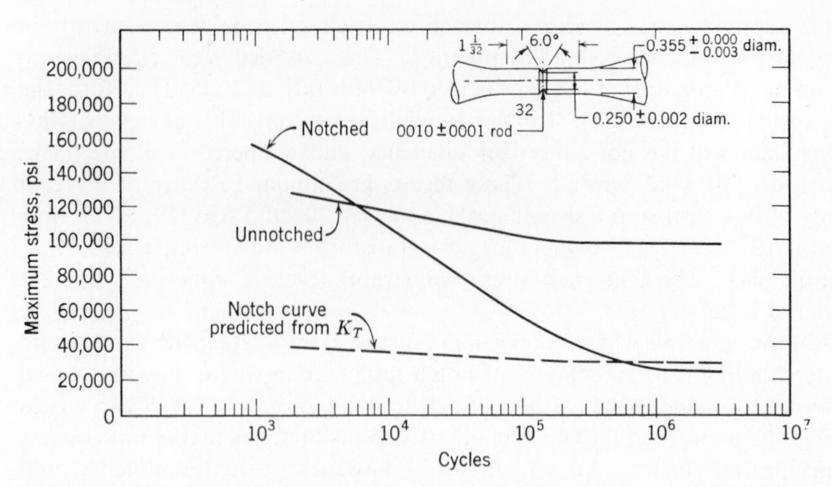

Figure 9.8 Comparison of experimental and K_T predicted notched fatigue curves for A-286 steel, $K_1 = 33$ [2].

Another problem in extrapolating notched fatigue data is that the notched properties of materials are affected by temperature. An example of this is shown in Figure 9.9 for a titanium alloy. The notched tensile strength of this alloy is shown to increase at low temperatures and to decrease at elevated temperatures. The ratio of notched to unnotched tensile strength, shown in the lower part of this figure, decreases at lower temperatures, although it remains constant at elevated temperatures. Therefore low-temperature notched fatigue data extrapolated from room temperature curves would result in premature failure.

9.1.2 Types of Testing to Determine Strength of Materials and Structures

Fatigue tests are conducted on simple specimens of materials, on small components, major assemblies, and on complete large structures. There are a number of commerical machines and innumerable "home-made" machines for testing small coupon-type material specimens and small components. Special fixtures and set-ups are required for tests of large components.

There are problems in applying the results of coupon type fatigue tests to the prediction of the fatigue life of large and complex structures. Some of these problems are discussed in Section 9.9 under "size effects." Other problems involving the interaction of complex biaxial and triaxial stresses are discussed in Section 9.10. In order to verify that the fatigue life of components and large structures will be adequate, such components may be tested under simulated service loads.

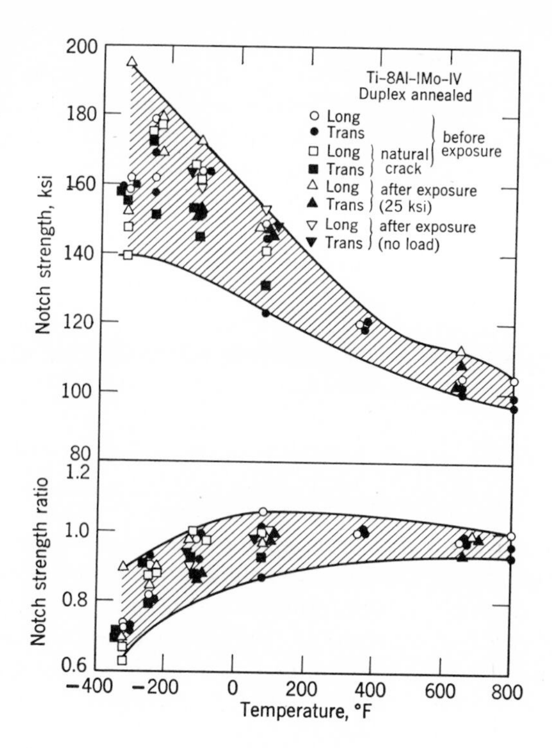

Figure 9.9 The effect of testing temperatures-on the notch strength and notch strength ratio of Ti-8Al-1MO-1V[3].

Figure 9.10 Special set-up for fatigue test of aircraft wing tank under combined fuel sloshing and vibration loads.

Special type fatigue tests include such service conditions as sloshing of fuel in tanks with superimposed vibration loads, Figure 9.10. Other types of fatigue testing involve use of vibration equipment which can be programmed to apply a spectrum of loads and frequencies from a tape which was recorded from strain gages attached to components during actual aircraft or missile flights. In some cases, special buildings are constructed to fatigue test very large structures such as complete full-size aircraft.

9.2 RECOMMENDATIONS FOR DESIGNS TO AVOID FATIGUE FAILURES

The designer can help to minimize the possibility of fatigue failure by proper design of structural components. A large proportion of fatigue failures can be attributed to lack of sufficient consideration of design details or a lack of appreciation of engineering principles. The engineering principles necessary for good design of structures subject to fatigue are well reported in the literature, but much of this information has been scattered through sources which may be relatively inaccessible to the designer who needs to know and utilize it. The following sections present a number of common problems of structural design which, if not recognized, frequently result in fatigue failure.

9.2.1 Sheet Metal Structures

It is frequently necessary to stiffen sheet metal panels by use of structural angles or beads. Such stiffening members must be designed to transmit the loads directly to adjacent structure, as shown in the preferred design of Figure 9.11.

The preferred design of Figure 9.11 has the stiffening angle overlying the adjacent structural members and firmly attached to them. The other configuration shown is not recommended because the stiffening angle must transmit its load through the sheet member to the adjacent structure. The resulting concentrated load in the sheet between the end of the stiffener and

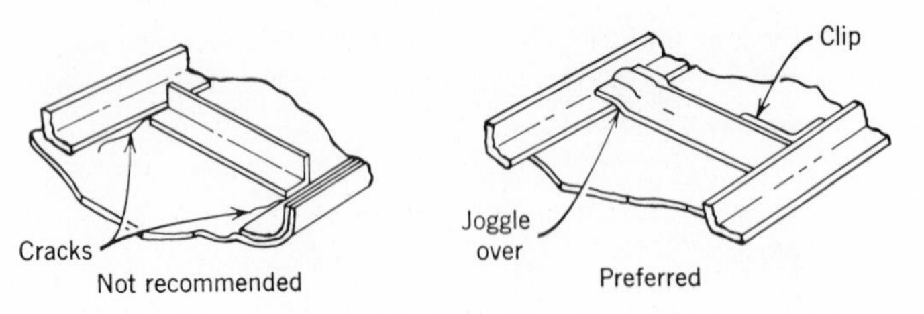

Figure 9.11 Stiffener ends must be tied to adjacent structure [4].

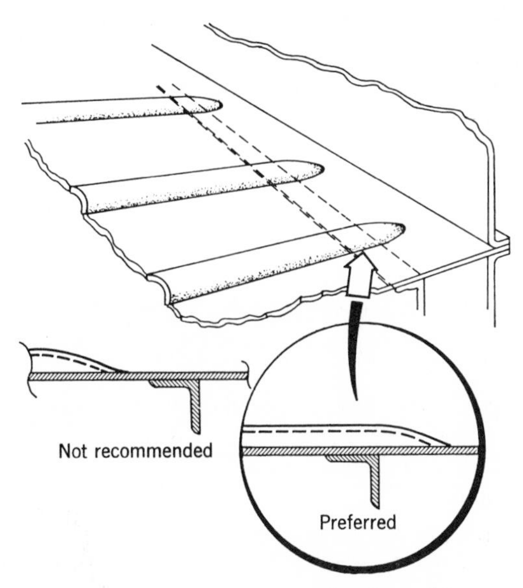

Figure 9.12 Overlap beads and stiffeners [4].

the other structure will cause failure of the sheet member such as the cracks shown. Similarly, the ends of beads which are used to stiffen flat panels should extend beyond adjacent structural members, as shown in Figure 9.12 to prevent failure because of flexing of the panels at the ends of the beads.

Cutouts for lightening holes or rain holes in sheet metal panels must be strengthened by stiffening members, as shown in Figure 9.13, or fatigue failure due to stress concentration will result.

When a beaded area of a sheet metal curved surface merges with another surface with a relatively small radius of curvature, as shown in Figure 9.14, the beads should stop at the beginning of the small radius curvature. All beads should fair into such curves with end radii not less than eight times the depth of the bead.

The cross section of a bead should be uniform or change only gradually throughout its length in order to prevent concentration of stress. Fatigue failure due to the abrupt change in contour of the center bead of the upper tank shown in Figure 9.15 was eliminated by a design change to the smoothly contoured bead shown in the lower tank.

Unsupported tubes extending from flat panels can cause fatigue failure of the panel, as shown in Figure 9.16, because of concentration of stresses induced by vibration of the unsupported mass of the tube.

An ideal shape for a tank is shown in Figure 9.17. There are no serious stress concentrations in the tank itself, and the protruding tubular members

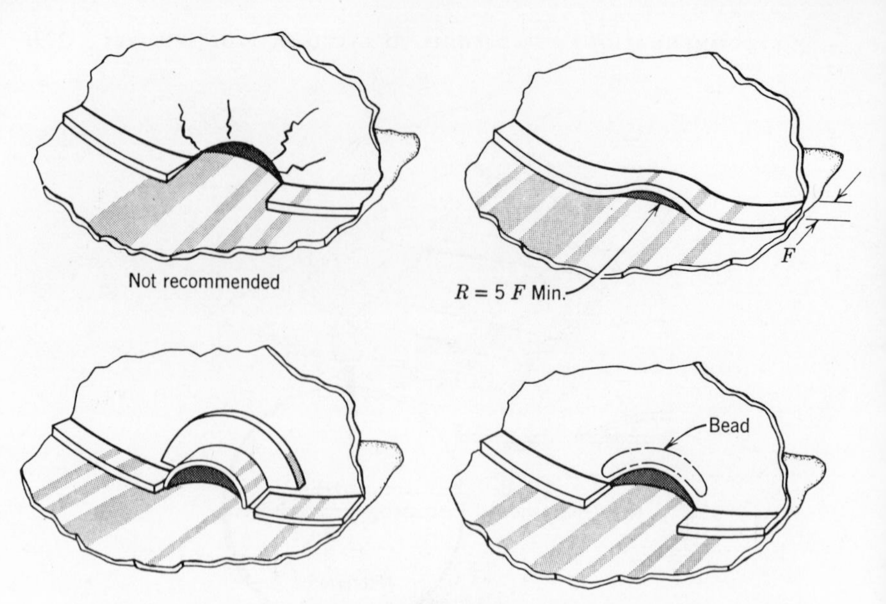

Figure 9.13 Cutouts for complete drainage where drilled holes cannot be used [4].

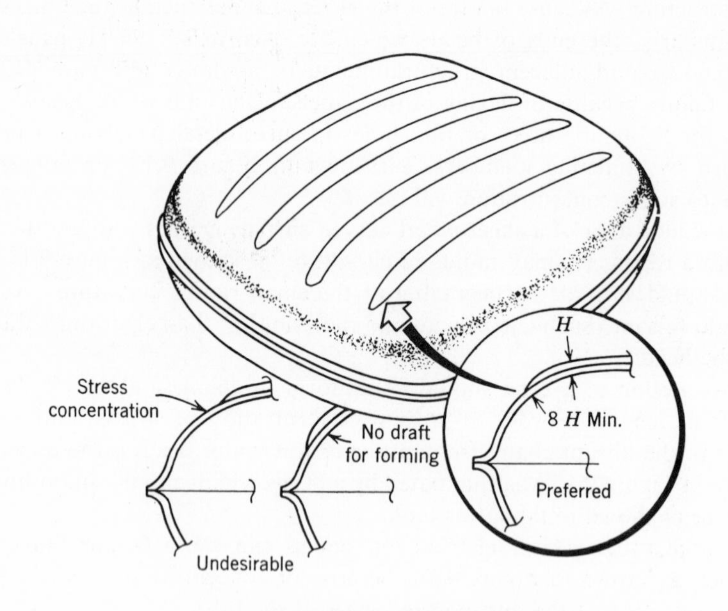

Figure 9.14 Preferred ending of bead [4].

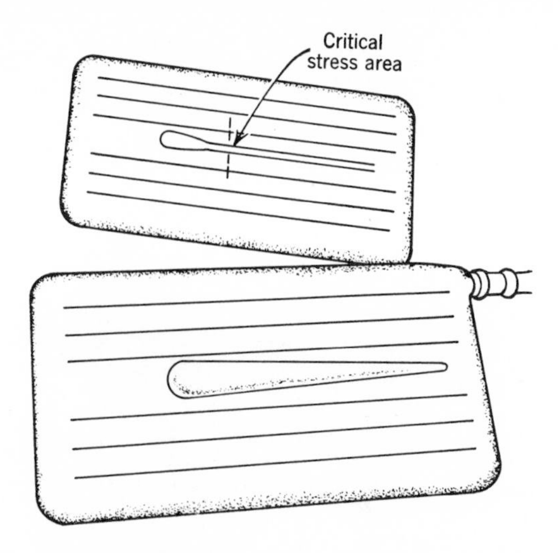

Figure 9.15 Gradual change of cross section reduces stress concentrations [4].

are adequately supported. However, like so many ideal things, such a spherical shape is not usually practical in aircraft design where fluid tanks must be fitted into available spaces. But, spherical tanks can be used in missile and spacecraft rocket propulsion systems.

The method of supporting component assemblies has an important effect on the fatigue life of the assembly and the support. A three-point support, with the three supports well-spaced on the component to transmit the stresses

Figure 9.16 Typical fatigue failure due to stress concentration caused by unsupported mass of tube [4].

Figure 9.17 An ideal shape for a tank assembly [4].

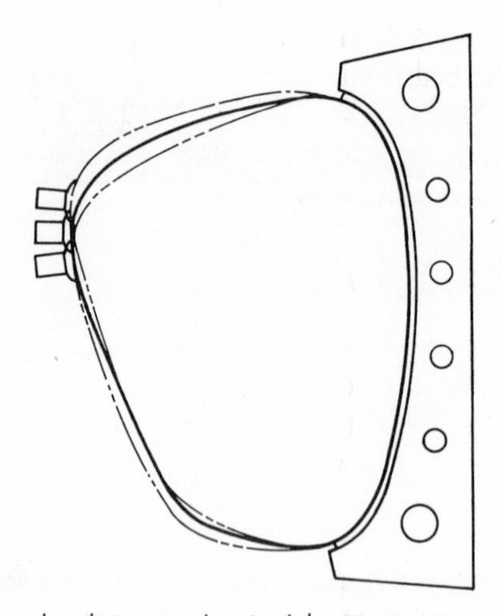

Figure 9.18 Irregular shapes require special support to prevent vibration [4].

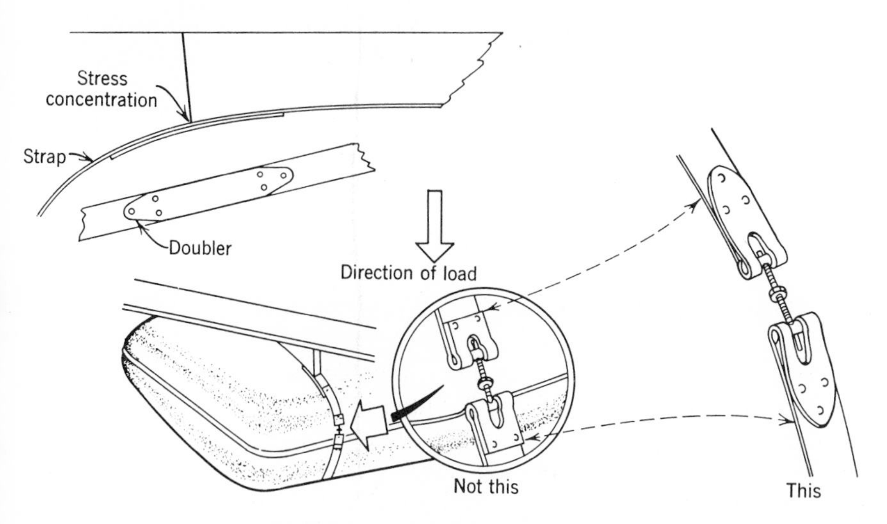

Figure 9.19 Use doubler to reduce stress concentration, attach straps with three or more rivets [4].

evenly, is desirable. Support should be secure but not so tight that the component is prestressed, nor should the component be required to carry any primary structural loads unless it has been specifically designed for this purpose. A tank supported as shown in Figure 9.18 is overbalanced and the mass of the tank and its contents will whip up and down.

Mounting straps require doublers to prevent stress concentrations in installations such as that shown in Figure 9.19. The doubler should always be located on the compression side of the strap relative to the load. Strap ends should have the bearing load distributed through three or more rivets. Strap and wear pad locations should be well defined. An improperly located strap or wear pad can cause failure due to excessive fretting or wear or because of an unanticipated stress concentration.

9.2.2 *Power Plant Sheet-Metal Components*

Fabrication defects can cause fatigue failure even in the most satisfactory designs. Failure can occur because of sharp edges, sharp corners, and burrs that are not removed. Very minor changes in design and close attention to manufacturing details have increased the service life of many parts from fifty to several hundred times.

Fatigue failures may also result from cyclic thermal stresses which cause buckling of inadequately stiffened sections. The intersection of two buckles can be a very severe stress concentration. Such stresses can also cause failure at attaching pads for fittings. All too frequently such pads are small in area and straight sided with sharp corners. When such pads are located in an otherwise unstiffened panel they form a concentration of stress because they stiffen a small area. The stresses from vibration or "oil canning" of the panel cause cracks to form at the corners or along the edges of such pads. Fatigue failure can be minimized by increasing the size of the attaching pad so as to spread out the stress, and the pad can be made up of several sheets, each smaller in size, to prevent an abrupt change of stiffness [5].

9.3 STRESS CONCENTRATIONS

A stress concentration is usually thought of as a change of section or size of a part which forces the stress paths closer together; that is, concentrates or increases the magnitude of stress in a localized volume of the part relative to adjacent volumes. Changes of section may be part of the design, such as fillets, shoulders, notches, holes, or joints between two parts of an assembly. A stress concentration may also be the result of accidental damage or poor workmanship; for example, a gouge or dent made by a screwdriver or hammer, or a burr left around the edge of a drilled or punched hole.

Less obvious stress concentrations which are frequently overlooked in a design review or failure analysis occur at the edge of an area of localized cold work, as where only a portion of a component surface is shot peened. The transition or interface between a hardened or plated surface and the core or basis metal can act as a stress concentration and initiate fatigue cracks under certain stressing conditions [35].

9.3.1 General Observations

The detrimental effects of concentrations of stress due to notches, screw threads, holes, and other changes in section required by the component design can be minimized by good design practice and careful attention by the designer to such details as fillet radii and avoidance of sharp edges from intersecting cuts.

9.3.2 Design Cautions

Two methods of minimizing stress concentrations at a sharp fillet radius are shown in Figure 9.20. If a shoulder-type stop is required on a shaft, but the stop does not itself have to resist much load, a narrow stop will cause a very small stress concentration compared to a wide stop or collar as shown in Figure 9.21. Figure 9.22 shows two methods of relieving stress concentrations where a wheel must be press-fitted on a shaft. Shallow grooves may be

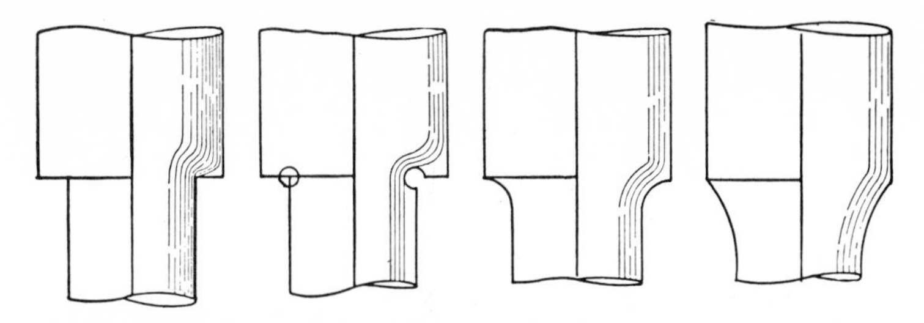

Figure 9.20 Effect of abrupt versus gradual change of section on stress concentration [6].

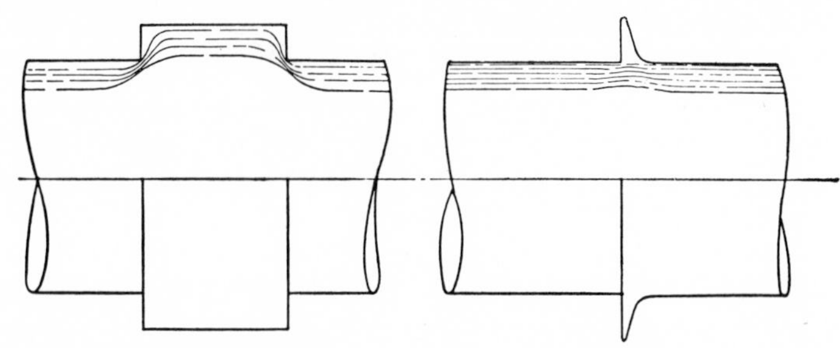

Figure 9.21 Comparison of stress concentrations with wide and narrow collars [5].

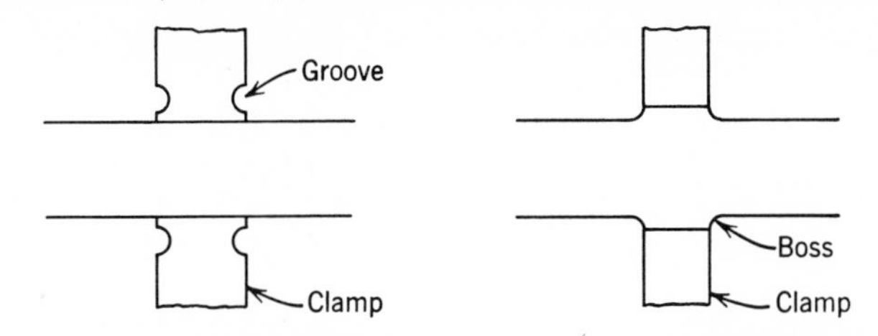

Figure 9.22 Acceptable methods of relieving stress concentrations in pressed fits [7].

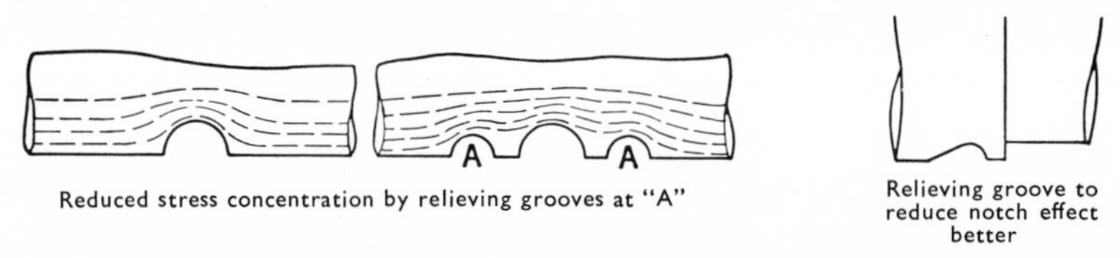

Figure 9.23 Devices for relieving stress concentration.

used to reduce the concentration of stress which a single deep groove or sharp fillet might produce as shown in Figure 9.23.

9.3.3 Rod Ends

Rod ends are frequently designed to fit into small spaces, and are very highly loaded. They are particularly subject to fatigue failure from stress

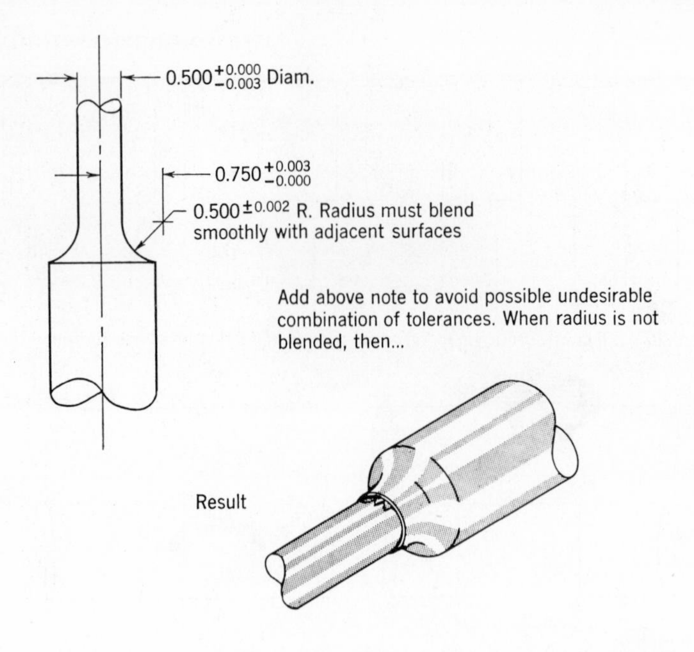

Figure 9.24 In service fatigue cracks developed at small sharp edge.

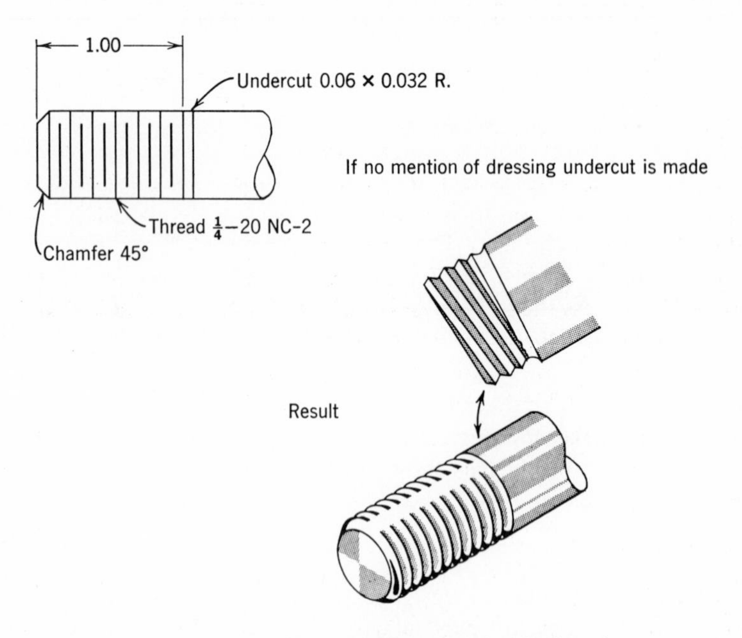

Figure 9.25 In service fatigue crack developed at feather edge.

Figure 9.26 Examples of redesign of rod ends to prevent fatigue failure.

concentrations because of their odd shapes. Fillet radii must blend smoothly into adjacent surfaces of rod ends and other highly loaded parts or fatigue failure will result. A note to this effect, as shown in Figure 9.24, should be a drawing requirement for such parts. The ends of threads are a common location for fatigue failures. Even the use of an undercut, or thread relief, may not help if a knife edge is produced on the last thread, as shown in Figure 9.25. Unless the knife edge is broken, or smoothed off; it can be the nucleus for fatigue failures. Examples of the results of unintended stress concentrations such as those noted are shown in Figure 9.26. This figure also shows the final design of each component which successfully met the fatigue life requirements.

9.3.4 Material Fabrication

The manner in which the material of a part is produced may also contribute to component fatigue failure, as shown in Figure 9.27. Fatigue failure originated at the flash line of the forging where there was an abrupt change in the direction of the flow lines of the material.

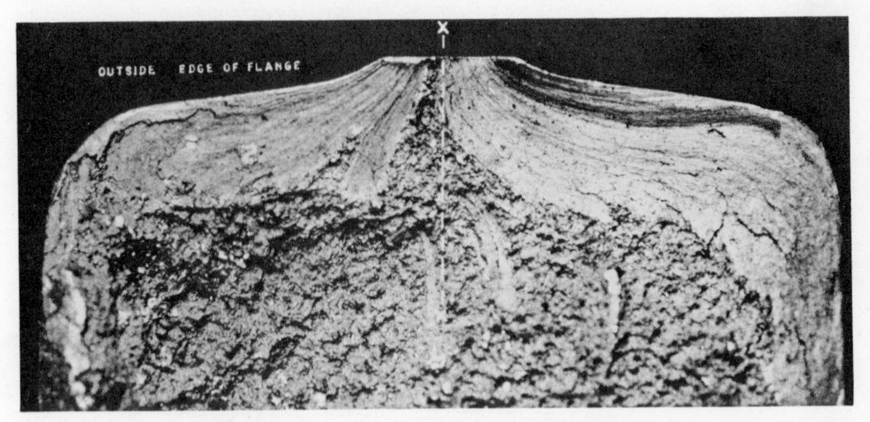

Figure 9.27 Enlargement of flap link. Note the extremely sharp divergence in the forging flow.

Another component in which material fabrication can cause fatigue failure is the multilayer bushing of lead-tin or lead-tin-copper case onto a steel backing which is widely used as a journal bearing. Such bearings may be subject to fatigue failure if the installation causes localized overloads, if the soft metal layer is too thick, or if the steel backing surface has been grooved or roughened to better anchor the layer of soft bearing metal.

In the case of local overload or when the soft metal layer is too thick the compression loads on the surface can induce high tension stresses in the interior of the soft bearing metal layer. Repeated surface loads then cause nucleation of a fatigue crack in the region of induced tensile stress. This crack may progress parallel to the bearing surface for some distance before finally turning to the surface [10]. Then the metal above the crack breaks off in a manner known as spalling, as shown in Figure 9.28.

In the case in which the steel backing surface has been roughened or grooved, the high points create notches in the bearing metal producing regions of stress concentration, Figure 9.29*a*, which can result in fatigue failure of the bearing as shown in Figure 9.29*b*.

9.4 FATIGUE PROPERTIES OF JOINTS

A joint, whether it be bolted, riveted, or welded, is a particular form of stress concentration. One could even go so far as to define a joint as a continuous defect in an otherwise perfectly sound material. It must be recognized that even the simplest joint can be, and too frequently is, the

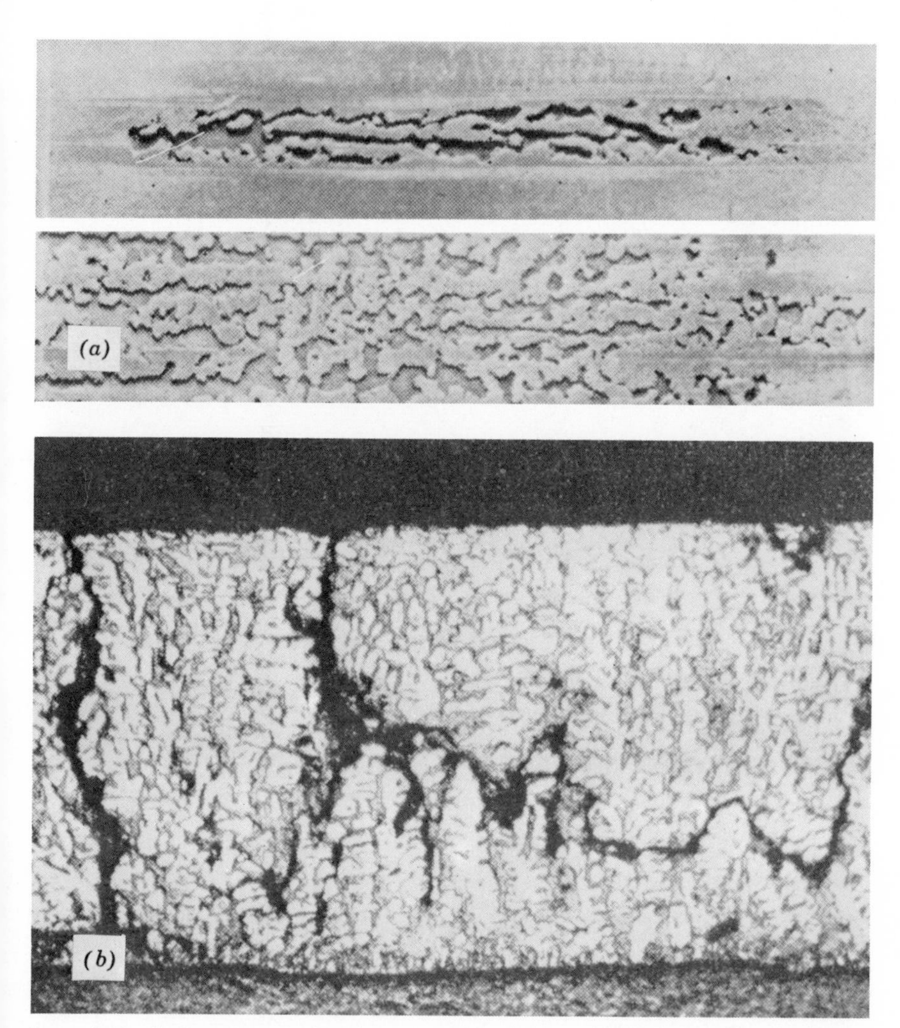

Figure 9.28 Spalling failures of bushings caused by fatigue: (a) disel main bearing showing fatigued central zone [11]; (b) photomicrograph showing fatigue in a copper-lead bearing [12].

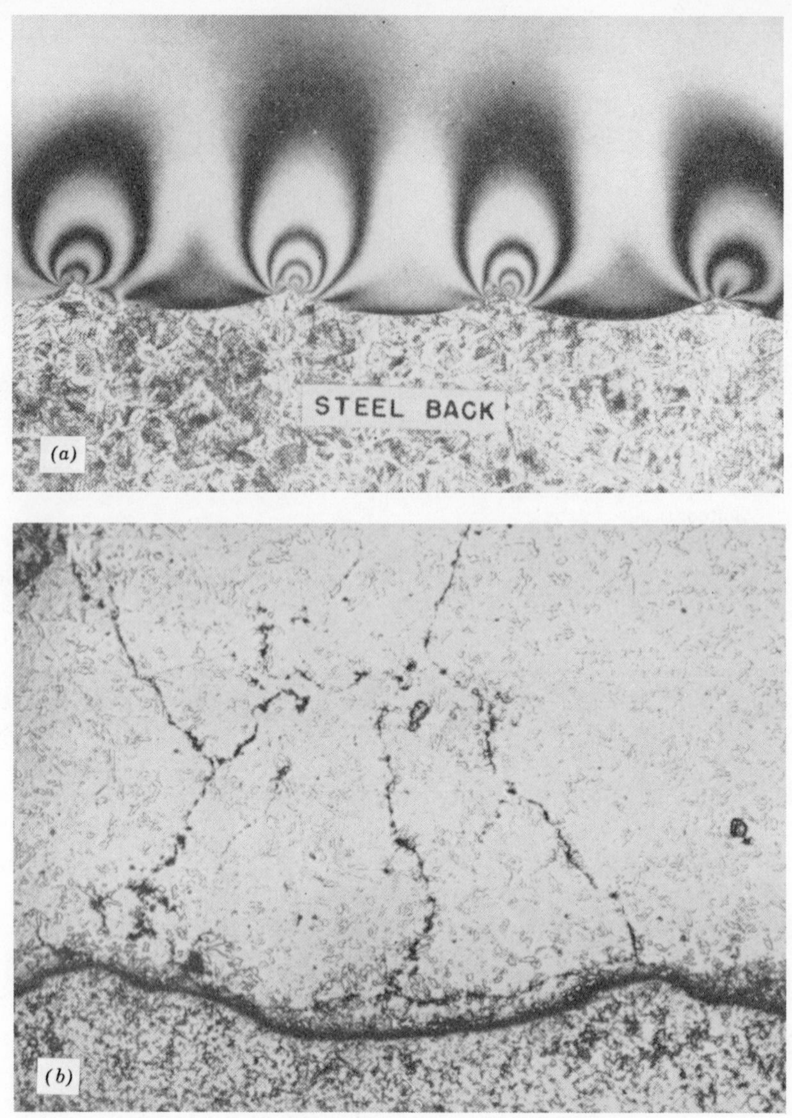

Figure 9.29 Fatigue failure of a bushing caused by roughened steel backing surface: (a) stress pattern in photoelastic polariscope showing enlarged model of bearing alloy cast onto a roughed steel backing; note how each peak on the steel back produces a stress concentration [11]; (b) cross section of Babbitt bearing failed by fatigue: Cracks are penetrating downward to the peaks of the rough-turned steel backing. Magnification 100X[11].

origin of a fatigue failure. Once this is understood it becomes possible to take preventive measures to minimize the detrimental effects on fatigue strength which are inherent in the different types of joints.

Design stresses must be kept within the capability of the joint to withstand fatigue failure. This will often involve fatigue tests of several variations of actual joint designs in order to determine the best practical design and to verify that there are no hidden problems or areas of critical load concentration which have been overlooked. A point to be emphasized is that static tests alone cannot determine the fatigue performance of a joint. Many times it has been shown that the design with the greatest static strength is not the strongest under fatigue loading conditions.

9.4.1 Bolted Joints

The tightness of a bolted joint will have a pronounced effect on the stress concentration factor for the joint. The fatigue stress concentration factor can be reduced appreciably by increasing bolting torque, as shown in Figure 9.30.

In many cases, stress concentrations in multi-bolt joints can be reduced by the use of oversize holes without adversely affecting the tightness of the joint. The use of such holes will reduce the residual stresses which would otherwise result from forcing bolts into mismatched holes. In any case, the residual stress pattern in multibolt joints, riveted joints, and spot-welded joints which exists immediately after assembly will change greatly during the first few cycles of repeated loading. The material around those fasteners which are carrying the load will be deformed, transferring load to the fasteners which were initially unloaded.

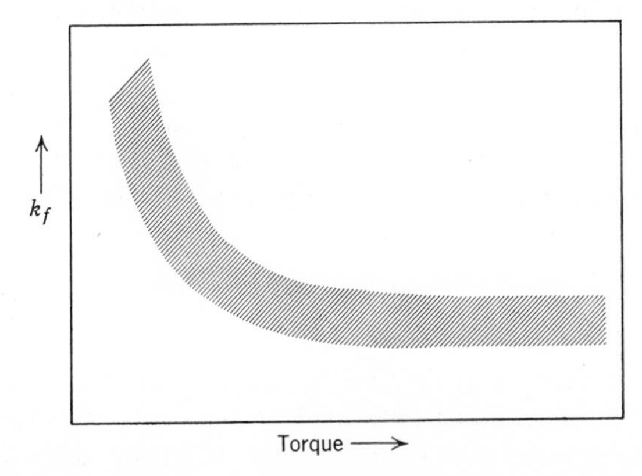

Figure 9.30 Effect of bolt torque on joint stress concentration factor [13].

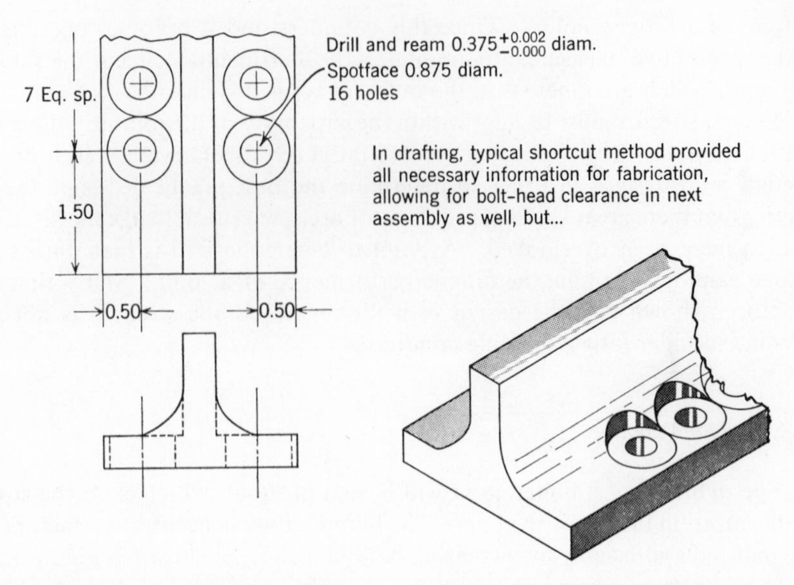

Figure 9.31 In service fatigue cracks developed at feather edges caused by intersecting spotfaces [14].

Designs for multibolt or multirivet joints frequently specify closely spaced fasteners. Machining of the seats for such fasteners may produce feather edges from which fatigue cracks can nucleate, as shown in Figure 9.31. Such edges must be broken or smoothed off, and this should be specified on the component drawing. A fracture of a multibolt joint, with fatigue markings indicating the points of origin, is shown in Figures 9.32 and 9.33.

Fatigue markings on the fracture surface of the broken bolt, shown in Figure 9.34, indicate that the failure was caused by repeated bending loads which concentrated the stress at the radius of the head-to-shank fillet.

Severe concentrations of load can also occur in the bolted part under the attaching bolt head, as shown in Figure 9.35. Failures such as the one shown can be prevented by the use of large washers which will increase the load bearing area and reduce the concentration of the applied load.

9.4.2 Riveted and Spot-Welded Joints

The effect of the number of rows of rivets on the stress concentration factor of a riveted joint is shown in Figure 9.36. A joint with two or three rows of fasteners will be stronger under static loads than a single-row joint. The stress concentration factor of such a joint, however, will still remain high, and the extra rows of fasteners will produce little increase in the fatigue strength.

Figure 9.32 *View of broken splice assembly [15].*

Figure 9.33 *Fracture surface and fatigue markings at failure origin [15].*

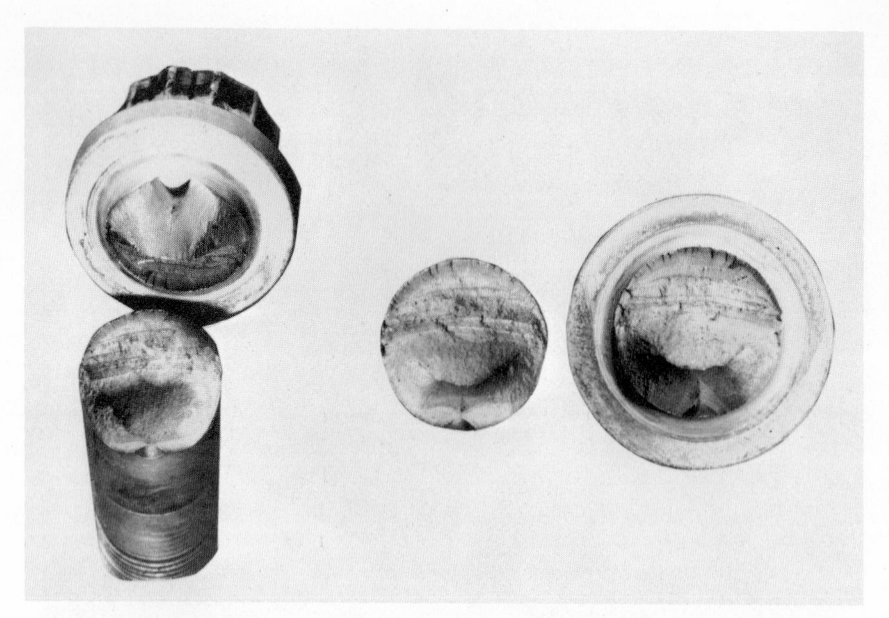

Figure 9.34 Fatigue failure of 12-point external wrenching EWB 18-8 bolt 0.500 in. diameter, 1-in. grip.

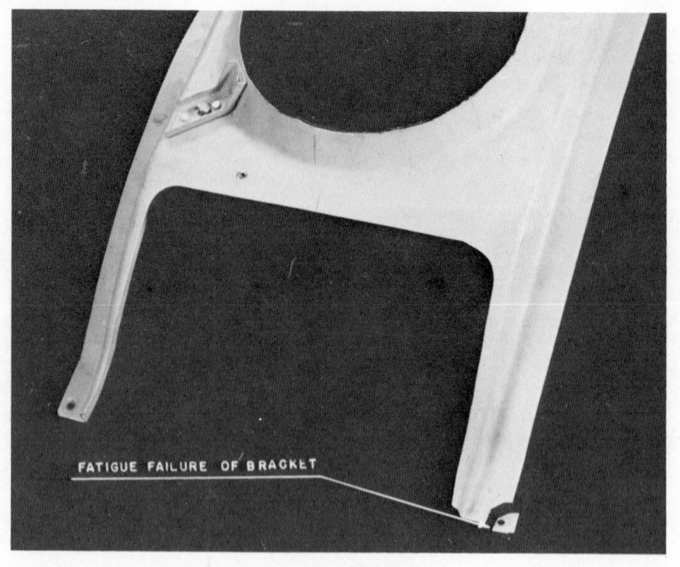

Figure 9.35 [16].

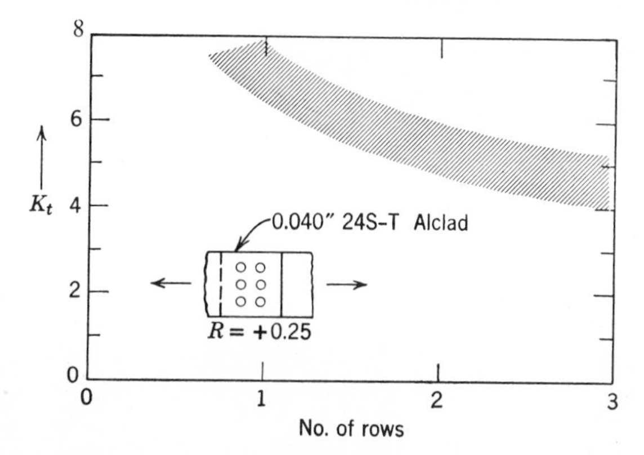

Figure 9.36 Variation of stress concentration factor with number of rows of flush dimple rivets [17].

The thickness of the materials being joined will have an effect on the stress concentration factor of the joint and, therefore, on the fatigue life. The variation of the stress concentration factor with material thickness is shown in Figure 9.37 for spot-welded joints.

Bolted, riveted, and spot-welded joints have different modes of failure under high and low loads. High loads that cause short-life failure will produce failure of the fastener, as shown in Figure 9.38 and Figure 9.39 for riveted and spot-welded joints, respectively. Fatigue failure at low loads and long lifetimes will occur by cracking of the sheet material along the edges of the fasteners, as shown in the same pictures. This is also illustrated in the *S-N* curves shown in Figures 9.40 and 9.41.

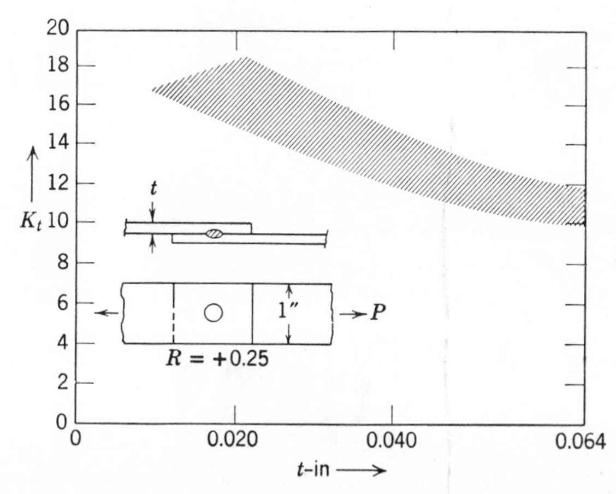

Figure 9.37 Variation of stress concentration factor with skin thickness for spot-welded lap joints [18].

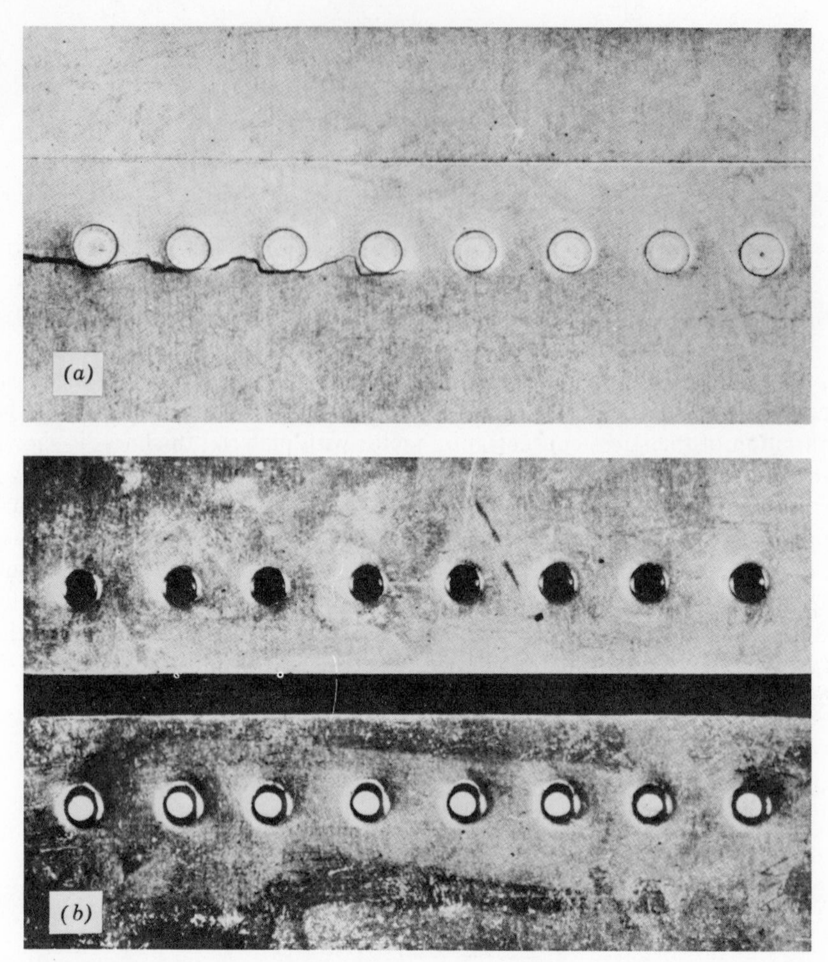

Figure 9.38 A "notch" effect in failures of a riveted lap joint in aluminum-alloy sheets [19]: (a) mode of fatigue failure at long life under low loads; (b) mode of fatigue failure at short life under high loads.

Design and manufacturing techniques have a pronounced influence on the fatigue life of riveted and bolted components. Figure 9.42 shows that the influence of the component material is overshadowed by the dimpling procedure used for flush-riveted joints. Even such a factor as whether dimpling is performed before or after heat treatment can have a large influence on the fatigue life of joints, as shown in Figure 9.43.

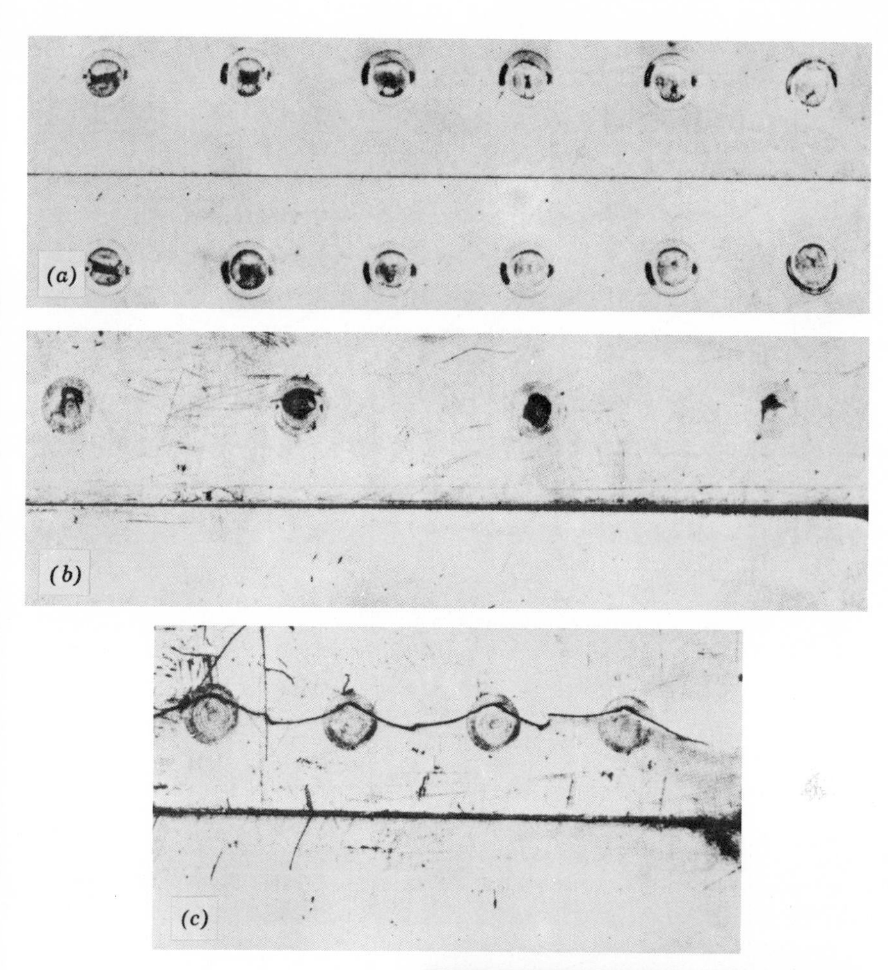

Figure 9.39 Typical failures of spot-welded lap joints [20]: (a) specimen showing shear type failure through spots; (b) specimen illustrating failure by "pulling a button"; (c) specimen showing propagation of fatigue cracks across sheet; (d) specimen showing section through spotweld that failed in fatigue.

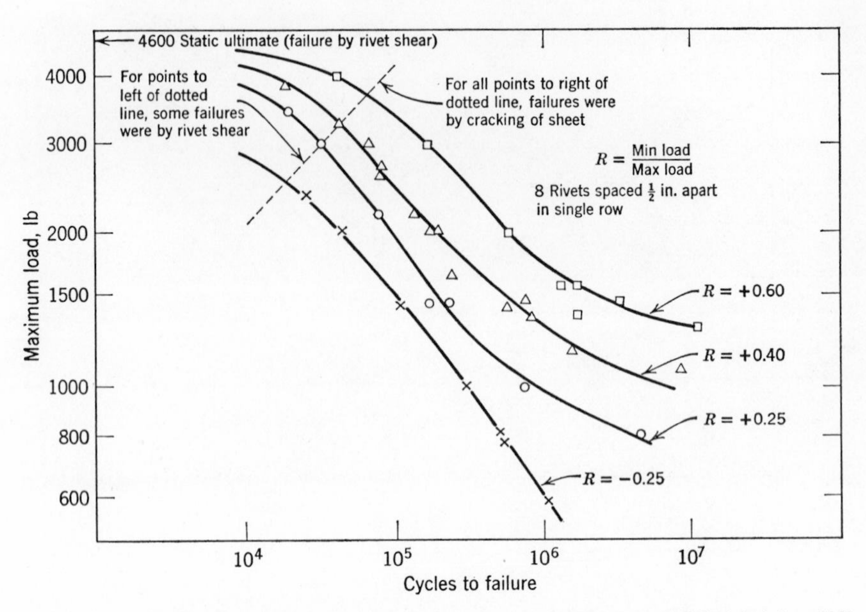

Figure 9.40 Direct-stress fatigue strengths of riveted lap joints of 0.040 in. 24S-T Alclad sheet [19].

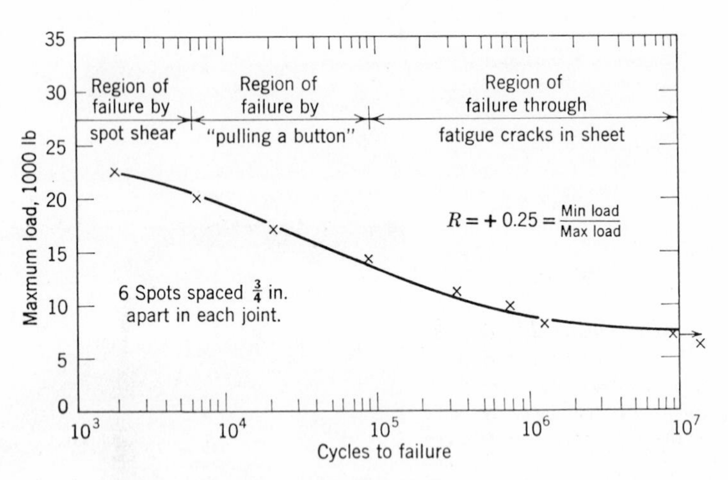

Figure 9.41 Load-life curve for spot-welded joint in 0.040 in. 24S-T Alclad sheet, showing approximate lifetime regions for each type of failure [20].

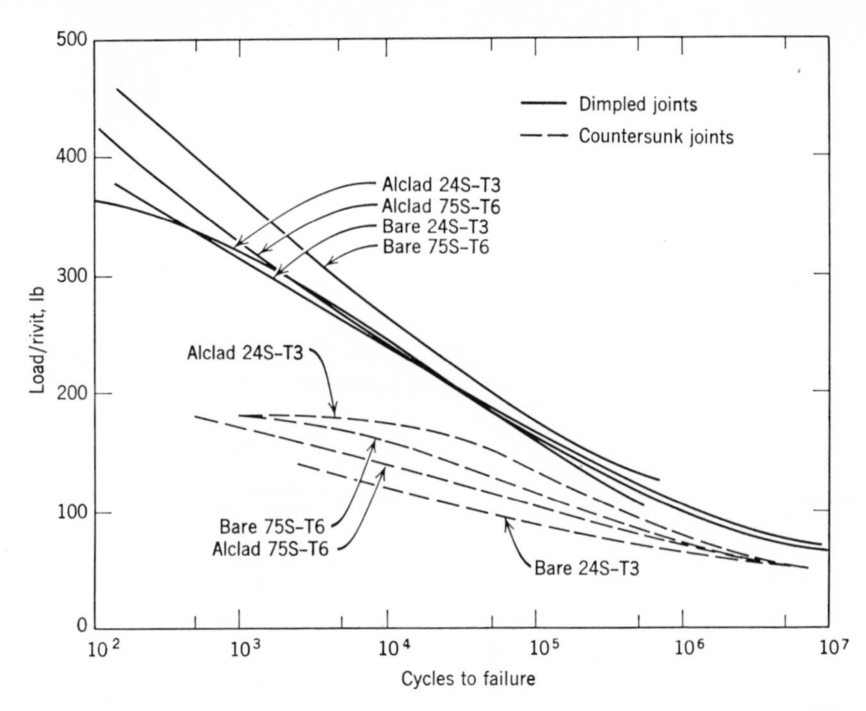

Figure 9.42 Comparison of lap joints of 24S-T3 and 75S-T6 strips 0.032 in. thick [21].

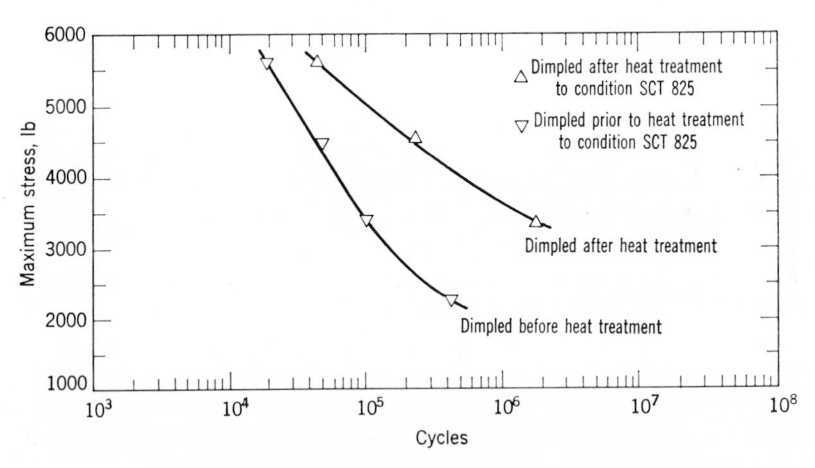

Figure 9.43 S-N curves for AM 350 riveted joint fatigue tests [2].

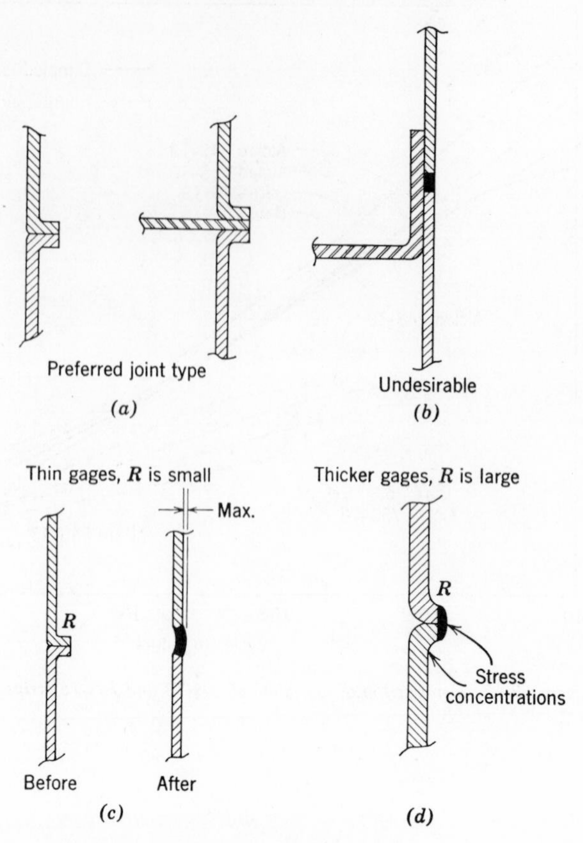

Figure 9.44 Welded joint details [4].

9.4.3 *Fusion-Welded Joints*

Fusion-welded joints will have stress concentrations because of discontinuities of component shape and also of the material itself. A butt-welded joint will have the least geometrical discontinuity; however, the difference between the parent material and that of the weld bead will frequently be sufficient to reduce the fatigue strength of the joint appreciably.

The influence of joint shape is illustrated by the burn-down flange weld joint shown in Figure 9.44. This type of joint can be used successfully with thin gauge sheet, but will create an undesirable stress concentration because of the larger flange radius when used with thick gauge sheet.

Carelessness in making a fusion weld can result in notches and other undesirable defects in the weld bead. Such defects can cause premature fatigue failure.

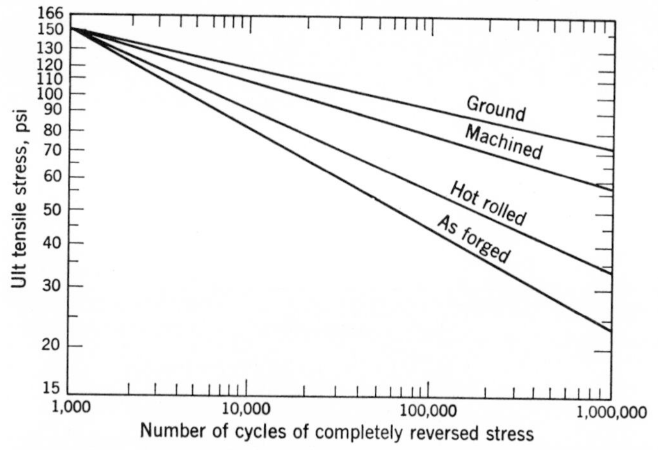

Figure 9.45 Surface roughness effect [9]. *Example of 166,000* UTS *steel.*

9.5 SURFACE FINISH

The roughness of the surface of a component and the metallurgical nature of the surface can have a considerable effect on the fatigue life. The effect of surface roughness, such as a smooth ground surface or a rougher machined surface, is shown in Figure 9.45. This figure also shows the detrimental effects of not removing the as-forged or as-rolled surface resulting from hot working or hot shaping of the material.

The reduced fatigue strength of as-forged and hot rolled surfaces is frequently due to oxidation or decarburization which will lower the strength of the surface material. Such surfaces should never be specified for highly loaded parts, but should be removed by machining. Machining and grinding operations may produce residual tensile stresses in the surface layers of a component. If such stresses are sufficiently high, premature fatigue failure may result. The effect of such detrimental stresses may be minimized or eliminated by shot peening or surface rolling [22]. The adverse effect of grinding and the beneficial effect of shot peening are shown in Figure 9.46.

A number of plated finishes are used to protect metal surfaces from corrosion. Fatigue failures due to corrosion may be eliminated or minimized by such platings, but in many cases the plating will themselves have an adverse effect on the fatigue properties of the material to which they are applied [22,24]. The effects of several electroplated coatings on the fatigue properties of a low alloy steel are shown in Figure 9.47. The detrimental effect of these coatings, like that of machining and grinding, can be reduced by shot peening or surface rolling, as shown in Figure 9.48.

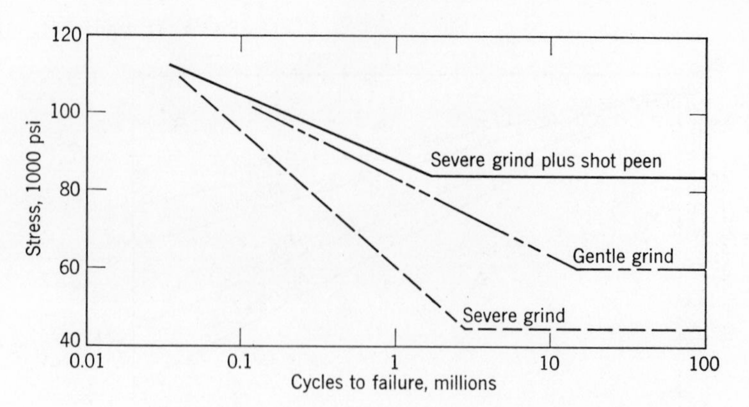

Figure 9.46 Effects of grinding and shot peening [23].

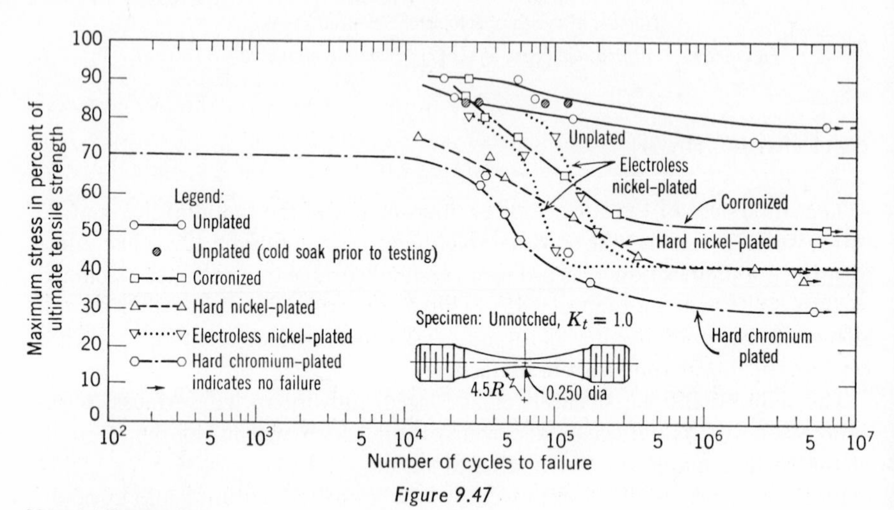

Figure 9.47

Material Properties:

Standard tensile test specimen	Static tension test of plated fatigue test spectrum
F_{tu} = 172,300 psi	Corronized = 177,700 psi
F_{ty} = 168,300 psi	Hard Ni plated = 176,100 psi
El. = 13.2 percent	Electroless Ni = 182,100 psi
R.A. = 53.9 percent	Chromium plated = 162,400 psi

Test Conditions:

Machine: Sonntag SF-20-U
Type of loading: axial
Stress ratio: $R = +0.02$
Speed: 1500 cpm
Temperature: room temperature

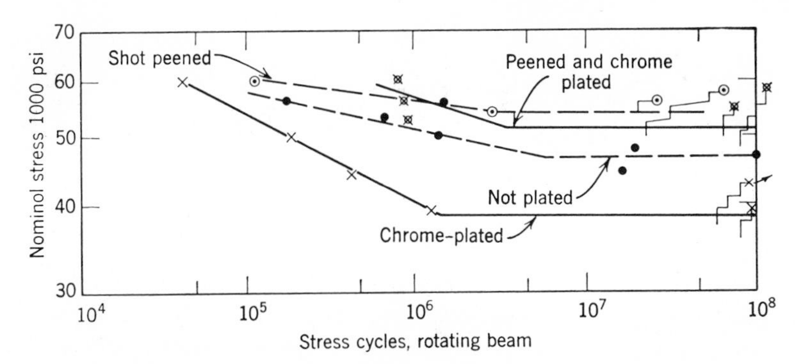

Figure 9.48 Fatigue strength of chromeplated steel, peened and unpeened [22,24].

9.6 ELEVATED TEMPERATURE EFFECTS

Elevated temperatures can produce dramatic effects on the fatigue properties of materials. These changes are due to a number of causes. The strength of the material is reduced at elevated temperatures. Metallurgical changes in the material may cause it to become soft or may embrittle it. The strengthening effects due to strain hardening become less as the temperature increases and disappear when the material is heated to approximately one-half of its absolute melting temperature. At this temperature dislocation barriers break down and diffusion processes become predominant in controlling material deformation and failure.

Elevated temperature testing is more difficult and involves more elaborate equipment than tests conducted at room temperature, and there are more variables which must be evaluated if useful information is to be obtained for design use. The rapid increase in aircraft speeds and the use of high-performance jet turbine and rocket engines have made it necessary to obtain fatigue information at increasingly higher temperatures. Aerodynamic heat of high speed aircraft, such as the hypersonic *X*-15 rocket research plane, produces temperatures exceeding 1200°F on the nose and other leading edges. Re-entry temperatures of spacecraft such as Gemini and Apollo will experience temperatures exceeding 2000°F. Fatigue testing at temperatures of 2000 to 4000°F requires the development of new techniques and equipment if the investigator is to obtain meaningful data.

Even simple static test fixtures for elevated temperature use, such as the shear and compression test fixture shown in Figure 9.49, require the use special heat resistant alloys in the construction of the test equipment, specially designed heating devices, and elaborate load, strain, and temperature sensing instrumentation. Elevated temperature tests of large components,

Figure 9.49

Figure 9.50 High temperature test of wing leading edge structure.

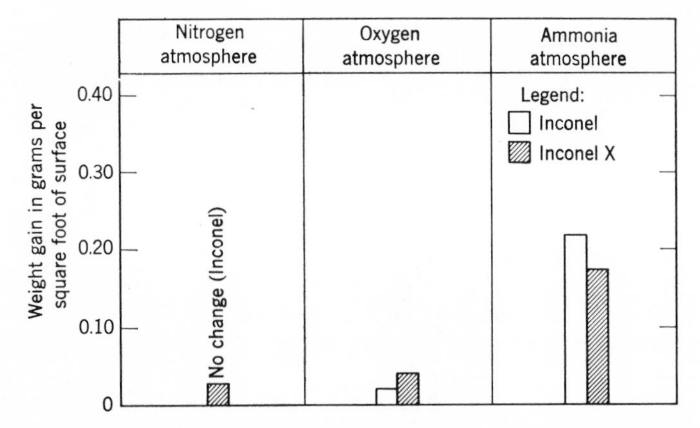

Figure 9.51 Weight change of Inconel and Inconel X due to exposure in various atmospheres for 1 hr at 1300°F [25].

such as the wing leading edge shown in Figure 9.50, require very elaborate heating and loading devices.

Factors which have little influence at room temperature can have a large effect on fatigue strength at elevated temperature. Some of these factors are corrosion due to accelerated chemical attack or oxidation of the heated metal, and chemical action of the rocket propellants and their combustion products. Figure 9.51 shows the effect of exposure of Inconel and Inconel X to various atmospheres at elevated temperature. In order to conduct tests at temperature and in such atmospheres, unusual apparatus may be required. Test set-ups for determining stress-rupture and fatigue properties at elevated temperature in ammonia atmosphere are shown in Figures 9.52 and 9.53, respectively.

The interaction of the variables which affect elevated temperature fatigue properties of structural components make extrapolation of information very difficult. Accuracy of data intended for design use must be verified by testing representative designs and joint configurations under anticipated service conditions. The effect of temperature on the fatigue strength of Inconel X sheet shown in Figure 9.54 would seem to indicate that the use of room temperature fatigue properties for higher temperature applications would be conservative. However, the results shown in Figure 9.55 for fatigue tests of fusion butt-welded Inconel X sheet demonstrate that such practice could result in premature fatigue failure at lifetimes less than 100,000 cycles.

The controlling phenomena in metal failure undergo a transition from strain-hardening controlled to diffusion controlled at temperatures near or above fifty percent of the absolute melting temperature of the material. Figure 9.56 shows that failure of butt-welded joints in Inconel X sheet at

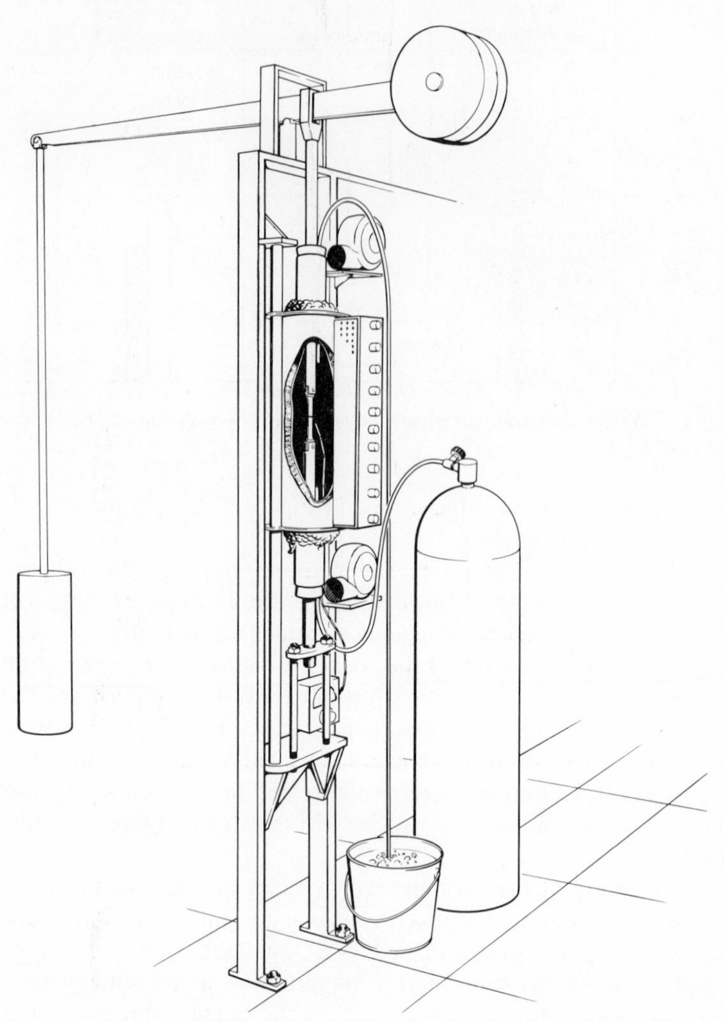

Figure 9.52 Arrangement of ammonia atmosphere equipment for stress-rupture tests [25].

1000°F is fatigue critical, but at 1330°F failure of Inconel X sheet material is critical in stress-rupture, not fatigue, as in Figure 9.57.

Another very important phenomenon in elevated temperature fatigue testing is the effect of speed of testing. This includes both the frequency and also the rate of loading. This is illustrated in Figure 9.58 which shows the reduction in fatigue strength at elevated temperature and also the further reduction in fatigue strength at slow testing speeds.

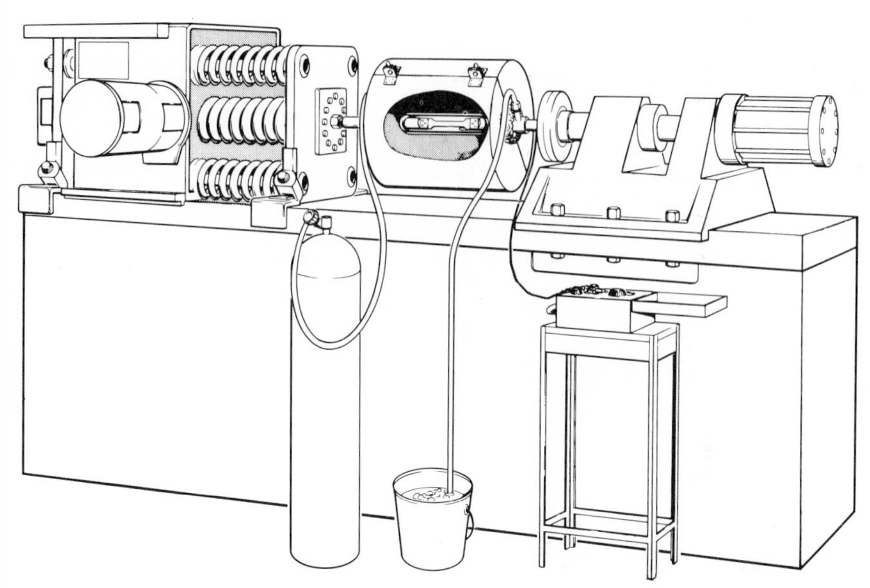

Figure 9.53 Arrangement of ammonia atmosphere equipment for fatigue tests [25].

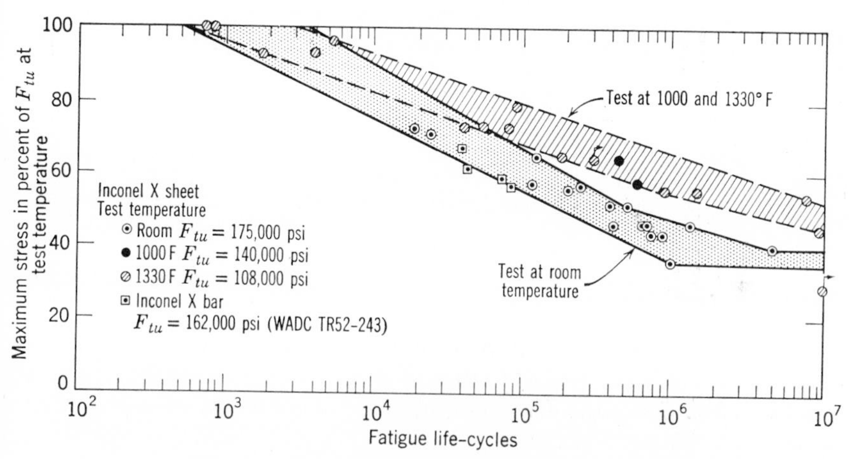

Figure 9.54 Effect temperature on the fatigue strength of Inconel X sheet [25].

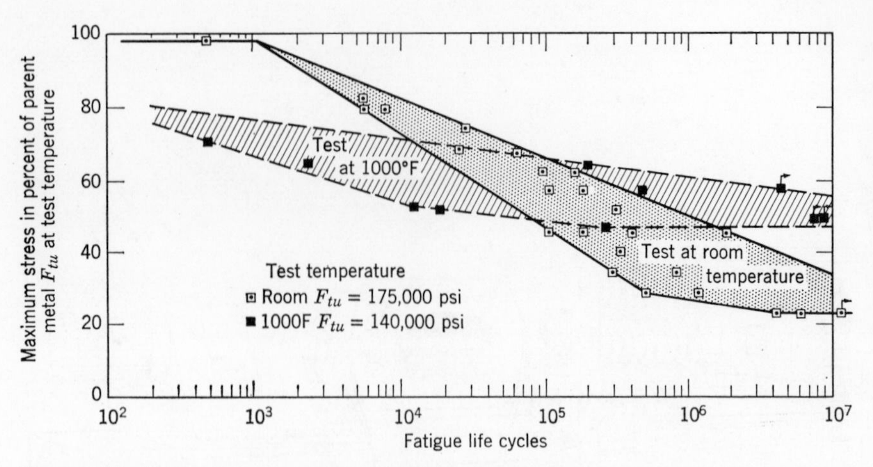

Figure 9.55 Effect of temperature on the fatigue strength of fusion butt-welded Inconel X sheet [25].

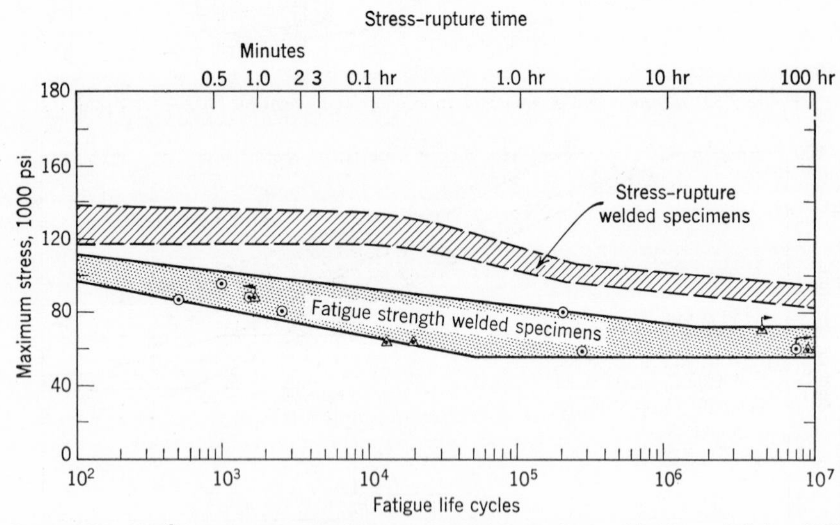

Figure 9.56 1000°F stress-rupture and fatigue strength fusion butt-welded Inconel X sheet [25].

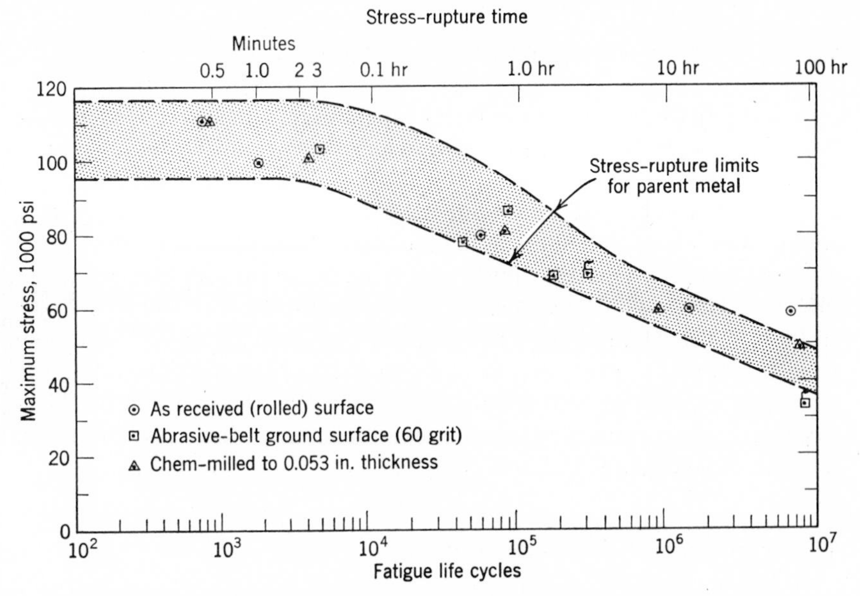

Figure 9.57 1330°F stress-rupture and fatigue strength of Inconel X sheet [25].

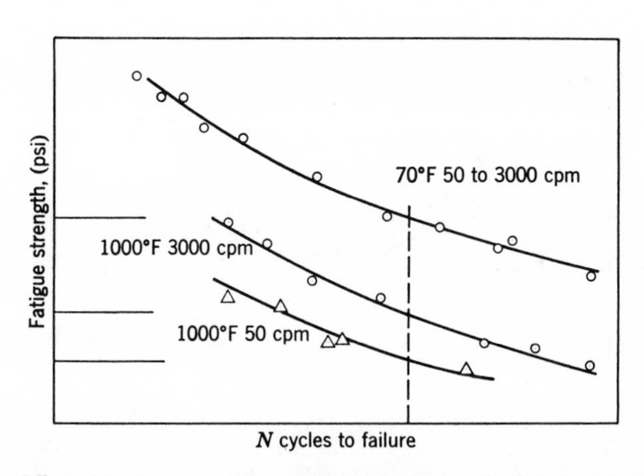

Figure 9.58 Effect of temperature and rate of cycle loading on the fatigue strength of materials [26].

9.7 SUPERSONIC TRANSPORT APPLICATIONS

In addition to the problems of high temperature, flight at speeds such as will be encountered in the supersonic transport create other types of fatigue problems. One of these problems is acoustic fatigue. The extremely powerful engines that will be used on the supersonic transport create intense levels of sound. During take-off and landing the energy from the sound waves will be reflected from the concrete runways onto the under surfaces of the wing and fuselage. These sound waves can set up very-high-frequency vibration in such surfaces, causing fatigue failure of the outer skin, or in the case of honeycomb panels the internal structure. In order to obtain adequate fatigue life, such structures exposed to high acoustic forces must be strengthened by thicker surfaces and properly positioned interior members which can absorb and dampen the induced vibrational energies [36].

9.8 EFFECT OF NOTCH SHARPNESS

Specimen configurations which are used for fatigue testing should be representative of the component designs for which the data are to be used. It has been common practice to use test bars with standard notch configurations. The sharpness of the notch, or severity of stress concentration, should compare with the degree of stress concentration which will be built into the

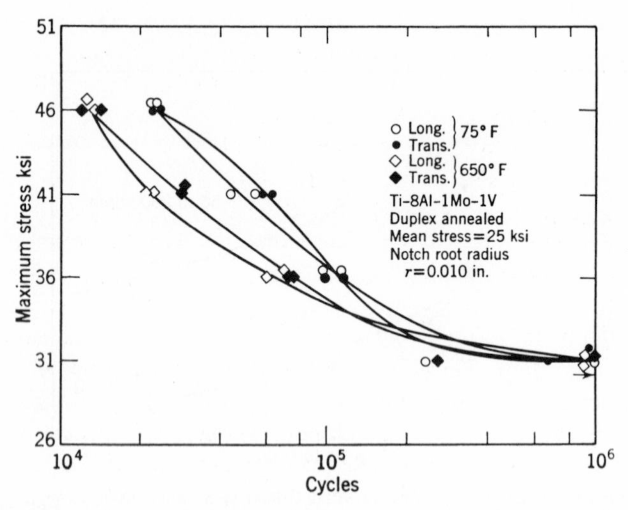

Figure 9.59 Effect of temperature and grain direction on the fatigue strength of notched titanium alloy specimens [3].

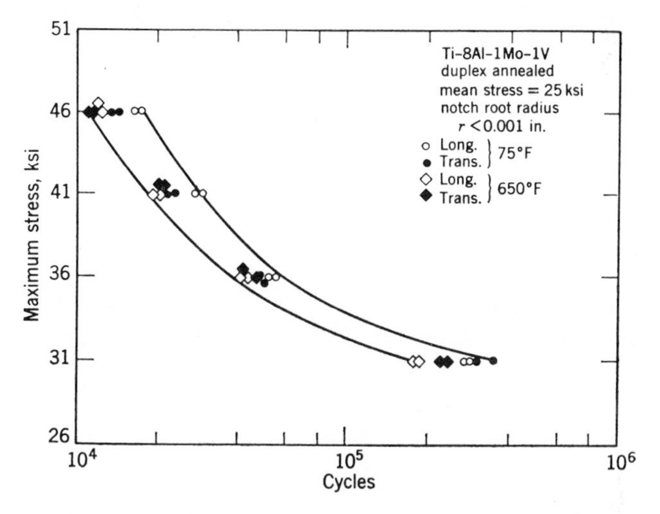

Figure 9.60 Sharp notch masks effects of grain direction and temperature on fatigue strength of notched titanium alloy specimens [3].

actual part being considered. If the fatigue test specimens have too sharp a notch with a very high stress concentration the effect of other important variables may be masked by that of the notch alone. Figure 9.59 shows notch fatigue data at room and at elevated temperature for specimens with a notch root radius of 0.010 in. The effects of longitudinal and transverse grain direction and of the two temperature conditions are clearly shown. When an extremely sharp notch with a root radius of less than 0.001 in. was used, as shown in Figure 9.60, the differences due to grain direction and temperature are not evident, being overshadowed by the effect of the very sharp notch.

Materials which have the same ultimate tensile strength will frequently have different fatigue properties. An example of this is shown in Figure 9.61, where the fatigue properties of a titanium alloy and a family of low-alloy steels heat treated to approximately the same strength level are compared. The two materials have greatly differing unnotched fatigue strengths. Their notched fatigue properties are much more comparable, and would more likely be the same if a sharper notch were to be used. A somewhat similar condition is shown in Figure 9.62 which compares the unnotched fatigue properties of two high strength aluminum alloys. When this data is plotted in terms of percent of ultimate tensile strength, the alloy with the high ultimate tensile strength will appear to have very much poorer fatigue properties. However, when the data is plotted as shown in terms of actual stress, it can be seen that, except at short lifetimes, the fatigue strengths of the materials are comparable. It is thus possible to take advantage of the higher tensile strength by use of the high-strength alloy without sacrifice of fatigue strength.

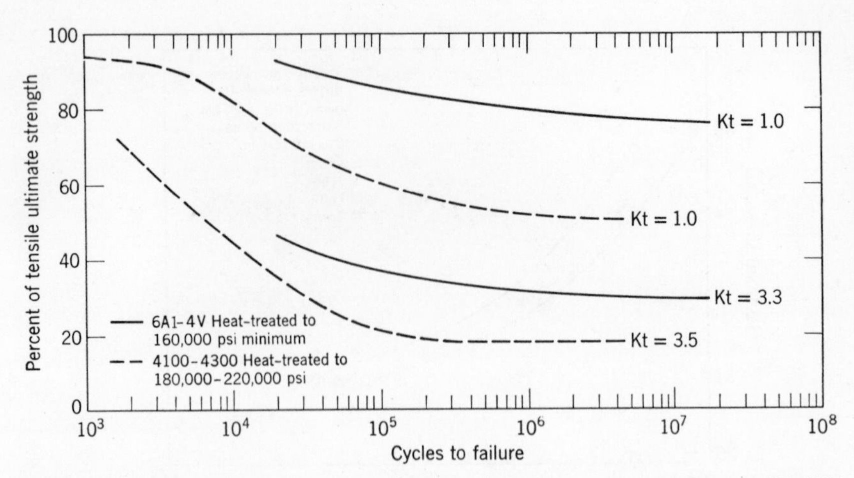

Figure 9.61 A comparison of the fatigue properties of 6Al-4V titanium alloy bar and SAE 4100-4300 low-alloy steel bar [27].

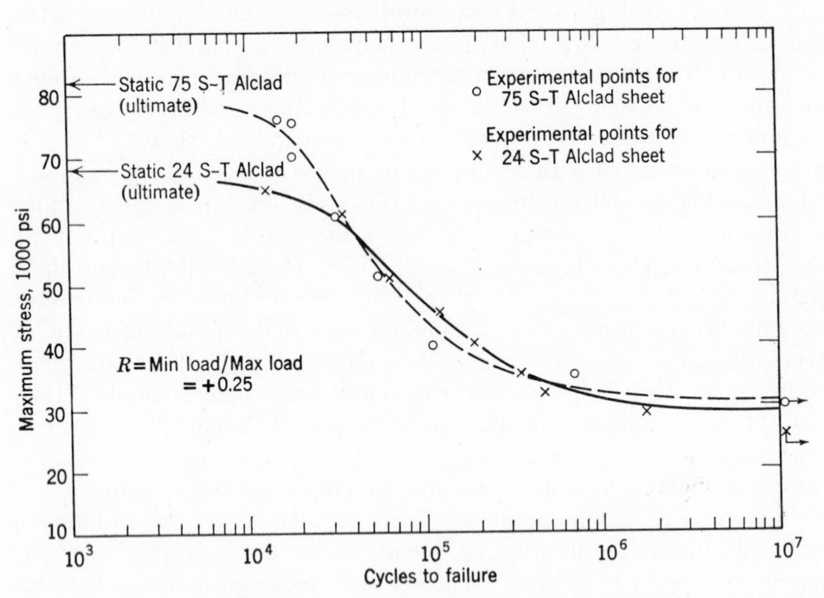

Figure 9.62 Axial-loading fatigue curves. Note that at 10^4 cycles the 75S-T appears to have higher fatigue strength than 24 S-T. At longer lifetimes the S-N curves coincide within experimental error of this test [28,29].

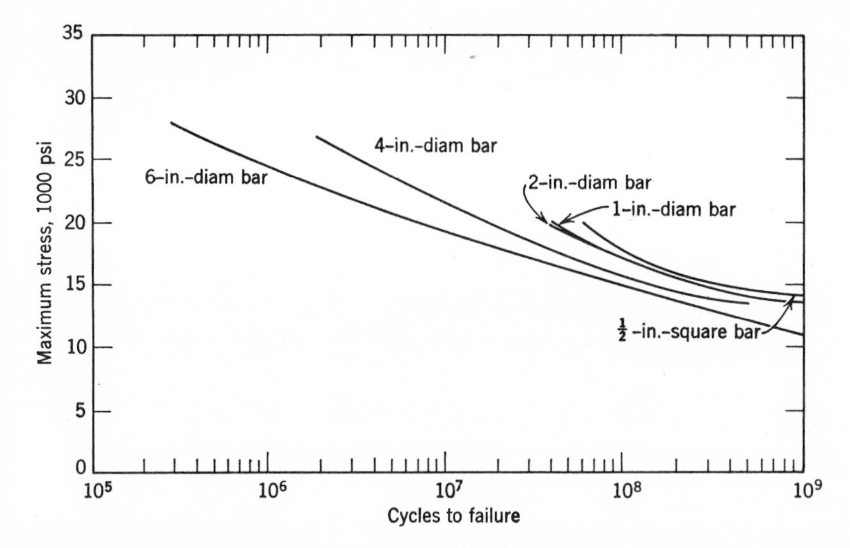

Figure 9.63 Rotating-beam fatigue strengths of specimens cut from bars forged to successively smaller diameters [30].

9.9 SIZE EFFECTS

Fatigue properties will vary with the size of a material. Small diameter bars will have much higher fatigue properties than bars of very large diameter as shown in Figure 9.63. The additional working of the material which is required to produce the smaller diameter pieces results in a finer grain structure and fragmentation and dispersion of inclusions which can reduce fatigue life. A similar condition is also shown in Figure 9.64 where specimens from the surface and from the center of a 9-in. billet have considerably different fatigue strengths. Fatigue data obtained with small-diameter laboratory coupon-type test specimens must be used with caution in the design of large components. Figure 9.65 shows that full-size structural members can have a very much reduced fatigue strength than small notched specimens which are machined from the same material.

9.10 SPECIAL TESTS

The design of large fuselage structures and pressurized tanks and large motor casings for solid propellant rockets has created the need for fatigue data obtained under multiaxis stress conditions. Specimens used in recently

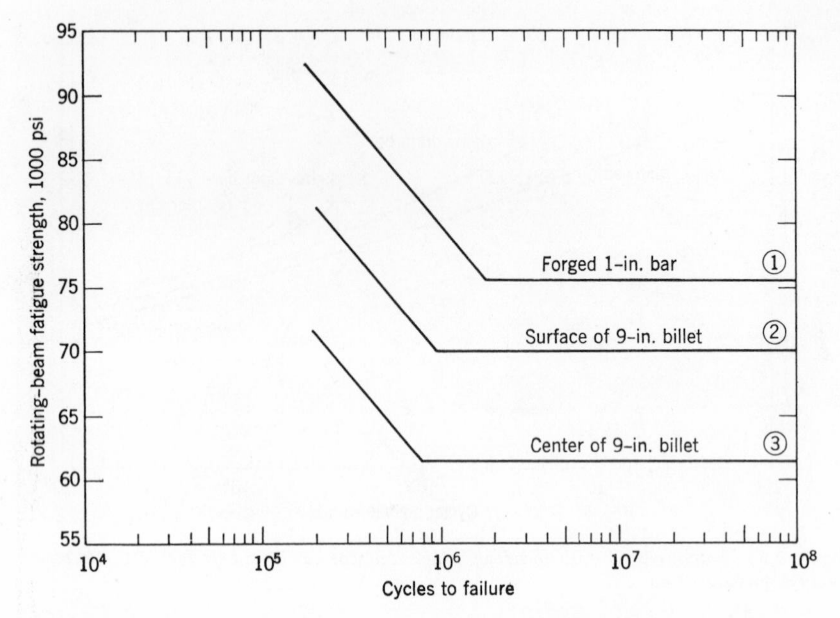

Figure 9.64 Fatigue strengths of specimens of SAE 4340 steel, cut from large billet [31].

SPEC.	UTS	YS
1	150	139
2	137	127
3	140	127

All specimens machined, oil quenched
from 150°F, drawn at 1200°F.

published tests conducted under triaxial fatigue stress conditions were made in the form of a six-armed cross with the arms mutually perpendicular to each other [33]. The results of triaxial tension fatigue tests with these specimens made of three constructional steels are shown in Figure 9.66.

If the results shown in Figure 9.66 are replotted in terms of the stress in percent of the ultimate tensile strength of the materials and are then compared with uniaxial tension fatigue data for similar materials, such as that in Figure 9.1, it will be seen that there is little difference between the triaxial tension fatigue data (which were for notched specimens) and the uniaxial tension fatigue data.

It was reported that the triaxial test specimens usually failed first in the short transverse direction of the material. When the specimens were then further tested under biaxial tension fatigue conditions, they were observed to fail in the long transverse direction of the material. The transverse fatigue properties of air-melted plain carbon and low-alloy steels can be as low as

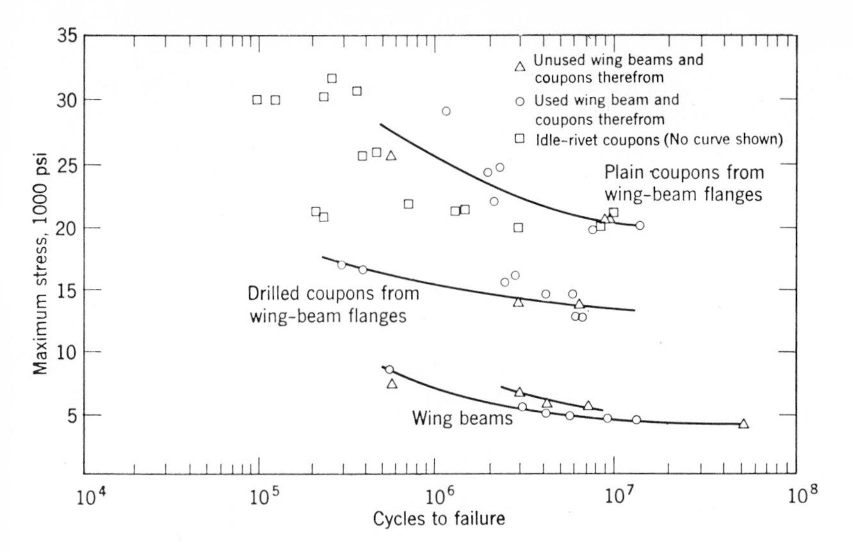

Figure 9.65 S-N curves for wing-beam specimens and coupons [32].

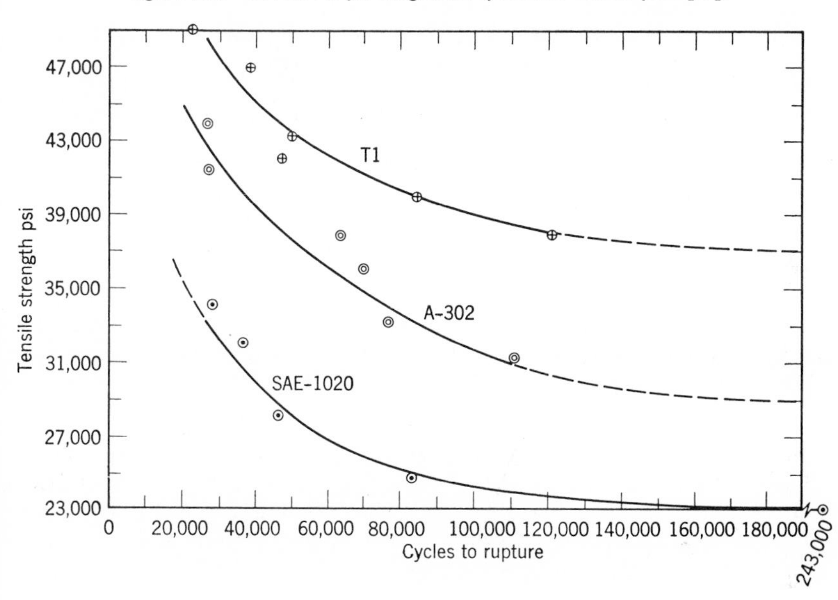

Figure 9.66 Triaxial tensile fatigue testing [33].

Material	Triaxial Fatigue limit (10^5 cycles) psi	Uniaxial Yield strength psi	Uniaxial Ultimate strength psi
SAE 1020	24,000	36,000	54,000–70,000
A-302	31,500	50,000	80,000–100,000
TI	39,200	90,000	105,000

60% of the longitudinal fatigue properties. These data would indicate that, for air-melt steels, the uniaxial fatigue properties may be used to predict fatigue failure under multiaxial stress conditions. This conclusion is also reported biaxial fatigue tests of similar materials where it was observed that a comparison with cantilever bar tests indicated that fatigue life under equibiaxial strains can be predicted from uniaxial fatigue tests [37].

REFERENCES

[1] M. H. Weisman and M. H. Kaplan, "The Fatigue Strength of Steel Through the Range from ½ to 30,000 Cycles of Stress," *Proc. ASTM*, **50**, 649 (1950).

[2] M. H. Weisman, J. Melill and T. Matsuda, "Uni-Directional Axial Tension Fatigue Tests of Beryllium Copper and Several Precipitation Hardening Corrosion-Resistant Steels, Fatigue of Aircraft Structures," *ASTM STP* **203**, 47 (1957).

[3] V. Weiss and A. Roy, "Further Material Evaluation for Supersonic Transport Aircraft," Syracuse University Research Institute Report No. MET. E. 873-6312-F, New York, December, 1963.

[4] M. H. Weisman, "Tank Design for Prevention of Fatigue Failure," North American Aviation, Inc., Report No. NA-49-68, Los Angeles, February 16, 1949.

[5] M. H. Weisman, "Investigation of Failures of Jet Engine Exhaust Cones and Tailpipes," North American Aviation, Inc., Report No. NA-48-237, Los Angeles, February 25, 1948.

[6] R. A. MacGregor, W. S. Burn and F. Bacon, "The Relation of Fatigue to Modern Engine Design," *Trans. North East Coast Inst. Engr. Shipbuilders*, **51**, 161, D100–D136 (1935).

[7] R. E. Peterson and A. M. Wahl, Fatigue of Shafts at Fitted Members, With a Related Photoelastic Analysis, *Trans. ASME*, **57**, 1935.

[8] H. J. Grover, S. A. Gordon and L. R. Jackson, "Fatigue of Metals and Structures," NAVAER 00-25-234, Bureau of Aeronautics, Department of the Navy (Superintendent of Documents), Washington, D.C., 1954.

[9] M. H. Weisman, "The Design of Rod Ends and Fittings for Actuating Cylinders," *Appl. Hydraulics*, **10**, 20 (1949).

[10] W. E. Duckworth and G. H. Walter, "Fatigue of Plain Bearings," Institute of Mechanical Engineers and ASME International Conference on Fatigue of Metals, London, September 10–14, 1956.

[11] H. W. Hayden, W. G. Moffatt, and J. Wulff, "The Structures and Properties of Materials, Vol. III, Mechanical Behavior," John Wiley, New York, 1965.

[12] A. J. Kennedy, "Processes of Creep and Fatigue in Metals," John Wiley, New York, 1963.

[13] E. H. Spaulding, "Detail Design for Fatigue in Aircraft Wing Structures," *Metal Fatigue*, McGraw-Hill, New York, 1959 pp. 325–354.

[14] H. J. Grover, "Allowance for Stress Concentration in Design to Prevent Fatigue," Institute of Mechanical Engineers and ASME International Conference on Fatigue of Metals, London, September 10–14, 1956.

[15] M. H. Weisman and G. Martin, Examination of a Lower L. H. Bolting Bar from the F-86 Wing Structural Fatigue Test Program, North American Aviation, Inc., Report No. TFD-64-370, Los Angeles, June 1, 1964.

[16] M. H. Weisman, Investigation of Current Fatigue Problems, North American Aviation, Inc., Report No. NA-47-504, Los Angeles, May 20, 1947.
[17] H. W. Russell, L. R. Jackson, H. J. Grover, and W. W. Beaver, "Fatigue Strength and Related Characteristics of Aircraft Joints, I. Comparison of Spot-Weld and Rivet Patterns in 24S-T Alclad Sheet—Comparison of 24S-T Alclad and 75S-T Alclad," NACA ARR 4F01, WR-W-56, December, 1944.
[18] W. F. Hess, R. A. Wyant, F. V. Winsor, and H. C. Cook, "An Investigation of the Fatigue Strength of Spot Welds in Aluminum Alloy Alclad 24S-T," NACA OCR 4H23, OCR Progress Report 15, August, 1944.
[19] H. W. Russell, L. R. Jackson, H. J. Grover, and W. W. Beaver, "Fatigue Strength and Related Characteristics of Aircraft Joints: II. Fatigue Characteristics of Sheet and Riveted Joints of 0.040-in. 24S-T, 75S-T, and R303-T275 Aluminum Alloys," NACA Tech. Note 1485, February, 1948.
[20] H. W. Russell, L. R. Jackson, H. J. Grover, and W. W. Beaver, "Fatigue Characteristics of Spot-Welded 24S-T Aluminum Alloy," NACA ARR No. 3F16, June, 1943.
[21] D. M. Howard and F. C. Smith, "Fatigue and Static Tests of Flush Riveted Joints," NACA Tech. Note 2709, 1952.
[22] H. O. Fuchs and E. R. Hutchinson, "Shot Peening," *Machine Design*, **30,** 116 (1958).
[23] L. P. Tarasov and H. J. Grover, "Effects of Grinding and Other Finishing Processes on the Fatigue Strength of Hardened Steel," *Proc. ASTM*, **50,** 668 (1950).
[24] J. O. Almen, "Fatigue Loss and Gain by Electroplating," *Product Eng.*, **22,** 109 (1951).
[25] M. H. Weisman, Elevated Temperature Stress-Rupture and Fatigue Properties of Inconel and Inconel X in Ammonia Atmosphere, Paper No. 61-AV-21, ASME Aviation Conference, Los Angeles, March, 1961.
[26] R. H. Christensen, Some Fatigue Design Requirements for Future Air and Space Vehicles, "Proceedings of the Symposium on Fatigue of Aircraft Structures, 11–13 August 1959," WADC Tech. Report TR-59-507, pp. 776–811, October, 1959.
[27] G. A. Fairbairn, An Appraisal of the Fatigue Characteristics of Materials for High Performance Air Vehicles, "Proceedings of the Symposium on Fatigue of Aircraft Structures," August 11–13, 1959, WADC Tech. Report TR-59-507, pp. 699–721, October, 1959.
[28] L. R. Jackson, H. J. Grover, and R. C. McMaster, Survey of Available Information on the Behavior of Aircraft Materials and Structures Under Repeated Loads, Report to The War Metallurgy Committee on Survey Project No. SP-27, OSRD, Washington, D.C., December 31, 1945.
[29] L. R. Jackson, H. J. Grover, and R. C. McMaster, Advisory Report on Fatigue Properties of Aircraft Materials and Structures, War Metallurgy Committee, OSRD No. 6600, Serial No. M-653, Office of Scientific Research and Development, Washington, D.C., March 1, 1946.
[30] T. T. Oberg and J. B. Johnson, Mechanical Properties of Forged Experimental Aluminum-Alloy Bars, AAF Materials Laboratory Report, M-56-2742, October 23, 1935.
[31] J. B. Johnson, *Airplane Welding and Materials*, The Goodheart Willcox Company, Inc., Chicago, 1941.
[32] W. C. Brueggeman, P. Krupen, and F. C. Roop, "Axial Fatigue Tests of Ten Airplane Wing-Beam Specimens by the Resonance Method," NACA Tech. Note 959, December, 1944.
[33] G. Welter and J. A. Choquet, "Triaxial Tensile Stress Fatigue Testing," *Welding J.* **42,** 565-s (1963).
[34] S. S. Manson and M. H. Hirschberg, "Fatigue Behavior in Strain Cycling in the Low- and Intermediate-Cycle Range," *Fatigue—An Interdisciplinary Approach*, Syracuse University Press, Syracuse, New York, 1964 pp. 133–178.

[35] L. J. Ebert, F. T. Krotine, and A. R. Troiano, "Why Case Hardened Components Fracture," *Metal Prog*. **90,** 61 (1966).

[36] W. J. Trapp and D. M. Forney, Jr., "A Review of Acoustical Fatigue," *Fatigue—An Interdisciplinary Approach*, Syracuse University Press, Syracuse, New York, 1964, pp. 261–283.

[37] K. D. Ives, L. F. Kooistra, and J. T. Tucker, Jr., "Equibiaxial Low-Cycle Fatigue Properties of Typical Pressure-Vessel Steels," Paper No. 65-MET-19, ASME Metals Engineering and Production Engineering Conference, Berkeley, Calif., June 9–11, 1965.

10

Fatigue Analyses

LOUIS YOUNG

The analytical and experimental techniques normally used to determine the service life of structural components on aircraft and helicopters are examined with two basic objectives. The first is the statistical levels of assurance involved in the evaluation of structural life. The other is the application of these techniques to provide the levels of safety, reliability and economic performance that are necessary to minimize the occurrence of fatigue cracking during the service life of a flight vehicle.

To achieve the first objective, a large body of fatigue test data was selected from the literature to evaluate the levels of assurance involved in calculating fatigue lives by conventional procedures. These data were also analyzed to provide statistical estimates of minimum test lives. Further investigations were made to show the effects on calculated life of progressive increases in the severity of the analytical definition used for the ground to air to ground transition between flight and ground loadings.

For the second objective concepts of design for fatigue reliability on rotating and nonrotating structural elements are discussed in terms of significant parameters, required fatigue test data, and interpretation of flight measured data in terms of this test data.

10.1 LEVELS OF ASSURANCE INVOLVED IN DETERMINING FATIGUE LIFE

Fatigue test data were selected from a literature survey [1–42] to determine the levels of assurance involved in the following analytical procedures.

1. Calculating the lives obtained by individual test specimens with conventional and reduced *S-N* data.

2. Statistically evaluating the minimum anticipated test life for a group of specimens tested under nominally similar conditions.

The application of these analytical procedures impose limitations on the selection of fatigue test data. For the first analytical procedure, both variable amplitude spectrum and constant amplitude *S-N* data were required for the same specimen configuration. In the analysis of minimum anticipated test lives, individual spectrum test groups had to have two or more specimens tested under essentially the same test conditions.

10.1.1 Test Data

For determining levels of assurance, 362 statistically independent groups of variable amplitude loading spectra tests were selected from the literature survey. These groups were distinguished by changes in specimen configurations, materials, test temperature, types of loading spectra, loading sequences or other test conditions. They contain individual results for 1809 coupon, joint and structural specimens. For 1245 of these specimens, both variable amplitude loading spectrum and constant amplitude *S-N* data were available. These particular specimens were used to determine the levels of assurance involved in calculating the test lives achieved with the concept of linearly cumulative fatigue life utilization. On the other hand, 338 groups of specimens were employed to statistically evaluate test life reduction factors.

Many of the selected spectrum test groups contained significant changes in mean or reference load level to represent the transition from ground to flight loadings and back to ground loadings in the test loading history. During some tests, this transition, commonly referred to as a ground-air-ground (GAG) cycle, was similated by a distinct test loading. In other tests, the GAG cycle occurred because of significant variations between the test applied flight and ground loadings. To account for the effects of GAG cycles the spectrum test results were separated into the following categories:

Category A: Contains 68 test groups with flight and ground loadings applied in significantly different stress ranges on a flight-by-flight basis. (See Figures 10.1*a* and 10.1*b*).

Category B: Contains 271 groups in which relatively simple variable amplitude loading with little or no variations in mean load were repeatedly applied by using a block loading sequence. (See Figure 10.1*c*).

Category C: Contains 23 test groups with flight and ground loadings periodically applied in blocks representing a large number of flights. The corresponding number of transition cycles between flight and ground loadings were generally grouped together at the end of each block. See Figure 10.1*d*.

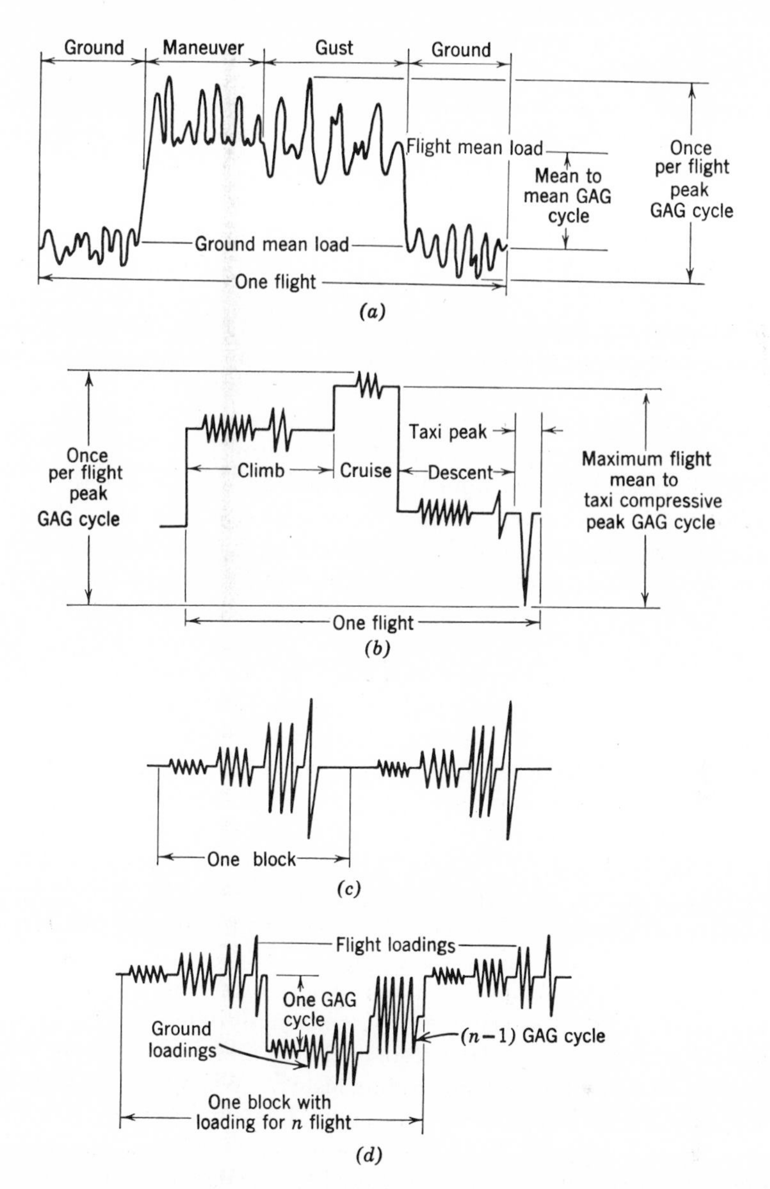

Figure 10.1 Typical spectrum test-loading sequences: (a) category A, flight-by-flight random loading sequence; (b) category A, flight-by-flight low-high ordered loading sequence; (c) category B, block loading sequence without GAG cycles; (d) category C, block loading sequence with GAG cycles.

10.1.2 Evaluation of Analytical Procedures

To provide a basis for evaluation of the significance of the analytical results obtained, some of the spectrum test data selected from the literature were employed to show the utility of life predictions using the conventional concept of linearly cumulative fatigue life utilization. As pointed out in [3], the other prediction methods investigated in the reference did not show any significant improvement over this concept when correlations were made between calculated and test lives. Such a concept is based on the use of *S-N* data obtained in constant amplitude loading tests. During this investigation the constant amplitude loading data were directly applicable to the specimen configurations used in the spectrum loading tests and were more completely defined than they generally are for the analytical evaluation of spectrum test results obtained on structural elements for flight vehicles.

Fatigue lives were calculated for the test applied loading histories by using the Palmgren-Miner concept of linear cumulative fatigue life utilization. *S-N* curves were derived during these calculations from the appropriate *S-N* diagram for the mean stress levels associated with a particular loading spectrum or multiple set of loading spectra used in the spectrum tests.

In addition, an assessment was made of the significance of making arbitrary reductions in the constant amplitude loading data to improve the reliability involved in using *S-N* curves to make conservative fatigue life predictions. During this assessment the varying stress levels for the appropriate *S-N* curves were reduced by 10, 20, 30, 40 and 50% before calculating fatigue lives in the conventional manner.

In all instances the amount of fatigue life utilization was calculated for all of the test applied loadings. When multiple sets of loading spectra with GAG cycles were analyzed, the largest calculated life utilizations came from the GAG cycles. Four types of GAG cycles were used to calculate test lives for category *A* tests where the GAG transition cycle occurs naturally on a flight-by-flight loading sequence without requiring an arbitrary definition of the GAG cycle for the test. However for the analysis of test results, arbitrary definitions are often made. In category *A*1, the arbitrary definition of the GAG cycle was based on the test loading history. When both the flight and ground loadings were applied at significantly different reference mean stress levels, as shown in Figure 10.1*a*, a mean-to-mean GAG cycle based on the maximum excursion in mean load was used to calculate life. A flight mean to once per flight taxi compressive loading peak, as shown in Figure 10.1*b*, was used to calculate category *A*1 lives for tests in which the flight loadings were cycled about fixed mean load levels with the ground loadings cycled as taxi compressive loading peaks.

For category *A*2, some of the calculated lives were based on the use of an

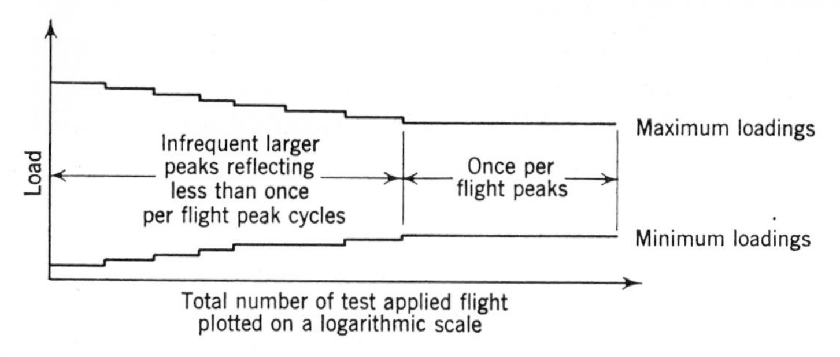

Figure 10.2 Spectrum of GAG cycles reflecting growth in loading magnitudes with test time.

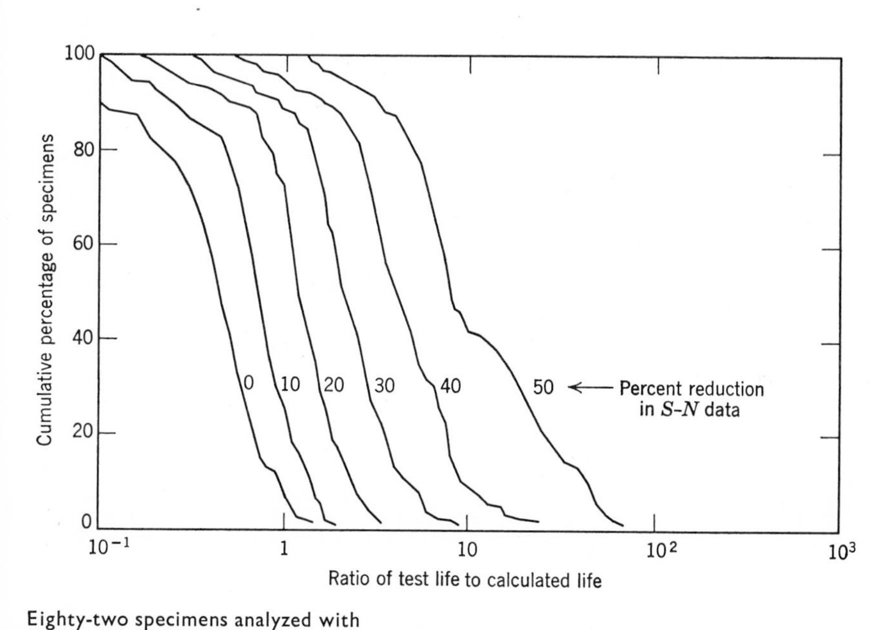

Eighty-two specimens analyzed with GAG cycle defined in terms of discrete flight mean to once per flight taxi compressive peak applied during tests; 36 specimens analyzed with arbitrary mean to mean GAG cycle; 79 specimens analyzed with arbitrary flight mean to once per flight taxi compressive peak GAG cycle.

Figure 10.3 Cumulative percentage of specimens equaling or exceeding specified ratios of test to calculated life-category AI-flight-by-flight loadings.

arbitrary once per flight peak GAG cycle of the type shown in Figures 10.1*a* and 10.1*b*. To complete category *A*2, these calculated lives were combined with similar calculations for tests conducted with a discrete test applied flight mean to one per flight taxi compressive peak GAG cycle.

A similar procedure was also followed in category *A*3 with a spectrum of GAG cycles being used when a discrete GAG cycle was not involved in the test loading history. The spectrum of GAG cycles reflects the magnitudes of all the infrequent less than once per flight loading peaks, as shown in Figure 10.2.

The fatigue lives calculated with both conventional and reduced *S-N* data were used to determine test to calculated life ratio for category *A*1, *A*2, *A*3, *B* and *C* tests. Cumulative distributions of test to calculated life are given on Figures 10.3–10.7 where conservatively calculated test lives are represented by test to calculated life ratios equal to or greater than unity. Comparisons

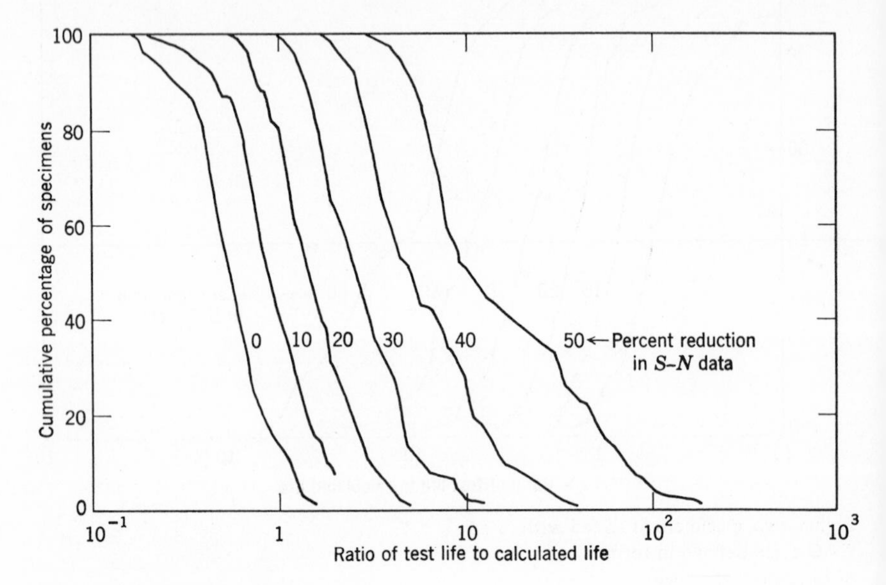

Eighty-two specimens analyzed with GAG cycle defined in terms of discrete flight mean to once per flight taxi compressive peak applied during tests; 115 specimens analyzed with arbitrary once per flight peak GAG cycle.

Figure 10.4 Cumulative percentage of specimens equaling or exceeding specified ratios of test to calculated life-category A2-flight-by-flight loadings.

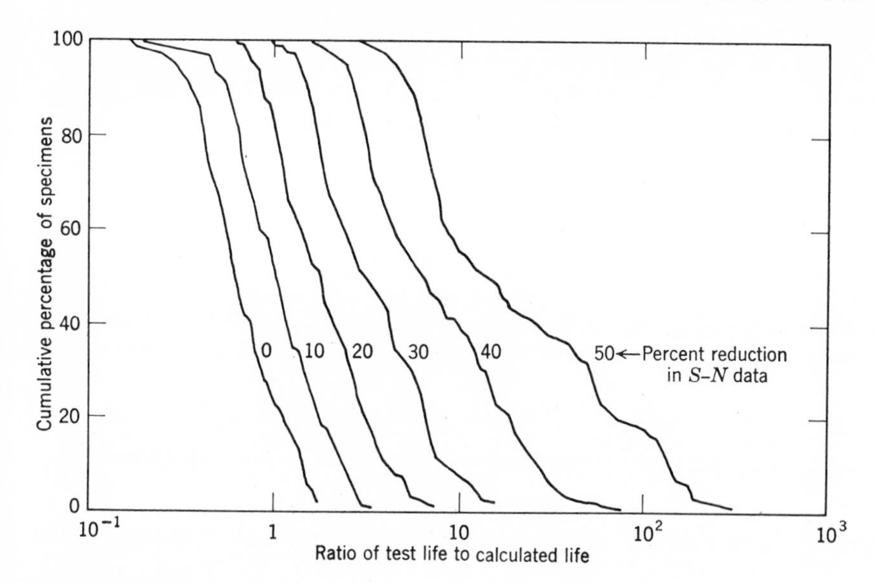

Eighty-two specimens analyzed with GAG cycle defined in terms of discrete flight mean to once per flight taxi compressive peak applied during tests; 115 specimens analyzed with arbitrary spectrum of GAG cycles.

Figure 10.5 Cumulative percentage of specimens equaling or exceeding specified ratios of test to calculated life-category A3-flight-by-flight loadings.

for category *B* and *C* tests in Figures 10.6 and 10.7 indicate 63 and 37% conservatism, respectively, for unreduced *S-N* data (zero percent reduction). When realistic flight-by-flight loading sequences are used, these percentages drop.

For the three category *A* classifications of GAG cycles in Figures 10.3 to 10.5, the amount of conservatism increases when the GAG cycles not defined by discrete test loadings are changed from a combination of mean-to-mean and mean to once per flight peak in category *A*1 to once per flight peak in category *A*2 and to GAG cycle spectrum in category *A*3. This growth in conservatism for conventional *S-N* data with no reduction from 7% for category *A*1 to 14% for category *A*2 and to 24% for category *A*3 is due to the increase in severity of the type of GAG cycle employed to determine fatigue life utilization when the GAG cycle was not a discrete test loading.

It is desirable to extrapolate the cumulative distributions in Figures 10.3 to

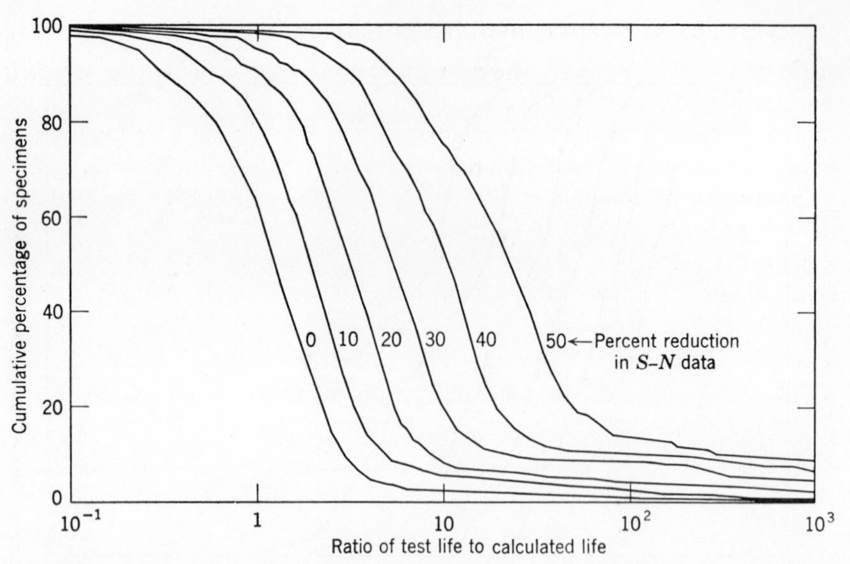

Figure 10.6 Cumulative percentage of specimens equaling or exceeding specified ratios of test to calculated life-category B-block loading without GAG *cycles—953 specimens.*

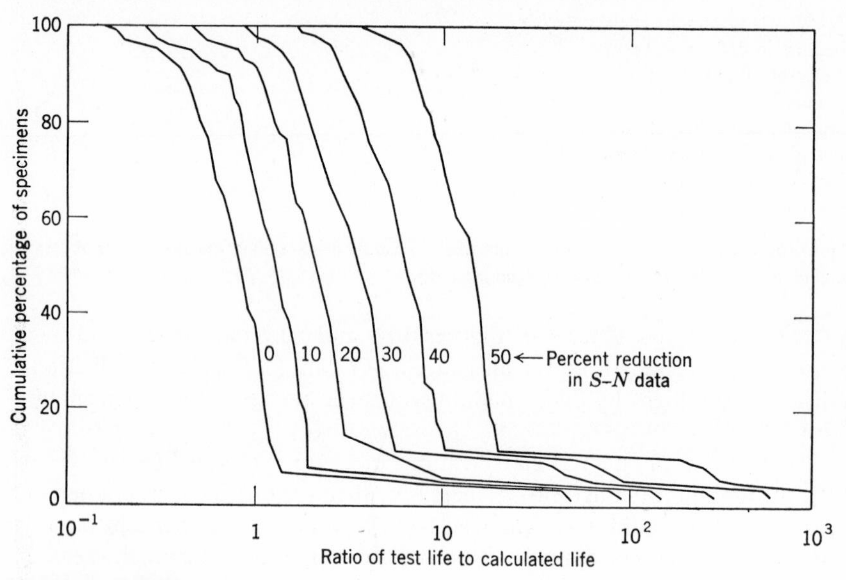

GAG cycle used to calculate life based on discrete definition of GAG cycle applied during tests.

Figure 10.7 Cumulative percentage of specimens equaling or exceeding specified ratios of test to calculated life-category C-block loading with GAG *cycles—95 specimens.*

10.7 to a probability basis. The percentage conservatisms could be accepted as probability estimates, with an error of $\frac{1}{2}\eta$ where η is the number of spectrum test results in the test category being considered. The cumulative percentage values were therefore reduced by a factor of $\frac{1}{2}\eta$ to provide a valid nonparametric best estimate of the probabilities for conservative predictions, which is presented in Figure 10.8.

While the groupings do not refer to the same number of individual specimens, Figure 10.9 gives further indication of the effects of growth in conservatism with increased severity in the analytical definition of the GAG cycle.

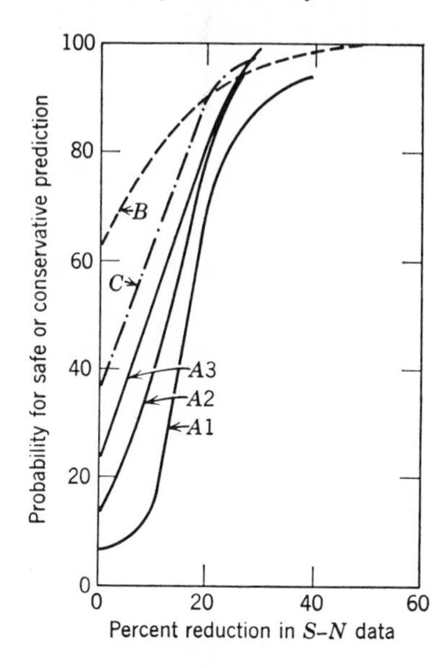

Test category	Derived from figure
A. Flight-by-flight with following analytical differences	
*A*1. Arbitrary mean to mean and mean to once per flight taxi peak GAG cycles	3
*A*2. Arbitrary once per flight peak GAG cycle	4
*A*3. Arbitrary spectrum of cycles	5
B. Block without GAG cycles	6
C. Block with GAG cycles	7

Figure 10.8 Probability for conservative prediction of spectrum test results.

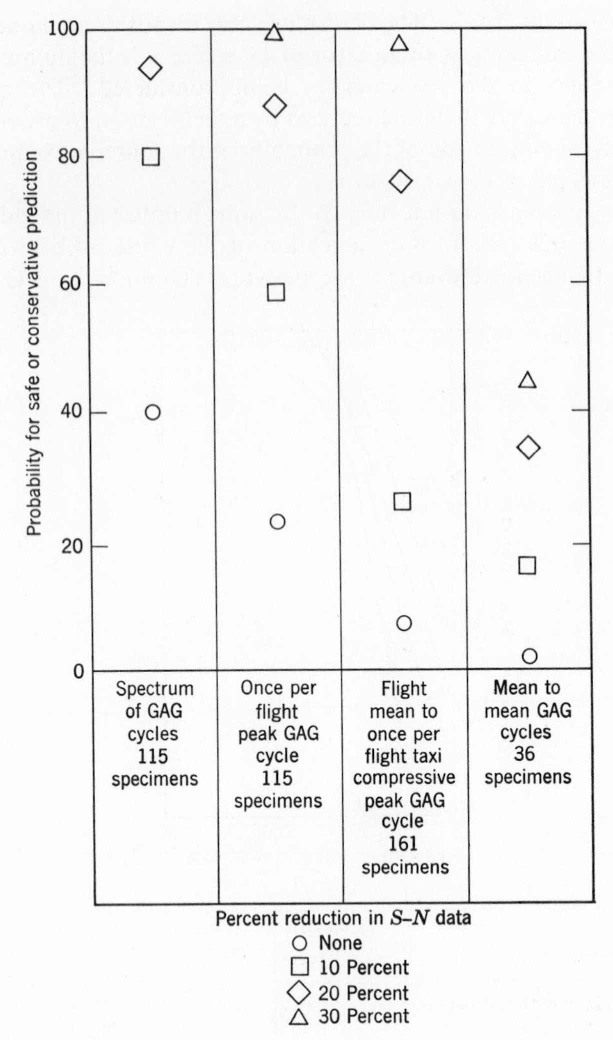

Figure 10.9 Probability for conservative prediction of flight-by-flight test result.

This growth in probability for making a conservative prediction with zero percent reduction in *S-N* data rises from an estimated probability of less than 2% for the mean-to-mean GAG cycle, to 7% for the flight mean to once per flight taxi compressive peak GAG cycle, to 23% for the once per flight GAG cycle and to 40% for the spectrum of GAG cycles. This figure also indicates that the use of a once per flight peak GAG with 20% reduced *S-N* data gives an 87% probability for conservatively predicting flight-by-flight test results.

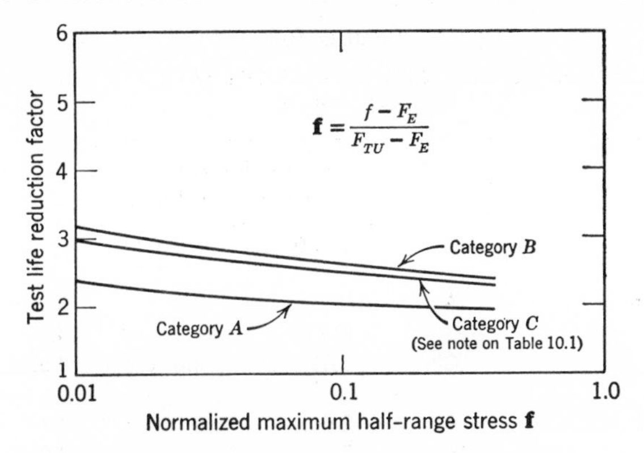

Figure 10.10 Variation in test life reduction factor for two nickel or steel specimens with maximum half-range stress.

10.1.3 Test Life Reduction Factors

In addition to evaluating the levels of assurance involved in calculating spectrum test life with conventional and reduced S-N data, the spectrum test data obtained from the literature survey were used to secure test life reduction factors. These factors are applicable to results of variable amplitude loading spectrum tests conducted on a few test specimens under nominally identical test conditions. When applicable, the derived life reduction factors are divided into the average spectrum test life to define the minimum anticipated life when additional specimens are tested. In deriving these test life reduction factors, which is more fully described in [43] and [44], they were found to be related to material and test variables.

The results obtained are shown in Table 10.1 and Figures 10.10 and 10.11

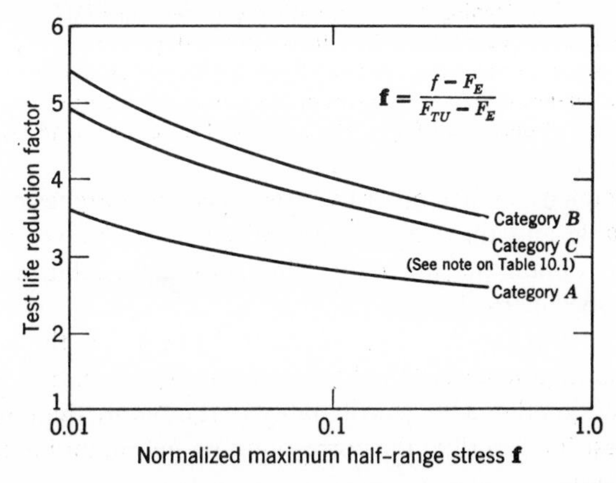

Figure 10.11 Variation in test life reduction factor for two aluminum specimens with maximum half-range stress.

TABLE 10.1

RECOMMENDED TEST LIFE REDUCTION FACTORS FOR TWO SPECIMENS

Material	Category	Factors for values of normalized maximum half-range stress $\mathbf{f}^{a}$		
		$\mathbf{f} = 0.18$	$\mathbf{f} = 0.06$	$\mathbf{f} - 0.02$
Aluminum	A	2.75	3.0	3.3
	B	3.8	4.3	4.9
	C[b]	3.55 × 2	4.0 × 2	4.55 × 2
Nickel or steel	A	2.0	2.1	2.25
	B	2.5	2.7	3.0
	C[b]	2.4 × 2	2.6 × 2	2.8 × 2

[a] $\mathbf{f} = \dfrac{f - F_E}{F_{TU} - F_E}$

where

$$f = \frac{1}{2}\left[f(\text{max})_{\substack{\text{tensile}\\ \text{test}\\ \text{peak}}} - (f\text{min})_{\substack{\text{compression}\\ \text{test}\\ \text{peak}}} \right]$$

F_E = endurance limit expressed in terms of the constant amplitude varying stress that produces failure at 10^7 cycles for the structural material,

and

F_{TU} = ultimate tensile strength of the structural material.

[b] For category C tests the test life reduction factors are multiplied by a factor of 2 to account for the relatively longer test lives that are obtained in using a block loading sequence with mean-to-mean GAG cycles as compared to the lives obtained by using a flight-by-flight loading sequence with the same set of loading spectra. If the GAG cycle used in the block loading sequence is defined in terms of the maximum stress expected on the average of once each flight, the additional factor is not necessary.

for two test specimens. When other than two specimens are tested, the values listed in this table and these figures are raised to the power indicated on Figure 10.12 to obtain the life reduction factor applicable to a different number of test specimens.

These test life reduction factors were derived to provide the same level of assurance. This assurance is that 85 times out of a hundred it would be correct to say that there is a 90% probability that no more than one specimen would have a test life less than the corresponding minimum life in 5,000,000 equivalent test hours.

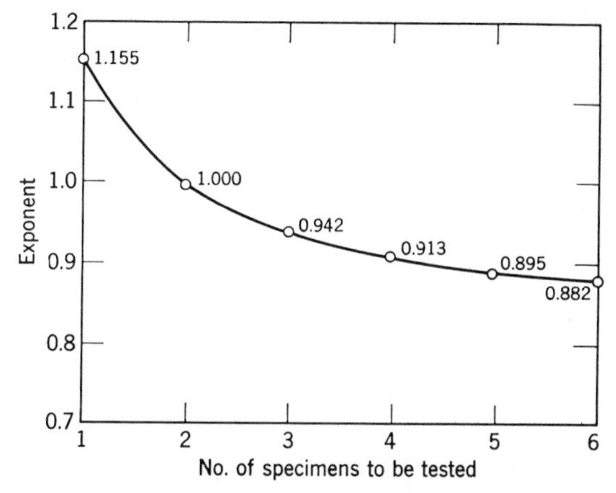

Figure 10.12 Adjustment in test life reduction factor when the number of tested specimens is different from two.

10.2 FATIGUE DESIGN

Having examined the levels of assurance involved in the use of life prediction methods and test life reduction factors, the next step is to look at the manner in which analytical and experimental techniques are applied to actual flight vehicles.

Since the designer has little or no control over the use or maintenance of a flight vehicle once it becomes operational, structural fatigue reliability on rotating and non-rotating structural elements can only be achieved by one of the three following methods.

1. Safe-life concept which requires the retirement and replacement of rotating parts after a definite period of time.
2. Fail-safe concept of damage tolerance with load carrying ability after moderate damage provides for unlimited life on nonrotating components when used with established maintenance and inspection procedures.
3. Combined safe-life and damage tolerance concept which provides longer replacement times in excess of those obtained for just safe-life on rotating structural components when adequate damage tolerance has been experimentally demonstrated.

For each of these concepts, the probability of catastrophic failure or excessive structural deformations must be extremely remote during the specified replacement times and inspection intervals. Within this framework, fail-safe structural elements are usually more economical to maintain.

Careful control is exercised during each phase of structural design to achieve minimum replacement times for rotating elements on helicopters that exceed the design life while nonrotating elements on both helicopters and airplanes are generally made fail safe to eliminate any need for a replacement time. The steps taken to provide such control are covered in the following section.

10.2.1 Design Procedures for Fatigue

To ensure an adequate fatigue life, the following design procedures are frequently used.

1. Materials are selected that have good fatigue characteristics.
2. Only processing and fabricating techniques are used that have little or no detrimental effects on fatigue.
3. The permissible design allowable gross area tensile stresses are established, when necessary, below the tensile strength capabilities of the structural material. The selected design allowable gross area stress is further reduced in areas subjected to biaxial loadings and in regions adjacent to large concentrated loads. Additional reductions are enforced on blind structure to prevent fatigue cracks from occurring in noninspectable areas. Any structure which cannot be inspected on at least one surface by direct means or by removal of cover plates, access doors or inspection panels is considered to be blind structure. A suitably reduced allowable tensile stress is also used in single load path components or vital structure other than landing gear structure. All of these allowable stresses must encompass the effect of fuel, air pressure and thermal loadings when applicable. Basically, design stress levels are selected to achieve low normal operational stress levels. The same objective can also be obtained by using fairly severe static design conditions that lead to low operational stresses.
4. The permissible gross area tensile stress is also limited for pressurized structure. A higher allowable stress is used for fuselage substructure, such as frames, than for the skin since the loading on the substructure is primarily uniaxial while the skin is subjected to biaxial stresses. In extremely hot or cold environments the allowable circumferential skin stresses must cover the combined effects of pressure loading and thermal strains, as well as primary flight, landing, and ground loadings.
5. Reduced plastic bending allowables are applied to landing gear and other components which are frequently loaded predominantly in bending to a high percentage of design limit loads. Such components, when not considered to be fail safe are required to experimentally demonstrate a specified safe life.
6. Fatigue tests are conducted on major joints, components, and assemblies. Such tests are sometimes conducted on the complete airframe of a flight vehicle. The nature, conduct, and choice of fatigue tests are described in

Section 10.4. When applicable, these tests must demonstrate that fatigue cracking will occur in visually inspectable areas rather than blind areas.

7. Structural arrangements that have previously exhibited a long fatigue life during service experience or fatigue testing are normally used in the design of new structure.

8. Load path discontinuities as well as stress and strain concentrations are minimized.

10.2.2 Design Life and Loading Spectra

Important elements in evaluating the effects of fatigue are the design life, mission profiles or condition lists, and significant fatigue loadings. The design or anticipated operational life and service loading spectra are given primary consideration in evaluating the effects of fatigue on the structural elements of flight vehicles.

For elements other than the power source that are limited by wear the design life represents an upper limit. For all other elements on helicopters or aircraft it represents a design goal. Design life is generally specified by contractual agreements or Federal Aviation Adminstration and Military specifications in terms of the number of flights, landings, cabin pressurizations and flight hours. Sometimes, the number of landings will exceed the number of flights to account for additional touch-and-go landings that may be used during training phases of flight. When the number of cabin pressurizations is not specified, the assumption is usually made that the number of pressurizations is equal to the number of flights to be performed at higher altitudes where the cabin and/or cockpit would normally be pressurized.

Mission profiles or condition lists are generally used to provide a basis for the selection of design stress levels, the choice of structural parts to be tested, the definition of the fatigue test spectra and for the evaluation of service life. Figures 10.13 and 10.14 contain typical mission profiles for transport and fighter aircraft. For helicopters, mission profiles are supplemented and sometimes replaced by condition lists of the type on Table 10.2. These mission profiles and/or condition lists must contain a detailed description of the activity versus time or distance applicable to each portion of flight. Also required are the flight configurations, anticipated changes in weight due to fuel usage or stores released, altitudes, speeds and any other parameter that could affect the fatigue loading spectra. All of these parameters normally reflect average operational conditions for aircraft. For helicopters, however, such parameters must also reflect critical flight and ground conditions.

The spectra used for analytical and experimental evaluation of fatigue should realistically represent the total operational loading history described by the mission profiles and/or condition lists. Depending on the structural

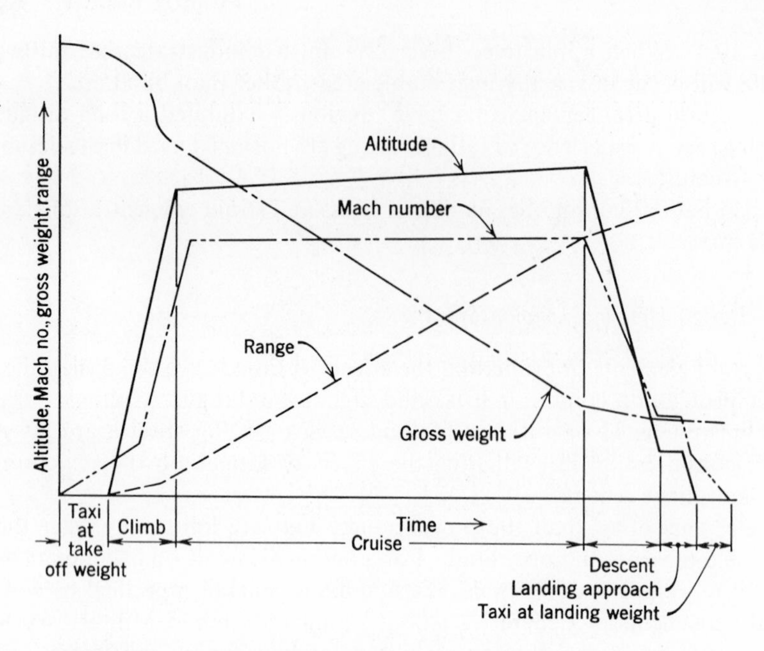

Figure 10.13 Typical mission or flight profile for transport or cargo airplane.

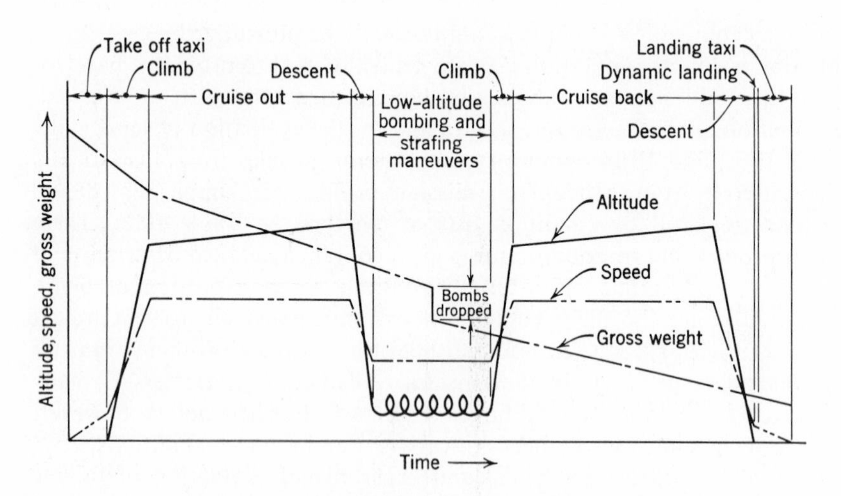

Figure 10.14 Mission or flight profile for fighter airplane.

TABLE 10.2

TYPICAL HELICOPTER CONDITION LIST FOR 1000 HOURS OF SERVICE

Type	Condition		Number of events in 1000 hours	Percent of flight time
Steady forward flights	Hover, air taxi	0–0.2 V_{NE}[c]		10
	Low speed (including transition)	0.2–0.6 V_{NE}	9600	21
	Low cruise	0.6–0.8 V_{NE}		25
	High cruise	0.8–0.9 V_{NE}		15
	Max. speed	0.9–1.0 V_{NE}		7.3
	Dive speed	1.0–1.11 V_{NE}		0.7
Steady flights (Others)	Nonforward level flight		3600	2
	Climb (T.O. power)		2800	2.4
	Vertical descent (partial power)		900	0.2
	Climb and descent at forward speed		5200	7
	Autorotation steady descent		750	1
Level ground conditions	Rotor on-off, on-idle-on events		6000	[0.2][a]
	Ground run time		5000	[25.0]
	Landing and take-off events (includes level autorotation loadings)		3510	[1.1]
Sloped ground conditions	Landing and take-off events (5° slope)		1400	[0.4]
	Landing and take-off events (10° slope)		90	[0.03]
Gust and maneuver conditions (power-on and power-off)	Pull-ups		19,000	3
	Flare		7900	1.7
	Turns		19,320	5.6
	Turn-on-the-spot, rudder reversals		3600	0.45
	Reversals (translation)		5150	0.30
	Sum of transitions to and from auto-rotation		4300	0.12
	Rapid recovery from autorotation		200	0.006
Special conditions	Lateral C.G. offset time			1
	RPM overspeed 7.5%		1250[b]	0.05
	RPM overspeed 13%		60[b]	0.0035

[a] Time in brackets [ ] not included in total flight time.
[b] No. of Flights.
[c] V_{NE} = never exceed velocity.

element, the significant fatigue loadings on nonrotating structural components arise from taxi, gust, maneuvers, landing impact and GAG cycles. For wing upper surfaces, the primary fatigue loadings are generally associated with taxi, landing impact and GAG cycles. Gust and GAG cycles produce the most important fatigue loadings on the wing lower surface of other than fighter aircraft where maneuver loadings are usually more significant. For pressurized structure, the primary source of fatigue is the number of pressurizations.

When loading spectra for taxi, landing impact, and GAG cycles are important, primary consideration is given to the total number of take-offs and landings in the mission profile or condition list. In defining loading spectra for gust sensitive elements, the time spent at various weights, altitudes and speeds is more important than the types and numbers of flights. For maneuver sensitive elements the types of flights or conditions would provide the most pertinent factors to be used for defining fatigue loading spectra.

Significant fatigue loadings on rotating structural components, such as the main and tail rotors on helicopters arise from reversals, turns-on-the-spot, and sloped ground conditions in the condition list of Table 10.2. For power train components, such as transmission shafts, couplings and gearboxes on helicopters, the primary sources of fatigue loadings are due to the magnitude and number of on-off cycles and fluctuations in power or torque.

10.3 FATIGUE ANALYSIS

Fatigue analyses are performed in two stages. The first is the design phase where spectra of the loadings throughout the airframe and/or rotor systems are analytically developed and used in linearly cumulative fatigue life utilization calculations to predict anticipated service lives. The results of these calculations provide guides to the selection of the fatigue quality and design stress levels, and to the requirements for loading spectra to be employed in development fatigue tests.

Because of mutual dependence, a simultaneous selection must be made of the fatigue quality or effective stress concentration and the design stress level required to provide an adequate structural life under a specified loading history. The best guide for the initial value of fatigue quality is obtained from the results of tests conducted with comparable specimen geometries and loading histories on structural specimens that have experienced satisfactory service on similar flight vehicles. For a particular structure, the fatigue quality is not known until the structure has been tested under anticipated loading histories. The fatigue quality is then expressed in terms of the stress concentration factor corresponding to a set of simple notched sheet coupon S-N data that exactly predicts the test life for initial cracking of the structure.

During the design phase of rotating components, the selected reference stress levels assist in the sizing of parts. These reference stresses reflect the maximum mean and cyclic stress levels anticipated during steady state flight, along with the maximum cyclic stress levels associated with the critical flight and ground conditions.

For nonrotating components the design phase calculations are used to establish the permissible fraction of the structural material's ultimate tensile strength that can be utilized at design ultimate loads.

The second phase of the fatigue analysis is the interpretation in terms of flight measured data of the results obtained from fatigue tests conducted on structural elements.

10.4 FATIGUE TESTS

The designer must produce an airframe or rotor structure with a good fatigue quality to resist cracking under repeated loads. This quality must account not only for geometric effects that result in local stress concentrations and changes in load path, but also for the magnitudes and frequencies of occurrence of fatigue loadings. The best initial guide for the fatigue quality of a new structure is provided by past satisfactory service or previous spectrum test results of similar structure under comparable loadings. A good design philosophy requires the demonstration by test of a fatigue resistance in joints and other design continuities that is slightly greater than that afforded by a simple hole-notched coupon. Only after the structure has been actually tested under anticipated loading histories, will the designer know whether or not an adequate fatigue quality has been designed and built into the structure. If the fatigue quality is not adequate, the structure is modified or redesigned and retested until the changes have been experimentally demonstrated to be adequate. At such a time the changes are incorporated in the production design.

10.4.1 Design of Test Specimen

In any test, a specimen is needed. The purpose in the fatigue testing of full-scale joints and critical areas of structure isolated from the complete airplane or helicopter is to economically determine the fatigue characteristics of the design when subjected to the best estimate of the environmental conditions possible to make in advance of actual service. In order to accomplish this, a number of things must be considered in designing the specimen, such as:

1. All details in the test region must be exactly representative of that portion of the airplane requiring test. The stress variability and lack of detailed knowledge of flexibility precludes compromises that sometimes may be taken in static tests. Dihedral and sweep angle in joints, rib flange

attachments and relative support flexibilities must be adequately simulated. Substitution of materials is not permissible without adequate corrections.

2. All construction details, clips, rivet holes, even the most innocent-appearing attachments (electrical wiring clips, etc., must be put into the specimen wherever they properly occur).

3. End attachments are very critical in specimen design. It is required that the load be introduced into the specimen at a lower stress concentration than appears anywhere in the specimen itself.

4. Location of the load axis throughout the specimen and especially at its end connections is important so that unwanted bending stress will not produce premature failure. Unless properly accounted for, deflections, especially in dihedral specimens, may build up restraining moments sufficient to cause premature failure.

10.4.2 Choice of Test Conditions

There is a growing realization that spectrum and *S-N* testing do not necessarily lead to the same conclusions. Therefore, both spectrum and *S-N* testing should be conducted to enhance the accuracy of material evaluations and subsequent decisions. The testing of large F-27 wing panels and complete C-46 wings (reported in References 45 and 46) respectively, indicated that the mode and location of failure was dependent on whether constant or variable amplitude loadings were used in structural testing.

During tests on the F-27 wing panels, a large number of cracks occurred in the constant amplitude loading tests that did not occur in the variable amplitude tests. These cracks may be attributed to the fact that the maximum loads employed in the constant amplitude test were less than those employed in the variable amplitude loading tests. The higher maximum loads in the variable amplitude tests could introduce favorable residual stresses at some structural locations that would not be present to the same degree during the constant amplitude tests. It was also noted that the same cracks occurred in the variable amplitude gust-type loading tests conducted with and without GAG cycles and with random flight-by-flight and ordered block loading sequences. For this reason the effects of load sequencing were not considered to lead to the systematic variation between the variable and constant amplitude tests. The change in fatigue sensitive elements between the variable and constant amplitude tests was felt to be due primarily to the effect induced by the higher load levels employed in the variable amplitude loading tests. The data in [46] also revealed that the location and mode of failure was not the same for variable and constant amplitude loaded fatigue tests conducted on C-46 wing specimens.

Another point to remember—structural failures are frequently associated with the rubbing of one part against another, which is referred to as fretting. Although the fretting mechanism is not fully understood, it seems probable that the fretting damage sustained in a constant amplitude test on a structural element is significantly different from that obtained when the load, and hence the relative slip, is varying from cycle to cycle. An indication of this was obtained from fatigue tests conducted on a bolted joint specimen [47]. The tests performed on specimens assembled both in greased and dry condition led to virtually identical *S-N* curves under constant amplitude loading with all failures originating from fretting damage. However, under block spectrum variable amplitude loading, the average life of the greased joints was approximately three times that of the dry joints.

Reference [48] in turn, has derived constant life diagrams based on variable amplitude loadings corresponding to a Gaussian distribution; *S-N* type curves were then defined in terms of the maximum stress applied in the spectrum. A plot of these *S-N* type curves for hole-notched sheet coupons, based on variable amplitude loading and plotted on a log stress–log cumulative cycle basis have revealed that there are differences when comparisons are made with similar plots for constant amplitude *S-N* test data.

For the constant amplitude *S-N* data, the slope on the log–log plot was a function of the ratio of minimum to maximum stress R, and the stress concentration factor K_t, as well as type of loading and type of the material being tested. On the other hand, the slope of the *S-N* type life curve for the variable amplitude loadings was not influenced by either the stress ratio R, the stress concentration factor K_t, or type of loading (rotating beam, bending, axial). The absence or presence of fretting, based on tests conducted on lugs, did not reveal any significant effects of fretting on the slope of the variable amplitude loading *S-N* type curves. However, the slope of these curves did vary with materials being slightly higher for steel than it was for aluminum.

These findings make it evident that the results of variable and constant amplitude loading tests are governed by different factors. They also indicate that the relationship between constant amplitude and variable amplitude tests was considerably more complex than is generally assumed.

10.4.3 Test Life Significance of the GAG *Cycle*

Similar to the differences between spectrum and *S-N* type fatigue tests, significant and consistent variations in fatigue life are introduced by the GAG cycle when the same loading spectra are applied in a flight-by-flight loading sequence and in a block loading sequence with periodically grouped GAG cycles.

The use of either the random or ordered flight-by-flight loading sequence of

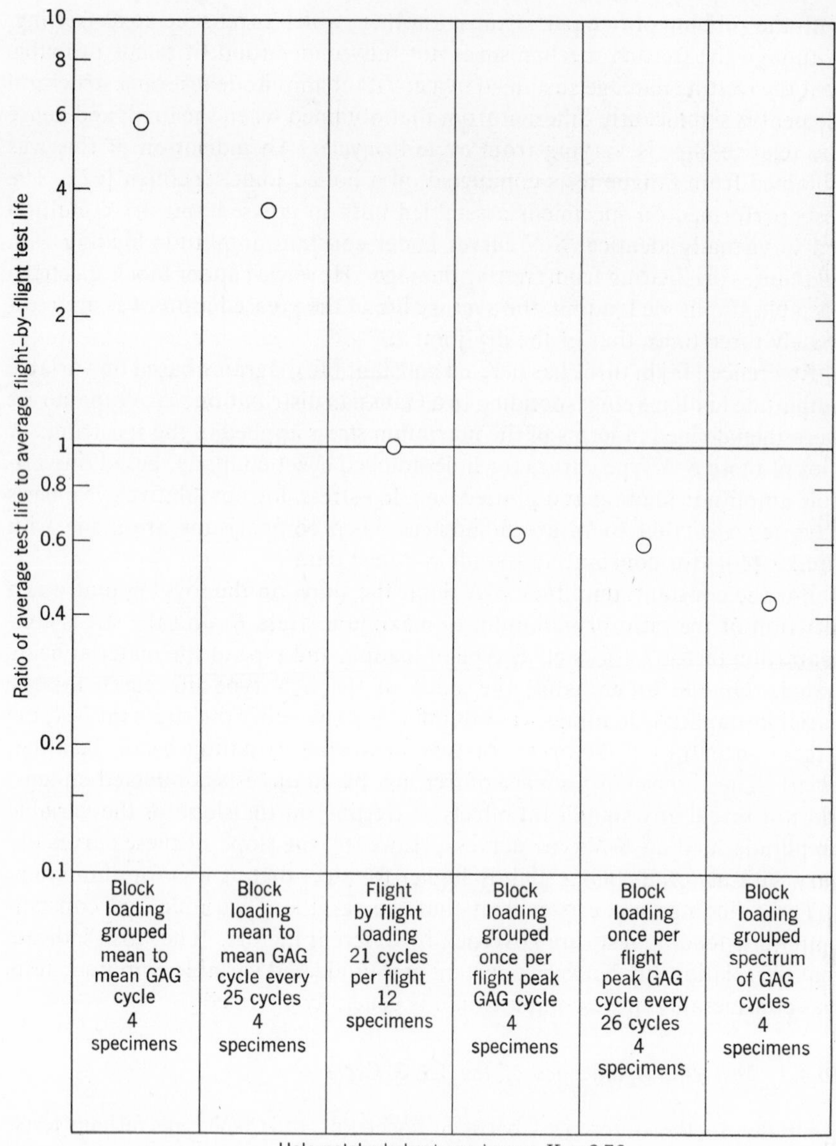

Figure 10.15 Effects of variations in magnitude and frequency of GAG *cycle on average test life.*

Figures 10.1*a* and 10.1*b* automatically introduces the loadings such that ground-air-ground transition cycle does not have to be specifically defined. The ground-air-ground transition cycle has two effects. One is the effect of the magnitude of the transition load itself. The other is its influence on the contribution to fatigue life utilization by successive loadings from other sources. The net result of these effects generally leads to a shorter test life than what would be obtained by complete omission of the transition cycle during testing.

For some full-scale panel tests, loadings due to air, ground and transition cycles are sometimes grouped together for an ordered spectral block loading that represents a large group of flights, such as shown in part *d* of Figure 10.1. The use of such groupings of loadings may lead to unconservative test results since they tend to produce longer test lives than those obtained by using a flight-by-flight loading sequence where the transition cycles are more frequently dispersed once each flight rather than being periodically grouped together.

Data from [4] indicates that the test lives obtained for notched sheet coupons under periodically grouped GAG cycles is approximately twice as long as the test lives obtained with a flight-by-flight loading sequence.

However, it is indicated in [49] that the use of periodically grouped GAG cycles could lead to a test life for notched sheet coupons of 6 to 13 times as long as the lives obtained with flight-by-flight sequence. This variation in effects of periodically grouped GAG cycles on flight-by-flight test life between the preceding references was attributed in [43] and [45] to the use of differences in loading sequences with the factor of 2 being obtained with a low-high loading sequence, whereas the factors of 6 to 13 were obtained with a low-high-low-loading sequence.

Figure 10.15 indicates that altering the test severity of the GAG cycle with its application either as a group or interspersed throughout a block loading sequence can lead to variations in average test life that range from less than half to six times the average test life obtained with a flight-by-flight loading sequence.

10.4.4 Development and Substantiation Fatigue Tests

Spectrum loading fatigue tests are usually conducted on all components that contain complex loadings and multiple or otherwise complex load paths while constant amplitude tests are employed for components with relatively simple loadings and load paths. Such tests are normally performed during two stages of design:

1. Development stage where fatigue tests are conducted early to get basic design data and to demonstrate attainment of design requirements. These

tests are conducted on material coupons, joints, rotating and non-rotating structural elements.

2. Substantiation stage where tests are performed late in the design stage to substantiate the fatigue life of individual components or the complete airframe. Such tests are normally conducted on full-scale elements that will appear on the production model.

Comparable types of structural specimens and test conditions during either the development and substantiation tests are illustrated below.

1. Safe-life components with simple external loadings. Constant amplitude loadings applied to helicopter pitch control links and rotor blades.
2. Safe-life built-up components. Spectrum loading with flight-by-flight sequence or block sequence with grouped GAG cycles applied to a helicopter transmission case or tension-torsion pack.
3. Repairable and nonrepairable fail-safe components. Spectrum loading with flight-by-flight sequence or block sequence with grouped GAG cycles applied to a panel structure (repairable) or a rotor hub (nonrepairable).

Inspections are normally conducted to detect fatigue damage during the tests. When cracks are discovered on damage tolerant structures, fail-safe tests are generally conducted to experimentally determine the amount of time for the crack to propagate to critical size. This time provides a valuable basis for recommended inspection and maintenance procedures during service.

10.5 FATIGUE LIFE EVALUATION

The basic concepts described in this section are generalized for aircraft and helicopters. These techniques are more fully described for helicopters in [44]. Basically, the factors used to evaluate the fatigue life of structural parts or components depend on whether or not they have been tested.

10.5.1 Untested Structural Components and Parts

For untested structural components, the following methods are used with conventional *S-N* data for fail-safe aircraft components and reduced *S-N* data for helicopter components:

1. For fail-safe aircraft components that have maximum operational below the endurance limit of the structural material at an appropriate stress concentration factor, tests and further analyses are generally not made. A similar procedure is used for rotating safe-life parts on helicopters when the operational stress levels are below a third of the material's endurance limit.

2. For some untested fail-safe aircraft parts, the method of linearly cumulative fatigue life utilization may be used to calculate an adequate life with conventional *S-N* data that is deemed appropriate to the part. For untested nonvital rotating parts on helicopters, a similar procedure is used with *S-N* curves reduced to one-third of the cyclic stress levels for the *S-N* data.

3. For untested fail-safe non-rotating aircraft shell structure, the anticipated crack-free fatigue life is calculated by using linearly cumulative fatigue life utilization with the applicable conventional *S-N* data. On similar helicopter structure, reduced *S-N* curves having 80% of the cyclic stress levels for the *S-N* data are used in making the analytical calculations.

10.5.2 Constant Amplitude Tested Parts

The data obtained from constant amplitude *S-N* tests on rotating parts and control system components of helicopters are used to derive *S-N* curves. These *S-N* curves are then reduced as shown for steel and nonferrous structural elements in Tables 10.3 and 10.4, respectively [44] and [50].

For the tested elements, the reduced *S-N* curves are used to either establish endurance limits for indefinite life or to calculate specific lives with the concept of linearly cumulative fatigue utilization.

10.5.3 Spectrum Tested Components

Test life reduction factors, as required by contractual agreement or government specifications, are used in analyzing the results of spectrum loading tests. The minimum test life reduction factors used for the safe-life structural components on a typical helicopter were the following:

Number of specimens tested	1 Specimen	2 Specimens
Aluminum	7	5
Steel and Nickel Alloys	6	4

When Table 10.1 called for higher factors, the higher factors were used.

For fail-safe or combined safe-life and damage tolerant structural components, a safe life equal to half the test life was permitted with adequate maintenance procedures and experimentally demonstrated inspection periods.

The loading spectra used during basic design and in setting up fatigue test loadings are based on analytically derived loads and extrapolations from previously recorded data for generally similar flight vehicles. Normally, the fatigue tests are in progress before flight test data becomes available for

TABLE 10.3

SAFE-LIFE STRESS REDUCTION FACTORS FOR CONSTANT AMPLITUDE *S-N* TESTED STEEL STRUCTURAL ELEMENTS

Types of *S-N* Test Results	Alternating Stress Reduction Ratio
Four or more specimens tested to 10^7 cycles	0.9 times average *S-N* curve for test results
Four or more specimens with two tested to 10^7 cycles	0.9 times minimum *S-N* curve for test results
Four or more specimens with one tested to 10^7 cycles	0.8 times average *S-N* curve for test results with the added requirement that all test points lie above the derived reduced *S-N* curve.
Four or more specimens with none tested to 10^7 cycles	0.75 times average *S-N* curve
Two specimens tested to 10^7 or more cycles	0.67 times alternating stress applied in test

TABLE 10.4

SAFE-LIFE STRESS REDUCTION FACTORS FOR CONSTANT AMPLITUDE *S-N* TESTED NONFERROUS STRUCTURAL ELEMENTS

Types of *S-N* Tests	Alternating Stress Reduction Ratio
Four or more specimens tested to $5(10)^7$ cycles	0.9 times average *S-N* curve for test results
Four or more specimens with two tested to $5(10)^7$ cycles	0.9 times minimum *S-N* curve for test results
Four or more specimens tested to 10^7 cycles	0.8 times average *S-N* curve for test results
Four or more specimens with two tested to 10^7 cycles	0.8 times minimum *S-N* curve for test results
Four or more specimens with one tested to $5(10)^7$ cycles	0.8 times average *S-N* curve for test results with the added requirement that all test points lie above the derived reduced *S-N* curve
Four or more specimens with none tested to 10^7 cycles	0.75 times average *S-N* curve
Two specimens tested to 10^7 or more cycles	0.67 times alternating stress applied in test

checking the accuracy involved in the analytical definition of test loading histories. The interpretation of fatigue test results must reflect comparisons between the test and flight-measured data or anticipated operational conditions when flight test measurements are not available. Any vital part in the control or rotor system on a helicopter, must have flight-measured data.

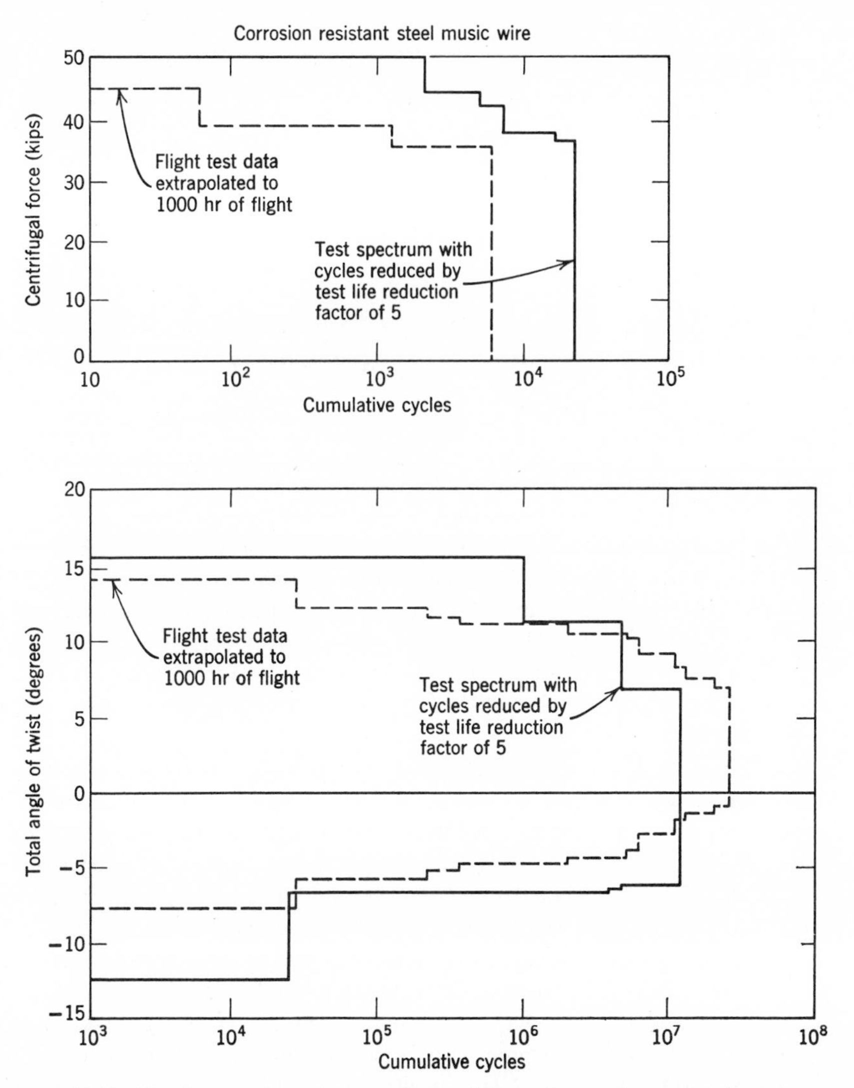

Figure 10.16 Visual comparison of test applied and flight measured spectrum for main rotor tension-torsion pack.

When flight recorded data is secured, the basis for comparison depends on whether the spectra deduced from flight data is similar or different from the test applied loading history. For similar shaped spectra a simple visual comparison on a graphical basis is used to evaluate the anticipated fatigue life from the test life. To illustrate, Figure 10.16 shows the centrifugal force and

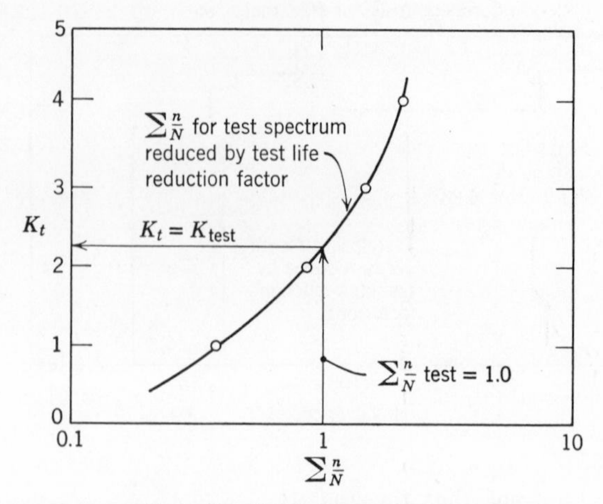

Figure 10.17 Evaluation of K_{test}.

angular twist spectra for a main rotor tension-torsion pack that was determined by extrapolating flight measured data to 1000 hours of service. The corresponding fatigue test spectra with the test applied cycles reduced by a test life reduction factor of 5 are also shown on this figure. Since more than 20 repetitions of the flight measured spectra would be adequately covered by the reduced test spectra, the tension-torsion pack is considered to have life of 20,000 hours.

When loading spectra are not directly comparable, correlations between flight measured or anticipated loading histories and the test applied spectral loadings are based on the use of linearly cumulative fatigue life utilization. One of the spectrum test results evaluated is the attainment of a particular level of K_t from *S-N* data for simple notched specimens made from the structural material that can be used to exactly predict the test life to crack initiation. To secure this value of K_t, the number of test applied cycles are reduced, when required, by the applicable test life reduction factor. Cumulative fatigue life utilization calculations are then made over a range of K_t values for the reduced test loading spectra. Normally, such calculations are made with conventional *S-N* data. The resulting fatigue life utilization ratios are crossplotted, as shown in Figure 10.17 to determine the value of $K_t = K_{\mathrm{TEST}}$ which produces a calculated fatigue life utilization ratio of unity for the reduced test life. The *S-N* curve corresponding to this value of K_{TEST} provides an experimentally derived fatigue life utilization boundary to be used for analytically determining the potential fatigue life of the test specimen under loading spectra that are moderately different from the test-loading history. Since it was experimentally derived from the results obtained for a complex loading history, a higher level of assurance is involved in using this boundary with the concept of linear cumulative fatigue life utilization than is otherwise possible.

10.6 SERVICE LIFE

The analysis of fatigue life is a continuing process of study and reevaluation in light of newly acquired data. The basic philosophy in analyzing fatigue is to provide the highest level of assurance for service that it is possible to do by using sound judgement and astute engineering knowledge with the currently available data. The acquistion of new service recorded loading and other environmental data can provide additional insight into the levels of assurance that exist in the evaluation of the service life of structural components on flight vehicles.

Feedback of loading and other environmental conditions from service is required for assurance that the analytical and experimental evaluation of fatigue life is fully adequate. Any significant variation between service recorded loading histories and related environmental conditions and the spectra used to determine fatigue life would require a reevaluation of anticipated structural life.

REFERENCES

[1] Investigation of Thermal Effects on Structural Fatigue, WADD-TR-60-410, Part I, August, 1960.

[2] Investigation of Thermal Effects on Structural Fatigue, WADD-TR-60-410, Part II, August, 1961.

[3] W. J. Crichlow, A. J. McCulloch, L. Young, and M. A. Melcon, "An Engineering Evaluation of Methods for the Prediction of Fatigue Life in Airframe Structures," ASD-TR-61-434, March, 1962.

[4] A. J. McCulloch, M. A. Melcon, W. J. Crichlow, H. W. Foster and R. Rebman, "Investigation of the Representation of Aircraft Service Loadings in Fatigue Tests," ASD-TR-61-435, January, 1962.

[5] "Effect of Compressive Loads on Structural Fatigue at Elevated Temperature," ASD-TR-62-448, October, 1962.

[6] A. J. McCulloch, M. A. Melcon, and L. Young, "Fatigue Behavior of Sheet Materials for the Supersonic Transport, Volume I, Summary and Analysis of Fatigue and Static Test Data," AFML-TR-64-399, Volume I, January, 1965.

[7] E. C. Naumann, H. F. Hardrath, and D. E. Guthrie, "Axial-Load Fatigue Tests of 2024-T3- and 7075-T6 Aluminum-Alloy Sheet Specimens under Constant and Variable Amplitude Loads," NASA TN D-212, December, 1959.

[8] E. C. Naumann and R. L. Schott, "Axial-Load Fatigue Tests Using Loading Schedules Based on Maneuver-Load Statistics," NASA TN D-1253, May, 1962.

[9] E. C. Naumann, "Variable-Amplitude Fatigue Tests with Particular Attention to the Effects of High and Low Loads," NASA TN D-1522, December, 1962.

[10] E. C. Naumann, "Evaluation of the Influence of Load Randomization and of Ground-Air-Ground Cycles on Fatigue Life," NASA TN D-1584, October, 1964.

[11] E. Gassner and K. F. Horstmann, "The Effect of the Ground to Air to Ground Cycle on the Life of Transport Aircraft Wings which are Subject to Gust Loads," Royal Aircraft Establishment Library Translation No. 933, February, 1961.

[12] J. Schijve and F. A. Jacobs, "Program-Fatigue Tests on Notched Light Alloy Specimens of 2024 and 7075 Material," National Aeronautical Research Institute, Amsterdam, NLL-TR M. 2070, 1960.

[13] J. Schijve and P. de Rijk, "Fatigue Lives Obtained in Random and Program Tests on Full-Scale Wing Center Section," National Aeronautical Research Institute, Amsterdam, NLR-TM.S.611, December 15, 1963.

[14] E. Gassner, "Review of Investigations on Aeronautical Fatigue in the Federal Republic of Germany," Laboratorium fur Betriebsfestigkeit LBF Report No. S-45, ICAF Lecture, 1963.

[15] E. Gassner and G. Jacoby, "Operational Strength Tests for the Determination of Permissible Design Stresses for the Lower Wing Surface of a Transport," translated from pages 6–19 in Luftahrttechnik-Raumfahrttechnik, Band 9, Heft 1, January, 1964.

[16] M. S. Rosenfeld, "Aircraft Structural Fatigue Research in the Navy, pages 216 to 238 in Symposium on Fatigue Tests of Aircraft Structures: Low-Cycle, Full-Scale, and Helicopters," ASTM Special Technical Publication No. 338, 1963.

[17] L. Mordfin and N. Halsey, "Programmed Maneuver-Spectrum Fatigue Tests of Aircraft Beam Specimens," pages 251 to 271 in ASTM Special Technical Publication No. 338, 1963.

[18] Unpublished Lockheed data.

[19] A. J. Berlioz, "Fatigue Tests of $\frac{1}{8}$ in. Countersunk and Dimpled Titanium Rivets in 0.030-in. Titanium Sheet," Lockheed Report No. 18484, December, 1964.

[20] Phase II-A Airframe Design Report-Volume VI A Supersonic Transport Development Phase II-A, Lockheed Report No. 18190, Volume VI A, June, 1964.

[21] "Design Concepts for Minimum Weight, High Performance Supersonic Aircraft Structures," Douglas Aircraft Company Report No. 31237, 1 April, 1963.

[22] *Proceedings of Full-Scale Fatigue Testing of Aircraft Structures, Symposium* (Amsterdam, June, 1959). Ed. F. J. Plantema E. J. Schijve, Pergamon, New York, 1961.

[23] R. H. Ketola, "Fatigue Test of Wing, (Nacelles), and Main Landing Gear (of P3A Airplane)," Lockheed Report No. 13645, April, 1964.

[24] E. W. Thrall, "A3D Wing Cycle Test," Douglas Aircraft Company Report No. 43134, September, 1962.

[25] R. H. Myers, "Repeat-Load Tests of Model F94F-4, −5 Wing Outer Panels," Report No. ASL NAM AD-279, Aeronautical Structures Laboratory, Naval Air Experimental Station, January, 1955.

[26] M. S. Rosenfeld and R. J. Zoudlik, "Determination of Fatigue Characteristics of a Typical (Fighter Model F-3A) Nose Landing Gear," Report No. NAEC-ASL-1079. U.S. Naval Air Engineering Center, Aeronautical Structures Laboratory. December, 1964.

[27] R. L. Schleicher, "Practical Aspects of Fatigue in Aircraft Structures, North American Aviation Inc.," paper presented at International Conference on Fatigue in Flight Structures, January–February, 1956.

[28] E. C. Naumann, "Fatigue Under Random and Programmed Loads," NASA TN D-2629, February, 1965.

[29] R. L. Lowe, "Exploratory Fatigue Test Model C-141," Lockheed-Georgia Company, Report ER 5992, April, 1963.

[30] G. V. Deheff, "Fatigue Prediction Study," Douglas Aircraft Company, WADD TR 61-153 (Preliminary Copy), May, 1961.

[31] E. Gassner, "Influence of Fretting Corrosion on the Fatigue Life of Notched Specimens of Al-CU-MG-2 Alloy," *Fatigue of Aircraft Structures Proceedings of Symposium at Paris, May 1962*. Macmillan, New York, 1963.
[32] G. Incarbone, "Fatigue Research on Specific Design Problems," *Fatigue of Aircraft Structures Proceedings of Symposium at Paris, May 1962*. Macmillan, New York, 1963.
[33] A. J. Troughton *et al.*, "Fatigue Lessons Learnt from the Argosy," *Fatigue of Aircraft Structures Proceedings of Symposium at Paris, May 1962*. Macmillan, New York, 1962.
[34] W. T. Kirby and P. R. Edwards, "Constant Amplitude or Variable Amplitude Tests as a Basis for Design Studies," preprint prepared for presentation at the ICAF Symposium on Fatigue Design Procedures, June 16–18, 1965, Munich.
[35] R. H. Meyers and M. S. Rosenfeld, "Constant Amplitude Fatigue Characteristics of the Wing of a Typical Fighter Airplane," U.S. Naval Air Engineering Center Report No. NAEC-ASL-1067, July 2, 1963.
[36] R. H. Meyers and M. S. Rosenfeld, "Variable Amplitude Characteristics of the Wing of a Typical Fighter Airplane," U.S. Naval Air Engineering Center Report No NAEC-ASL-1087, June 23, 1965.
[37] R. P. Swartz and M. S. Rosenfeld, "Variable Amplitude Fatigue Characteristics of a Slab Horizontal Tail for a Typical Fighter Airplane," U.S. Naval Air Material Center Report No. NAMATCEN-ASL-1023, Part II, September 18, 1961.
[38] C. R. Smith, "Linear Strain Theory and the Smith Method of Predicting Fatigue Life of Structures for Spectrum Type Loading," Aerospace Research Laboratories Report No. ARL 64-55, April, 1964.
[39] R. E. Whaley, "Fatigue Investigation of Full-Scale Transport-Airplane Wings," NACA TN 4132, November, 1957.
[40] C. A. Patching and J. Y. Mann, "Comparison of a 2L.65 Aluminum Structure with Notched Specimens Under Program and Random Fatigue Loading Sequences," preliminary preprint of Paper, presented at ICAF Symposium in Munich, June 16–18, 1965.
[41] R. P. Swartz and M. S. Rosenfeld, "Constant Amplitude Fatigue Characteristics of a Slab Horizontal Tail for a Typical Fighter Airplane," U.S. Naval Air Material Center Report No. NAMATCEN-ASL-1023 Part I, January 25, 1960.
[42] G. S. Jost, "The Fatigue of 24S-T Aluminum Alloy Wings Under Asymmetric Spectrum Loading," Structures and Materials Report 295, Dept. of Supply, Australian Defense Scientific Service Aeronautical Research Laboratories, February, 1964.
[43] C. S. Davis and L. Young, "A Comparison of Fatigue Life and Reliability from Constant and Variable Amplitude Loading Tests," *J. Am. Helicopter Soc.*, **12,** 65, (1967).
[44] W. J. Crichlow, C. J. Buzzetti, and J. Fairchild, "The Fatigue and Fail-Safe Program for the Certification of the Lockheed Model 286 Rigid Rotor Helicopter," paper presented at 5th ICAF Symposium in Melbourne, Australia May 22–24, 1967.
[45] J. Schijve, D. Broek, P. DeRijk, A. Nederveer and P. J. Sevenhuysen, "Fatigue Tests with Random and Programmed Load Sequences With and Without Ground-to-Air Cycles, A Comparative Study on Full-Scale Wing Center Sections," National Aerospace Laboratory, Amsterdam, The Netherlands, NLR-TR S.613, December, 1965.
[46] W. B. Huston, "Comparison of Constant-Level and Randomized-Step Tests of Full-Scale Structures as Indicators of Fatigue-Critical Components," ICAF-AGARD Symposium on Full-Scale Fatigue Testing of Aircraft Structures in Amsterdam, The Netherlands, June 5–11, 1959.
[47] W. T. Kirby and P. R. Edwards, "Constant Amplitude or Variable Amplitude Tests

as a Basis for Design Studies," preprint presented at ICAF Symposium on Fatigue Design Procedures, June 16–18, 1965, Munich.

[48] E. Gassner, Review of Investigations on Aeronautical Fatigue in the Federal Republic of Germany, LBF Report S-45, prepared for presentation at ICAF Symposium in Rome, April 19, 1963.

[49] E. Gassner and W. Shutz, "Assessment of the Allowable Design Stresses and the Corresponding Fatigue Life," paper presented at ICAF Symposium on Fatigue in Munich, Germany, June 1965.

[50] *Rotorcraft Airworthiness*, Normal Category, Civil Aeronautics Manual 6, Federal Aviation Agency, Appendix A, Revision January 15, 1963.

11

Fatigue Considerations in Helicopter Design and Service

HENRY G. SMITH

Fatigue considerations in helicopter design and service may be thought of as similar in many respects to those of fixed-wing aircraft. However, the helicopter is subject to many additional fatigue considerations not found in fixed-wing design and service. For instance, the helicopter experiences significant oscillatory loads even in straight and level flight due to the harmonic content of aerodynamic loads from the combined rotational and translational blade motion through the air. Also, the need for both steady and cyclic pitch control of the helicopter rotor provides additional sources of oscillatory loads. The complex power transmission system of the helicopter is yet another unique source of oscillatory loads. Because of the wide range of operating rpm's during flight and the many vibratory modes of the complex helicopter structure that may be excited, resonant amplification of the forcing loads is often unavoidable. Add the rotor start-stop cycles, the transient maneuver loads, the gust loads, the coriolis loads, and the transient power (torque) loads, and we perceive a rather formidable picture of the fatigue loads which are unique to the helicopter in normal operating service.

Furthermore, because of the empty weight penalties inherent in the helicopter as compared with an airplane (of equivalent gross weight) due to the complex rotor, control, and power transmission systems and to the greater required horsepower per pound of gross weight, the helicopter designer is extremely hard-pressed to keep the structural weight to the

Figure 11.1 Turbine-powered light helicopter.

absolute minimum required for safety in order to provide a machine with a militarily and commercially useful payload. (Figure 11.1.)

Unlike its fixed-wing aircraft counterpart, the helicopter structure, in many areas, is determined not by static strength but rather by fatigue considerations and by vibratory characteristics. For these reasons, it is currently commonplace in the helicopter industry to use special design and fabrication techniques to improve the fatigue-to-static-strength ratio of important components and to detune or damp vibration. For instance, adhesive bonding fabrication of metal rotor blades is now standard practice, and fiberglass-reinforced plastics are being used more extensively because of their unique fatigue as well as stiffness characteristics. Special defect limitations are specified, such as vacuum-melt steels for gears, bearings, etc. Extremely tight fabrication and process quality control is the rule on critical fatigue loaded components.

In view of the unique problems associated with the helicopter, it is no wonder that its successful development and use followed that of fixed-wing aircraft by many years in spite of its early inventive consideration, going back to flying models in the eighteenth century [1]. We will proceed then to consider this particular type of flight vehicle, which has been variously referred to as a "chopper," a "shaker," a "vibrator," and even a "flying fatigue test machine."

11.1 DESIGN CONCEPTS AND PHILOSOPHY

The helicopter is, in principle, best described by how it differs from fixed-wing aircraft, viz.:

1. It is not dependent on forward speed to impart lift producing momentum flux in the surrounding mass of air. Vertical take-off, landing, and flight are thus an important characteristic.
2. Forward speed is not essential for control of the vehicle in flight.
3. Propulsion in the horizontal plane is obtained by components of the resultant lift produced by the rotor(s) rather than by separate propulsion devices.
4. Power-off flight, even down to zero forward speed, is obtained through autorotation of the lifting rotors(s).

From the fatigue viewpoint, helicopters are best classified as to rotor configurations, rather than overall vehicle configuration, inasmuch as the predominant fatigue considerations generally are derived from the rotor systems. The present industry trend is toward improved safety and reliability, easier and less maintenance, longer fatigue service lives, more fool-proof and fail-safe, reduced vibration, increased speed, and improved stability and control characteristics. (Figure 11.2).

11.1.1 Rotor Configurations

Three principal types of helicopter rotor configurations are of importance in structural considerations, viz.:

1. Fully-articulated—with feathering, flapping, and lag hinges.
2. Semirigid—with feathering and teetering hinges (or feathering and gimbaled-hub hinges, feathering and flapping hinges, or feathering, flapping, and gimbaled-hub hinges for rotors of three or more blades).
3. Rigid—with feathering hinges only (rotors with elastically restrained articulation in the flapping and lag directions have also been popularly referred to as "rigid" rotors).

The above classification of helicopter rotor configurations is more important structurally than that of the rotor arrangement on the air-frame, i.e., single main rotor plus tail rotor, two tandem rotors, two side by side rotors, synchropter (two rotors side by side with angular high intermesh), and coaxial (two counterrotating rotors vertically displaced). Therefore, for the remainder of this paper, examples, theory, and comparisons will in general

Figure 11.2 Helicopter primary trainer with articulated rotor.

Figure 11.3 Experimental helicopter with tip driven rotor.

refer to the single main lifting rotor plus single tail rotor configuration. The ducted, tip-driven rotor, Figure 11.3, which is in the advanced experimental development stage at this time, will no doubt be operational in the near future in the heavy lift (that is, high payload ratio at high gross weight) and high-speed (stopped rotor convertiplane) categories, and its unique fatigue considerations will also be touched upon in the subsequent paragraphs.

11.1.2 Propulsion Systems

Because of the higher induced power requirements per pound of lift in order to hover, the helicopter requires a higher power-to-weight ratio engine and transmission system than does a comparable airplane. The combination of turbine engines and improvements in mechanical power transmission systems have therefore made the over-all propulsion system a most important factor in permitting the high state of helicopter development as we know it today. Another interesting propulsion development showing considerable promise today is the hot-cycle tip-driven rotor, which utilizes the direct turbojet engine exhaust that has been ducted through the blades, thus almost completely eliminating the usual mechanical drive system.

11.1.3 Control

The methods that have been used for controlling helicopters in flight are almost as diverse as the rotor arrangements, which have already been mentioned. For instance, control tabs, direct hub control, cam systems, crank systems, gyro systems, and servo rotors, as well as the more universal tilting swashplate, have been tried at one time or another in order to control the longitudinal or lateral tilt of the rotor plane or the pitching or rolling moments of the helicopter.

For lift and power control, the present trend, particularly for turbine-powered helicopters, is to use a governed, nearly constant, rotor speed, along with a collective pitch system functioning as a simultaneous lift and engine power control. A most important reason for using collective pitch rather than rotor speed control alone is to permit autorotation of the helicopter under power-off conditions.

Yaw control of helicopters has also been obtained by various schemes, such as differential torque (two rotors), vane in slipstream, differential rotor tilt (two rotors), and the antitorque or tail rotor (the latter being by far the most common).

Thus control of all six degrees of freedom is obtained in hovering flight as well as horizontal flight. It may also be mentioned at this time that teetering or gimbaled semirigid rotors can obtain pitching or rolling control-power only through the pendulum action of the suspended fuselage (or in forward flight through fuselage control surfaces linked to the rotor controls), and the rigid rotors can obtain lateral or longitudinal forces only by tilting or banking the fuselage along with the rotor. Fully articulated rotors obtain control-power in a manner similar to that in the case of the semirigid rotors, except that additional control-power may be obtained from rotor centrifugal forces by means of offset flapping hinges.

11.1.4 Structural Concepts and Philosophy

As previously mentioned, there is great pressure on the helicopter manufacturer, both from competitors and from the customer, to enhance the performance and useful load characteristics of the helicopter. This requirement is passed on to the structural designer, along with such additional constraints as: (a) make it safer; (b) make it more producible; and (c) make it easier to maintain.

To accomplish this he seeks more accurate prediction of loads, either by rational analyses utilizing electronic computers or through extrapolation of actual flight test data, or both. He applies fail-safe principles to the design of important structure wherever possible, considers Murphy's law, utilizes design ingenuity to convert a fatigue loading to a steady loading, simplifies the structure by reducing the number of critical joints or components, provides the shortest, most direct, load paths, and utilizes the most appropriate materials and processes for each component.

11.1.5 Present and Future Trends

A present trend toward higher top speed and better altitude performance of the helicopter is well advanced. The future will also see further development of hybrid rotary-wing aircraft, such as the compound helicopter, which would utilize auxiliary lifting surfaces as well as additional forward propulsion means. Speeds on the order of 300 knots may be achieved by the compound helicopter. Vibratory loadings on the structure will be maintained within the present limits by unloading the rotor at the top speeds [11]. Also, speeds above 300 knots, even supersonic, are technically feasible with stopped rotor convertiplanes.

The present trend is also toward longer-lived structural components. Long-service-life components are a definite competitive asset to the helicopter of today. The future trend is toward "good for the life of the vehicle" structural components (i.e., $\geq$5000 hr). Fail-safe structure is receiving strong emphasis. Gains in crashworthiness are significant.

Greater attention is being given to the reduction of the vibratory loads and accelerations encountered by the helicopter. This is because of the trend to higher speeds and the increased customer emphasis on reducing pilot fatigue. Rotors with three or more blades and with full or elastically restrained articulation will become almost standard in the near future. Optimally tuned rotor and airframe components are receiving more attention from all helicopter manufacturers. Special dampers or vibration detuning systems are also being applied to advantage on the newer designs. The advent of the armed helicopter will also require design attention to the repeated (fatigue) loadings caused by the use of armament.

11.2 ARTICULATED ROTOR CONSIDERATIONS

In the fully articulated rotor system, the flapping hinge is used to force an equalization of lift at all rotor azimuths and under all flight conditions and to minimize blade bending moments by forcing an inflection point near the important root attachment. The reason for a blade lag hinge in addition to the flapping hinge is primarily structural, but also to reduce vibration. First, the lag hinge permits rotor torque to be resisted by centrifugal force components rather than by blade chordwise bending moments. Second, a convenient radial placement of the lag hinge provides the blades with a low natural frequency in the rotor plane ($\simeq\frac{1}{3}$ per revolution), thus permitting harmonic aerodynamic forces as well as coriolis forces (due to coning and flapping) to be isolated from the rest of the vehicle. A more recent development of the articulated rotor is the incorporation of moderately offset flapping hinges, which, through centrifugal force components, provide hub moments for improved control power and rotor damping. Another development which is used primarily on tail rotors (which do not have cyclic pitch control) of both articulated and semirigid types, and occasionally on main rotors, is the δ_3 hinge, which amounts to pitch-flap coupling. This has been found advantageous in reducing blade flapping under transient conditions. Even the steady lift forces of the fully articulated rotor are resisted by components of centrifugal force resulting from the coning of the blades. In short, the fully articulated rotor system could almost be referred to as the "centrifugally restrained rotor system."

11.2.1 Advantages of the Fully Articulated Rotor System

The notable advantages of the modern fully articulated rotor system are (a) lowest steady and vibratory loads in the rotor system, (b) lowest transmitted vibration, (c) lightest weight rotor for a given gross lifted weight,

(d) adequate control power and rotor damping, (e) lowest airframe fatigue loads, (f) horizontal forces as well as moments can be applied at the hub, and (g) simplest control systems.

11.2.2 Disadvantages of the Fully Articulated Rotor System

The following disadvantages of the fully articulated rotor system are listed: 1. Usually more rotor components. 2. Requires blade lag dampers and landing gear dampers. 3. Less rotor kinetic energy versus gross weight for the same tip speed unless tip weight is added.

11.2.3 Fatigue Problem Areas in Articulated Rotor Systems

As previously stated, the articulated rotor system is subjected to the lowest steady and vibratory loadings of all the types of rotor systems currently used on helicopters. Thus, most present day helicopters using fully articulated rotor systems would exhibit unlimited fatigue lives for all rotor, control, and airframe components were it not for maneuvers, start-stop, transient operating conditions, and resonant amplification of loads.

On the rotor blades themselves the following potential fatigue problem areas may be encountered:

1. Root fitting attachment area of the blades.
2. Blade root fitting at hinge lugs or at blade attach bolt holes.
3. Blade folding fittings and bolts.
4. Chordwise blade bending at the lag damper attachment or outboard.
5. Constant blade section between blade root and tip-flapwise bending.
6. Attachment areas of trailing edge pockets (if used).
7. Tip weight attachment.

The blade section at the root fitting attachment will normally be critical at the outboard bolt hole if a bolted attachment is used. Special processing to minimize the additional stress concentrations due to fabrication defects or due to fretting is normally applied to this area by the manufacturer; nevertheless, reduction in the fatigue allowable in this area is on the order of 3-to-1, as compared with the smooth continuous portion of the blade. Frequently adhesive-bonded fittings and/or doubler reinforcements are added in the root area of the blade in order to offset this reduction in the fatigue allowable and to provide an improved blade service life. Since very stringent process and quality control measures are required in order to assure extreme reliability of the adhesive attachment, the present design trend is to provide a "fail-safe" feature. Welded and brazed attachments are normally avoided, because of poor fatigue strength as well as poor reliability.

The hub and blade retention fittings normally are subjected to a certain amount of vibratory loading derived from the blades. If antifriction bearings are used at all articulated joints, both the bearings and the interconnecting fittings are subjected to the full centrifugal and vibratory loads. Special attention to material quality, stress concentrations, processing, and fabrication is usual for these critical parts of the rotor retention system. The hinge bearings, on the other hand, fail in fatigue by gradual pitting of the race surfaces due to fretting and/or to subsurface fatigue, and are sufficiently fail-safe that frequent inspection is depended upon almost entirely for the retirement of the deteriorated parts in service. The use of tension-torsion type strap flexures to remove the blade centrifugal load from the feathering bearings has been successfully practiced in all three types of rotors. A recent innovation has been the use of a special strap-type flexure to transfer the primary blade loads and to permit all hinge motions (feathering and flapping) inboard of the lag hinge on a fully-articulated rotor system. Such a strap system has proven in tests to be fail-safe and eliminates the primary centrifugal and fatigue loads from the remaining hub-retention components. The strap flexures are normally composed of many thin high strength stainless steel laminations installed as a pack. The hub-to-rotor shaft connection may also require consideration as a potential fatigue problem area, even though vibratory loads from the individual blades have largely been either cancelled or converted to nonrotating steady shears or moments (and lift) within the hub ([9] or Section 11.6.2) for more detailed explanation. The fatigue problem occurs mainly due to the hub and shaft rotating relative to these comparatively steady flight loads and due to torque transients or vibration which may be simultaneously applied during the various maneuvers. Although fully-cantilevered rotor shafts have been widely used in the past in conjunction with articulated rotors as well as with semirigid or rigid rotors, the latest design trend is to transfer these nonrotating, relatively steady load vectors into the nonrotating structure as close to the load source as possible. A recent successful design separates the rotor torque loads and the rotor lift and control loads within the hub itself, the torque going to the rotating system (which does not change the fatigue loads) and the primary flight loads going directly through the hub bearings to a nonrotating structural mast that is directly attached to the airframe. This latter approach, in addition to eliminating all unnecessary exposure of critical rotor support structure to fatigue loading, provides a fail-safe feature whereby power-transmission system components such as the main gearbox or the main rotor shaft may fail completely and yet leave the main rotor, its support structure, and its control system intact for a comparatively safe autorotation landing. On the other hand, where flight loads are carried entirely by the rotor shaft and on out through a structural gearbox, shaft failure would be catastrophic in nature.

The control system for the fully-articulated rotor system is normally simpler and lighter than for the other types of helicopter rotor systems we shall discuss. This is because when three or more blades are used (as is required for the fully-articulated rotor) the major harmonics of the control loads cancel or compensate at the swashplate. (The swashplate is the mechanical device whereby loads and motions in the nonrotating control system are transferred to the rotating control system). Current practice is to locate the swashplate as close to the rotor as possible, thus exposing a minimum of primary flight control components to major fatigue loads. This inherent control stability and load compensation eliminates fatigue considerations that would otherwise also be required for any added compensation, feedback, or damping systems. The inherently good control power of the articulated rotor with offset flapping hinges has also normally avoided the need for movable control surfaces on the fuselage.

11.3 SEMIRIGID ROTOR CONSIDERATIONS

The advantages of the semirigid rotor system from a fatigue standpoint are summarized as: (a) fewer rotor components, (b) elimination of centrifugally loaded rotor hinge bearings (assuming use of tension-torsion member), and (c) avoids the ground resonance phenomenon without the need for either blade or landing gear dampers.

The disadvantages of this system from a fatigue standpoint, on the other hand, are (a) no relief from even harmonic flapwise loads—thus higher steady and fatigue loads in the flapwise direction; and (b) little relief of chordwise loads—steady or cyclic (all major harmonics).

Fatigue problem areas are numerous with this type of rotor system, because of the much greater severity of both steady and vibratory rotor, control, and drive system loads as compared to a fully-articulated system. Because of the absence of blade hinges (flapping and lag) near the blade root, the fatigue loads for the lowest elastic modes are quite high. Steady plus second harmonic flapwise bending and steady plus first and second harmonic chordwise bending predominate in the blade-root, hub, and retention region. Heavier blade root reinforcements and attachments, as well as heavier hub-retention components, are thus required as compared with the articulated rotor system in order to provide reasonable component service life. The rotor shaft receives major fatigue loads in the form of combined steady plus second harmonic torque and steady plus second harmonic nonrotating, cantilever bending. The rotor controls are also subject to two-per-revolution (nonrotating) loads, which may be alleviated by gyro-bars or servo-rotor as well as by irreversible mechanisms or power control units. Even though the pilot is relieved of most of these vibratory control loads, the upper control system, including the alleviation devices noted above, is subjected to these fatigue loads. Thus the rotor control system components are a source of potential fatigue problems to a much greater degree than is the case for an articulated rotor. Finally, the rotor-engine-transmission suspension system, as well as the entire airframe, is subjected to greater vibratory loads by the two-bladed semirigid rotor than would be the case for an articulated rotor, although alleviated by a special vibration isolating suspension. Thus airframe fatigue problems would be expected to be more numerous for helicopters using this type of rotor system.

11.4 RIGID ROTOR CONSIDERATIONS

The attempts to develop rigid rotor systems for helicopters have been almost synonymous with the development history of the helicopter itself. The only modern rigid rotor developments that have met with reasonable success use three or more blades and employ a certain amount of elastic flexibility in the hub-retention area of the rotor. Thus, the only successful rigid rotor systems so far developed might more accurately be referred to as "elastically restrained" rotor systems [11]. This trend is most interesting in view of the present trend in articulated rotors toward increased flapping hinge restraint through a flexure hinge that is offset from the center of rotation. Therefore, the fundamental structural differences between the blade root-hub-retention area of the modern rigid rotor and the modern restrained articulated rotor (with offset flapping) appears to lie mainly in the degree of flapping restraint and in whether or not a lag hinge (bearing or flexure) is used. Tests by NASA [12], as well as theory, have shown that a very rigid rotor will have the following undesirable flight characteristics:

1. Excessive rotor damping.
2. Gyroscopic coupling of pitch-roll.
3. Poor control sensitivity.
4. Inability to tilt the lift vector relative to the airframe.

In order to overcome these objectionable flight characteristics, flap-feather coupling, as well as gyro coupling, plus additional special devices, must be included in the rotor control system, thus increasing complexity over that required for a fully-articulated rotor. The introduction of flapwise flexibility at the rotor hub-retention alleviates, to some extent, the steady and cyclic flapwise bending moments in forward flight. Also, the control coupling and feedback schemes reduce the transient flapwise bending moments due to pitching or rolling velocities or accelerations of the helicopter. (For gyroscopic effects, including spring restraint and transients, see [12].) Nevertheless, the steady and cyclic flapwise bending moments at the blade root, hub, and retention area are more severe than for the equivalent articulated rotor [7] and [11]. Also, the vibratory loads in the plane of the rotor are comparatively high, except as relieved by an isolating suspension. (For vibration levels see [12].) In short, it now appears that the optimum rotor system for the high-speed turbine-powered helicopter of tomorrow will contain a minimum of three blades and either a "rigid" rotor with a certain amount of flexibility or an "articulated" rotor with a certain amount of flapping offset and restraint. The advantages of one system over the other will lie in which provides the simplest, lightest, most reliable system with the lowest vibration levels and the best flight and landing characteristics. As of today, the "rigid" rotor system [8] requires considerably more structural weight in proportion to gross weight than does a well designed articulated rotor; however, it certainly does appear competitive in this respect relative to the two-bladed, semirigid rotor systems, and no doubt will improve with further refinement and development.

11.4.1 Advantages of Rigid Rotor System

The advantages of the modern rigid rotor system may be summarized as (a) potentially excellent control power, (b) mechanical simplicity of the rotor itself, (c) excellent static stability, (d) relative insensitivity to stall or turbulence, (e) potential aerodynamic cleanliness of rotor hub-retention area, (f) no blade or landing gear dampers required.

11.4.2 Disadvantages of Rigid Rotor System

The disadvantages of this rotor system, on the other hand, are (a) control system complexity, (b) higher steady and fatigue loads cause increased rotor system weight, (c) poor inherent vibration characteristics in comparison with the articulated rotor, (d) excessive static stability may restrict the maneuverability; and (e) high rotor loads under transient conditions.

11.4.3 Fatigue Problem Areas of Rigid Rotor Systems

The fatigue problem areas of the rigid rotor lie primarily in the blade root area, its attachment to the hub, the rotor hub, and its attachment to the rotor shaft, including any flexure elements. In many respects, the rigid rotor fatigue problem considerations are similar to those of the semirigid rotor. The rotor shaft, transmission, and suspension problems are also reminiscent of those for the semirigid rotor, except that the forcing loads are lower in magnitude (due to better harmonic compensation with three or more blades) and higher in frequency (integer multiples of the number of blades). Likewise, the airframe vibratory loads should not create as severe a problem as in the case of the semirigid rotor. The modern "rigid" rotor also shares, along with the articulated rotor, an inherently lessened sensitivity to blade stall at high forward speeds, since a smaller fraction of total vehicle lift is carried by the stalled blade.

11.5 POWER TRANSMISSION SYSTEM CONSIDERATIONS

Fatigue considerations of power transmission system components involve not only imposed transient and vibratory loads from the engine or rotors, but also cyclic stressing as a result of rotation under even steady external loads or deflections. Also the usual mechanical transmission system constitutes a multiple degree of freedom vibratory system which may amplify the fatigue loading under certain conditions of operation. The current trend toward higher system rotating speeds produces marked effects on the fatigue considerations for the power transmission system.

11.5.1 Typical Power Transmission System Components

The modern helicopter employs essentially every major technical advance in the field of mechanical power transmission (Figures 11.4 and 11.5):

1. Precision involute gears of carburized and heat treated alloy steel.
2. Precision thru hardened or carburized alloy steel antifriction bearings.
3. High-strength alloy steel shafts.
4. Constant speed mechanical or flexure couplings to provide for small shaft misalignments and structural deflections at high rpm.
5. E. P. gear lubricants.
6. Super-critical speed shafting.
7. Mg alloy gearbox housings.
8. Al alloy shafting and fittings where speeds are high and loads are light.
9. High-capacity, one-way, overrunning clutches.
10. Fail-safe warnings such as noise change, temperature indicators, and electric chip detectors.
11. High-capacity involute splines.
12. Shot peening, etc., for improved fatigue strength.

Other systems of power transmission which have been used in automotive vehicles, such as hydraulics, pneumatics, or electricity, have been ruled out of

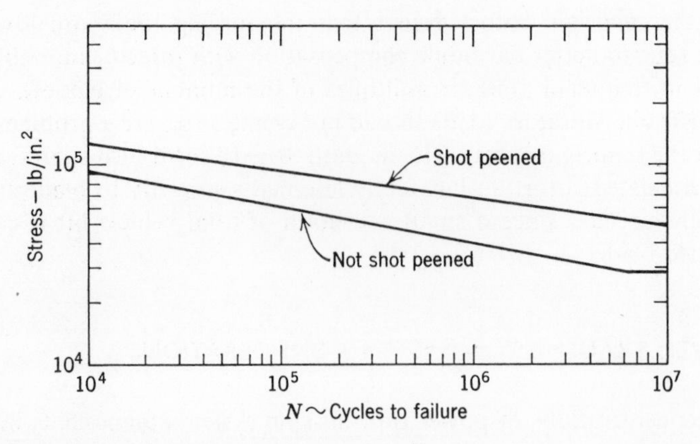

Figure 11.4 Comparative fatigue strengths—shot peened gears versus not shot peened (A.S.M.E. Paper No. 51-S-21).

helicopter applications because of weight and efficiency penalties. (Inefficiency of power transmission is actually viewed as a weight penalty on the basis of approximately 9-lb loss in payload for each 1-hp loss.) The present advanced state-of-the-art permits engine power to be mechanically transmitted to the rotors at over 98% efficiency and for a weight of approximately $\frac{1}{3}$ lb per hp, and with excellent service reliability. Disassembly joints in the power transmission system are normally provided by means of involute splines for maximum static and fatigue strength under the primary torque loading. Bolted connections, where used, are conservatively designed so that the vibratory torque in normal maneuvers is transmitted entirely by the friction between the clamped surfaces. The heavily loaded, low-speed, main rotor

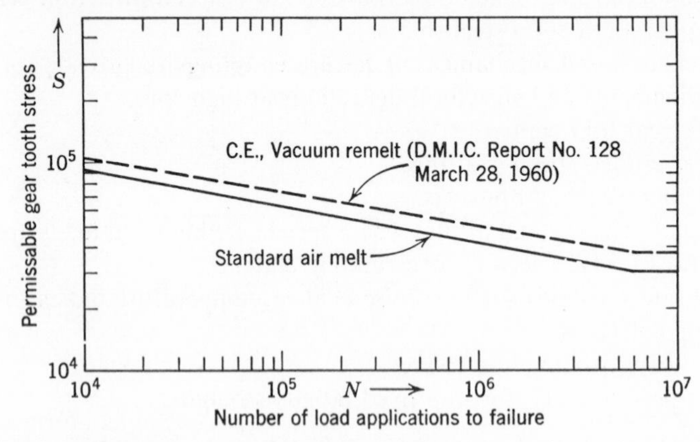

Figure 11.5 Comparative gear tooth fatigue strength (S-N) standard air melt versus C.E., vacuum remelt carburized steel.

drive shaft is normally designed to operate below its first critical speed (especially if it doubles as a cantilevered supporting beam for the main rotor) and is fabricated of heat treated alloy steel. The lightly loaded, high speed, tail rotor drive shaft may be fabricated of thin-walled aluminum alloy tubing and designed to operate at several times the first critical speed (this trend permits added strength and stiffness for the long tail rotor drive shaft, reduces weight, and eliminates most—if not all—shaft bearings and associated supports). The main engine drive shaft usually turns at comparatively high rpm, but is also usually of short length, so that it is usually designed to operate below its first critical speed. It may be fabricated of either steel or Al alloy, depending upon the torque vs. length ratio and upon the torsional stiffness requirements.

Although Cardan couplings (*U*-joints) have been used in the past, especially in tail rotor shafting, they offer many disadvantages and are avoided as much as possible in present advanced designs. They are comparatively heavy, introduce two-per-shaft-revolution torque and speed fluctuation (which increases approximately with the square of the misalignment angle), impose cyclic bending moments on the connected shafts, and present added problems of resonant response in the shafting and supports. Thus, a potential fatigue problem is avoided and weight is saved by merely not using this type of coupling. A very common and reliable light-weight coupling for use where misalignment angles are small and speeds are not excessive, is the crowned (or spherical) gear coupling. Flexure couplings of either the Thomas type (annular pack of stainless steel laminations) or the Bendix type (single-load-path radially-tapered stainless steel discs) (Figures 11.6 and 11.7), are being used increasingly at the present time. The metallic flexure coupling can be designed to withstand torsional fatigue loads much better than a gear coupling, and for a given torque and misalignment angle, can be operated at a rotational speed well beyond that at which a gear coupling would overheat.

The design of helicopter gearboxes is continually trending toward more science and less art. Although there are variations in heat treatments and gear finishing techniques between manufacturers, the helicopter gearbox of today must operate with a consistent, even, polished tooth pattern under load. With the usual involute gearing, sliding occurs at every point on the tooth face except at the pitch line, and lubrication and cooling is thus essential. Tooth contact stresses are estimated based on elasticity formulae developed by Hertz, and bending stresses at the root of the teeth are calculated by methods in [12] and [13]. It should be recognized that dynamic elements such as gears and bearings, as well as shafts and couplings, normally are subjected to repeated loading primarily as a result of rotation, even though the net torque load transmitted may be relatively steady. Thus, power

Figure 11.6 Fatigue test specimen of a metal dish flexible coupling.

transmission gears and bearings in helicopter applications will always be critical for fatigue, rather than static strength. Gear shafts also are usually critical for rotating beam fatigue loading in combination with the steady plus oscillatory torque loading.

11.5.2 Vibratory Loads

The complete helicopter power transmission system constitutes a complex, multiple degree of freedom, vibration system. The major natural frequencies of the system are estimated early in the design stage and corrections to avoid resonance with the obvious forcing frequencies are made either by changes in stiffness (shafts mainly) or mass (flywheels mainly) or by dampers. A few of the more important sources of vibratory problems are (a) pilot excitation—especially through the tail rotor control; (b) transients due to sudden power chop; (c) transients due to sudden power recovery, possibly also involving the sudden reengagement of the overrunning clutch; (d) excitation at NP from either the main or tail rotor (where N = No. of blades of the rotor, and

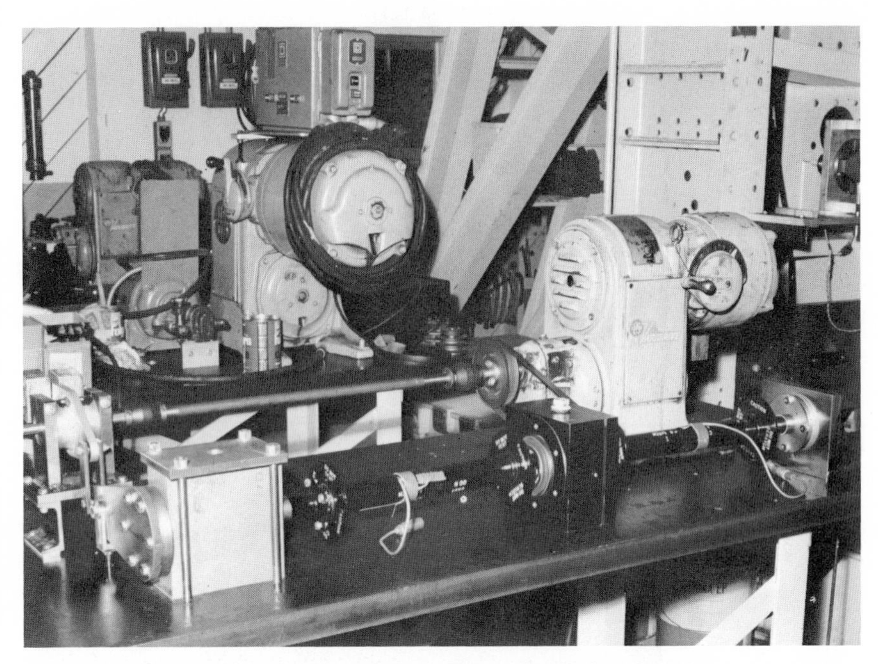

Figure 11.7 Fatigue test of a helicopter driveshaft and coupling.

P is the rotor rotational frequency); and (e) excitation due to engine power fluctuations (particularly for reciprocating engines).

Pilot excitation is generally avoided by keeping the natural frequency of the affected modes at or above approximately 5 cps. Transients due to power chop and power recovery may be the most severe fatigue loadings for portions of the system where a reciprocating engine is employed, but are considerably alleviated with the slower responding turbine engines. Excitation at NP from the rotors may be reduced by blade lag dampers or by other damping devices. Gear tooth engagement frequencies, while detectable, normally do not constitute a fatigue problem (though possibly a noise problem).

11.5.3 Fatigue Considerations of Power Transmission Components

Gear teeth may fatigue due to either excessive Hertzian stresses or excessive tooth bending stresses. Figure 11.8 and [12] and [13], show typical allowable operating stresses which are derived from a mass of test data. The statistical limits shown are conservative because of recent improvements in materials and processes, manufacturing, and quality control. Thus the 5% failure line actually amounts to less than 1% in present aircraft practice. The current trend is to use vacuum melt alloy steels, employ maximum tooth-root fillets,

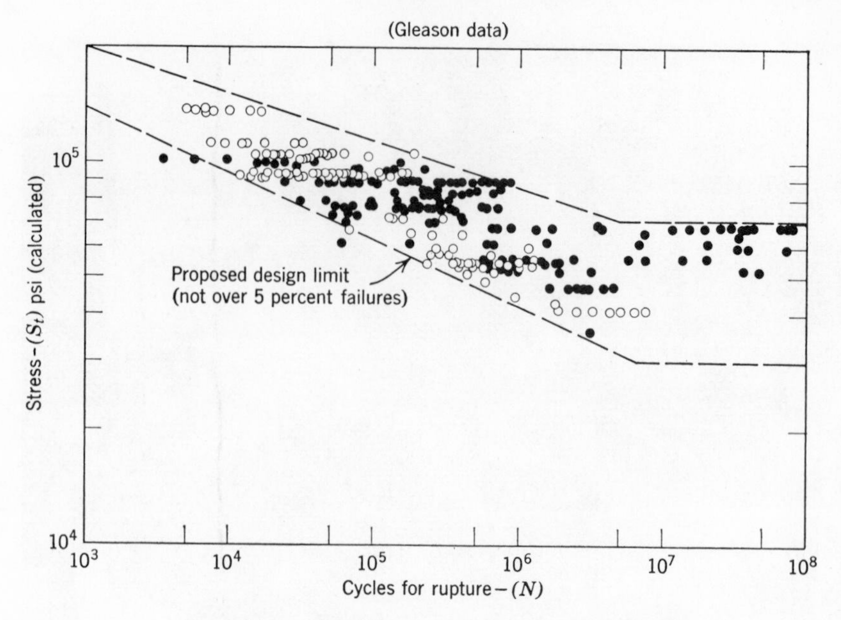

Figure 11.8 Allowable gear tooth bending stress [12,13].

shot-peen, etch inspect for grinding cracks, employ careful metallurgical control of carburizing and heat treating, magnetically inspect, and avoid designed-in stress raisers. Anti-friction bearings, though conventionally rated on a B-10 basis (i.e., 10% failure), actually perform with less than 3% failure, and even these failures are detectable before they become serious. In addition to change in noise level, the pilot is warned of bearing or gear surface fatigue failures by oil temperature indication and by electric chip indication in the cockpit of recent advanced helicopters. Gear shafts and rotor shafts are trending toward vacuum melt steels, generous fillets, heavy preload at fits and attachments, as well as special processing, such as shot peening. Gear shaft failures are thus extremely rare in service. The present trend is also toward use of higher strength steels, such as Hy-Tuf, which have shown improved fatigue, as well as static, strength in shaft applications (refer to Figure 11.9). Gear couplings are usually fabricated with a combination of nitrided and carburized steel mating surfaces (hardness in the range R_c 60–68). These couplings will withstand heavy steady torque and moderately heavy vibratory torque ($\pm\frac{1}{6}$ of the steady torque). However, flexure couplings are employed successfully at even more severe vibratory torque levels ($\pm\frac{1}{4}$ of the steady torque). Gearbox housings (usually of Mg alloy castings) are usually critical more for rigidity than for fatigue stresses, except where the housing serves also to support the main or tail rotor to the airframe. In these cases redundant

Figure 11.9 Fatigue test specimen of a gear coupling.

attachments are now normally made between the gearbox and airframe so that the installation is "fail-safe" to at least limit load.

The future trend toward higher rotational speeds, and the resulting lighter, yet more rigid, systems, will call for even greater consideration of fatigue loadings; for example, at 4000 to 10,000 rpm, those items which see $1X$ (where X is the rotational frequency) such as rotating beams (shafts), couplings and flexures, and gear teeth, and which, in some particular maneuver(s), operate at equal stresses (above the endurance limit), will have $\frac{1}{4}$ to $\frac{1}{10}$, respectively, of the service life had the rotational frequency been 1000 rpm. By similar logic, those items which are critical for stress cycles applied at frequencies of $2X$ and higher will have their service lives reduced. The gist of this is: the higher the frequency of cyclic load application on a component, such as due to higher rotational speed, the greater is the portion of the maneuver spectrum which must be kept within the component endurance limits in order to provide long service life. Also, the more important become the infrequent extreme maneuver loads or the transient conditions. Should the operating frequency be increased without limit, the only possibility of achieving a finite service life would be to have even the peak cyclic loads from extreme maneuvers or transients lie within the component endurance limits. This phenomenon is not actually as bad as it may appear at first blush, since the steady loads are reduced as the rpm is increased for the same transmitted power. In fact, a very real safety advantage occurs as a result of this design trend, since more of the most frequently occurring maneuvers are kept within the component endurance limits, and only the infrequently encountered conditions cause fatigue damage. Thus, there is a much improved chance of detecting, by inspection, the rare cull components before catastrophic failure occurs.

The previously mentioned trend toward "fail-safe" design also applies to the helicopter power transmission system. However, the concept may be applied somewhat differently than for the rotor or airframe structures. For example, incipient failure may be detected by unbalance (shake), noise, or change in vibration level. Safe landings after failure of tail rotor shafts, and engine shafts have actually been made (although the credit may be due to an alert pilot). For the helicopter design (previously mentioned) which does not utilize the main rotor shaft to transmit flight loads, it should be possible to make a safe landing even after failure of the main rotor shaft.

11.6 LOAD SPECTRA CONSIDERATIONS

Reference [4] summarizes typical methods by which the fatigue strength of helicopter structural components may be substantiated to the F.A.A. at the present time. These methods, though not perfect by present advanced theoretical standards, nevertheless have in fact filled a practical need for the substantiation of the safe service lives of helicopter fatigue loaded components. This document requires consideration of three basic factors affecting any rational determination of the fatigue life of a structural component:

1. The stresses or loads associated with the expected maneuvers and operating conditions in service.
2. The frequency of occurrence of these maneuvers and conditions, so as to arrive at an over-all operating stress or load spectrum.
3. The actual fatigue strength characteristics of the structure.

Although great progress is being made in the analytical prediction of helicopter loads and stresses, and these methods may be adequate for initial engineering design purposes, the final fatigue lives must nevertheless be based on strain and other data obtained from a carefully controlled instrumented flight test program. Typical maneuvers and occurrence percentages are shown in Table I of [4]. Instrumentation is applied in areas determined, from stress analysis or previous experience, to be of interest from the fatigue standpoint. Then each instrumented component is carefully load-calibrated. Such associated information as acceleration (in g's), rpm, airspeed, engine torque, control motions, and blade motions are also recorded so that the adequacy of each flight maneuver can be assessed before the strain data is used. This flight loads program is always extremely thorough, investigating the extremes of rpm, airspeed, control inputs, power inputs, center of gravity, altitude, etc.

The stress cycles within each maneuver are counted and the time to initiate and complete each maneuver is determined. Then, using Table I, [4], as a guide, the number of maneuvers of each type per hour of flight is determined, and a complete over-all stress or load spectrum per flight hour is thus arrived at for each critical component (Figure 11.10). Often, a more detailed maneuver spectrum breakdown is used in order to provide for severity of maneuvers (i.e., mild maneuvers will occur with much greater frequency than extreme maneuvers), variations in gross weight, speed, altitude, rotor rpm, and so forth. In the end, the maneuver points of major interest in the fatigue load spectrum are usually repeated a number of times, to be sure the data obtained is repeatable and representative. The maneuver severity distributions and frequency of occurrence are also reviewed to see that they are both reasonable and conservative for the particular helicopter design being investigated. Figure 11.12 schematically illustrates the over-all fatigue loads determination. Reference [4] suggests that the load spectrum be based upon flying a representatively severe mission profile. The military services have actually recorded flight loads data during their normal service usage for certain types of helicopters. In some cases these data show the F.A.A. spectrum to be quite conservative (Figures 11.10 and 11.11). The statistical nature of helicopter loads may also require further consideration [2]. For instance, piloting procedures, manufacturing variations, maintenance adjustments, and wear may alter the loads encountered.

The vibratory loads that are of interest from the fatigue viewpoint occur on any particular rotor blade as a combination of integer multiples of the rotor frequency, and may be expressed in Fourier series form; namely,

$$\sigma = \sum_{0}^{n} |\sigma_{A_i} \sin i\Omega t + \sigma_{B_i} \cos i\Omega t|.$$

Of course, for transient flight conditions such as maneuvers, both the steady stresses and oscillatory stresses will vary in different amounts throughout the maneuver, until a steady-state condition is again reached. These loads may be counted and classified as to cyclic stress level by actually measuring the counting from representative oscillograph records. More recently, various systems of automatically counting and classifying the oscillatory loads data have been studied. The most common of these systems are (a) mean crossing peak count; (b) simple range count; and (c) simple interval crossing count [17].

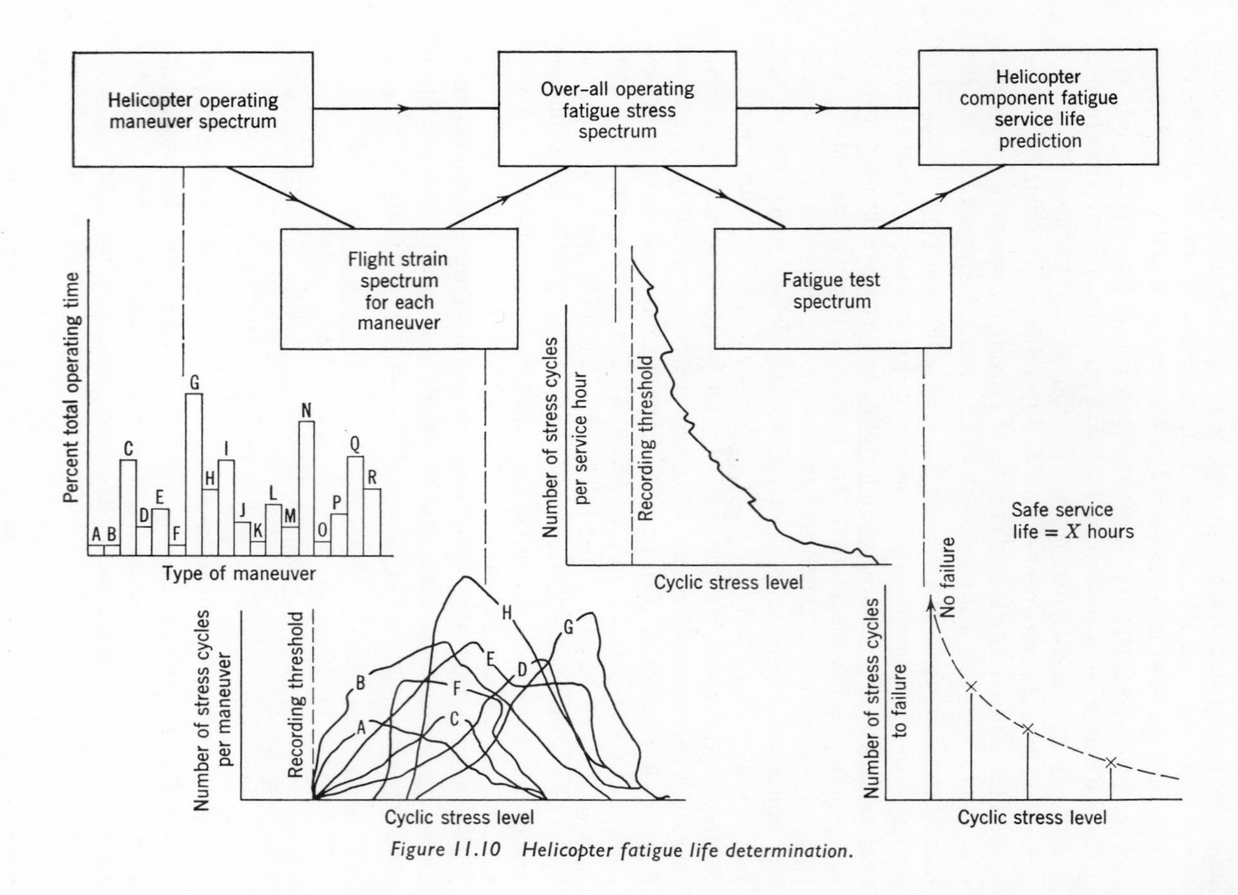

Figure 11.10 Helicopter fatigue life determination.

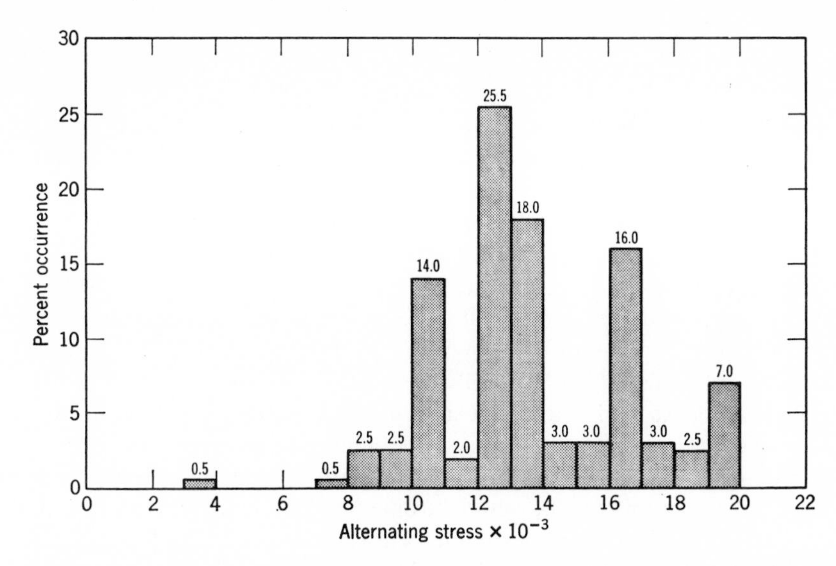

Figure 11.11 Typical helicopter rotor blade spectrum A. CAM 6, appendix A distribution.

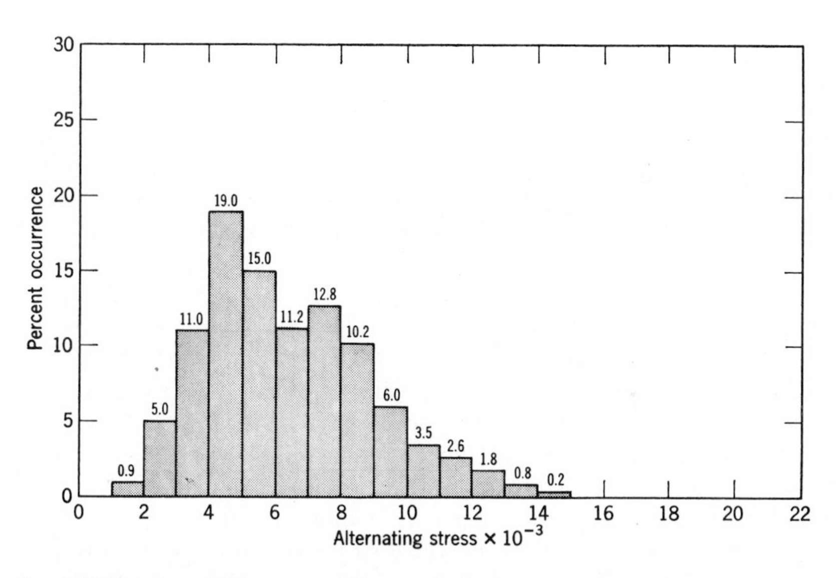

Figure 11.12 Typical helicopter rotor blade spectrum B. Service strain survey distribution (data from 120 flight hours—3163 observations).

11.6.1 Rotor Loads

Rotor loads are obtained primarily from strain gages located at likely points of interest, such as flapwise bending at the blade root attachment area and at additional spanwise locations, and one or more chordwise bending locations (at least at the blade root). The flapwise bending, pitch control, and chordwise bending loads are scrutinized to detect any unwarranted or detrimental resonances, coupling, flutter, or other interaction. Oscillatory loads are normally low during hover, but increase significantly during transition to forward flight, because of changing flow conditions. In forward flight, the oscillatory blade loads are primarily due to the cyclic $1P$ variation in blade element airspeeds and pitch, but also due to nonuniform downwash (or inflow), rotor coning, higher harmonic flapping, and so forth. The effects of gross weight, altitude, and air-speed variations require more complex operating restrictions for the newer turbine-powered machines, inasmuch as these variables are no longer negligible. The severity of blade stall encountered will also greatly affect the rotor blade loads experienced. More detail data on rotor loads are presented in [11], [8], and [7]. A last, but very important, fatigue loading for portions of the rotor system is that resulting from repeated centrifugal force applications due to the start-stop cycles.

11.6.2 Hub, Pylon, Control Loads

The methods by which individual blade or control loads combine and either cancel or add are discussed in [9]. We will summarize by stating that it is predominantly the integer multiples of NP (where $N =$ number of blades and $P =$ speed of rotation) that are transmitted to nonrotating structure. It is important not to overlook the fact that the effective frequency of most of the oscillatory loads changes by $\pm 1P$ when changing the frame of reference from rotating to nonrotating, or vice versa. For instance, a three-bladed rotor will pass $3P$, $6P$, $9P$, and so forth, loads to the nonrotating elements, and these would derive from $2P$ and $4P$, $5P$ and $7P$, and $8P$ and $10P$ loads, respectively, in the rotating system. As a class problem, you may prove this for the lowest passed frequency mentioned above (that is, $3P$). In general, the pylon and rotor support structure in the fuselage will not be critical in fatigue for the vibratory loads transmitted from the rotor in normal flight (assuming the vibratory accelerations are within the military specifications for crew comfort levels). However, these structures may still be critical in fatigue for cyclic loads incurred during maneuvers. Many load cycles of varying severity will usually occur during each type of maneuver, the peak cyclic stress being that determined by the one-per-maneuver combined variation in the steady stress and the peak cyclic stress (assuming they occur simultaneously—as they usually do).

11.6.3 *Tail Boom and Stabilizers*

These components, in addition to being loaded by rotor vibration of the fuselage, are loaded in fatigue by such maneuvers as pedal reversals and by the turbulent rotor wake, as well as by gusty air. Occasionally these items may respond as tuned vibration absorbers for main or tail rotor unbalance or out of track conditions, and thus be unpredictably affected in regard to fatigue loading. For this reason, fail-safe design of these components is highly desirable [21].

11.6.4 *Power Transmission System*

Except for rotor- and engine-induced vibratory loads, the main fatigue loads in the power transmission system are the result of excitation of the system natural frequencies by maneuver or power (torque) transients. Also, as previously noted, even steady system loads may constitute fatigue loading on rotating components.

11.7 DETAIL DESIGN AND FABRICATION EFFECTS

Quality of detail design and fabrication markedly influences the fatigue strength of helicopter structural components. Generally, the added cost and weight required to eliminate unnecessary local discontinuities and stress concentrations are negligible. Similarly, the avoidance of detrimental defects caused by careless fabrication or handling is merely a matter of indoctrination of the workers and foremen involved and the provision of the proper handling and storage facilities during fabrication. Favorable processing methods in regard to fatigue must be specified during the engineering design phase, as also must the adequate detailed inspection of finished parts.

11.7.1 *Stress Concentrations and Discontinuities*

Probably the most common area of stress concentration and discontinuity in helicopter structural components is the bolt or rivet hole and the associated mechanical joint area. First, the finish on the inside of the hole is important; reamer marks, galled areas, and other tool marks must be eliminated. Second, the edges of the hole (intersection of the hole with the surfaces of the part) must be particularly free of defects inasmuch as higher stresses may occur at the surface due to unavoidable eccentricity of loading. Thus burrs, nicks, rough or torn edges, or any other tool mark or surface defect must be avoided. Normally, the edge of the hole is carefully "broken" or rounded to remove burrs and torn metal from the machining operations. Shot peening and/or burnishing or bearingizing are often used in order to leave a favorable residual compressive stress at the hole. The hole fits are also important. Frequently removed bolts (in service) are usually provided with bushed holes to minimize damage in the critical fatigue area. Fretting, both in the holes and at the faying surfaces of the joint, can seriously reduce the fatigue life; however, this may be minimized by nonmetallic liners or coatings (even zinc chromate is good), adhesive bonding, and by high frictional restraints due to heavy clamping preload of the fasteners.

11.7.2 Materials and Process Effects

References [16], [18], and [19] show the deleterious effects on fatigue strength of various metal platings, decarburization (steels), brazing, etc. The various alloys and heat treats of the same metal will exhibit different "perfect specimen" fatigue strengths as well as different notch sensitivities. For instance, 7075*T*6 offers no increase in fatigue strength over 2024*T*4 for most fabricated structures for low-stress, high-cycle, types of loading. The beneficial effects of certain case hardening processes for steel, such as nitriding and carburizing, are also indicated. Figure 11.4 shows the beneficial effect of shot-peening for one particular application. Figure 11.5 shows the advantage of vacuum melt steels in respect to fatigue strength. Special quality control to eliminate surface defects is needed to obtain maximum fatigue strength, particularly in the case of very high strength materials. This may be accomplished by magnification, reflected light techniques, solvent and dye techniques, etc. Such metal joining processes as welding, brazing, and adhesive bonding are of interest. Fusion welding is avoided except where vibratory stresses are very low (on the order of ± 5000 psi for steel). Brazing will also reduce the safe fatigue strength of steel appreciably (to the order of ± 9000 psi for stainless steel) based on unpublished data. Adhesive bonding, however, detracts hardly at all from the fatigue strength of the basic metal, be it aluminum alloy or steel, and is therefore a preferred joining method for fatigue applications.

11.7.3 Environmental and Human Factors

Protection against corrosion pitting and stress-corrosion cracking are necessary for fatigue loaded components. Platings, coatings, paints, anodic treatments, etc., are used. Accidental damage due to careless handling during manufacture and maintenance is also guarded against for critical fatigue loaded components.

11.7.4 Detail Design Factors

Careful control of alignments, fits, and tolerances are required, especially at joints or fittings. Provision of gradual transition in cross sections, generous fillets, and adequate preloads are made by the designer. Eccentricities in loading are avoided wherever possible. Rotating beam loadings are kept as low as possible throughout the helicopter, by placing support bearings as close to the load points as possible and by transfer of the rotating beam loads as directly as possible to adjacent non-rotating structure. Structural stiffness is tailored to the desired natural frequency of each main system. Attention is also given to "fail-safe" characteristics of important structural areas, including ease of inspectability.

11.8 VIBRATORY RESONANCE CONSIDERATIONS

A vibration is basically any repeating periodic motion [10] and [7]. The period of the vibration is expressed in time units. Also, the number of periods (or repetitions) per unit of time is called the frequency of the vibration. The simplest type of periodic motion is harmonic motion, i.e., of the form $x = x_{0_j} \sin j\omega t$; and, more generally, any complex vibration of constant amplitude and period may be expressed by a Fourier series of the form

$$x = \sum_{0}^{n} x_{0_j} \sin j\omega t.$$

In mechanical systems, we are interested primarily in mass (inertial) characteristics and in elasticity characteristics, as well as damping characteristics, since each resists motion in its own unique manner. A mass (weight) reacts to the accelerations accompanying the motion; an elastic element (structure) reacts to the deflections of the motion; and the damping reacts to the velocity of the motion. The relations of these types of vibratory forces for the simplest harmonic motion thus would be

$$\text{inertia, } F_i = -m\ddot{x} = +mx_0\omega^2 \sin \omega t$$
$$\text{elastic, } F_e = -kx = Kx_0 \sin \omega t$$
$$\text{damping, } F_d = -c\dot{x} = -cx_0\omega \cos \omega t,$$

where m = mass,
k = stiffness,
c = damping coefficient,
ω = angular frequency,
t = time,
x = motion coordinate,
ω_n = natural frequency (angular).

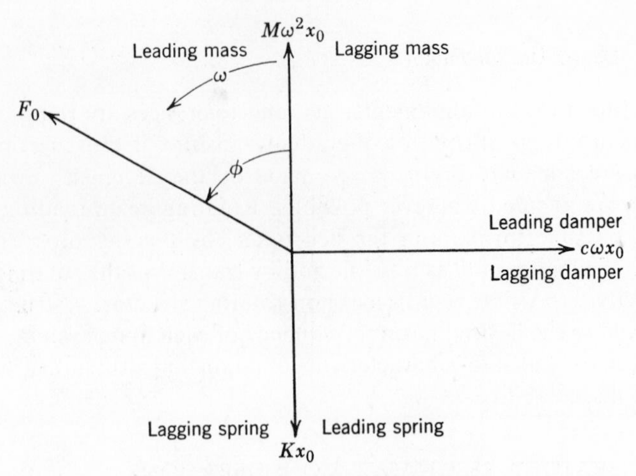

Figure 11.13 Mass, spring, and damping forces as vectors.

It may be seen from the above equations that the inertia and elastic forces are 180° out of phase and thus oppose each other. Also, the damping force lags the motion by 90° (Figure 11.13). At this point, the concept of natural frequency (undamped) should also be given some thought. Physically, it is the frequency at which the inertial and elastic forces are in exact balance, i.e., $F_i + F_e = 0$, or $\omega_n = \sqrt{K/m}$ (Figure 11.14).

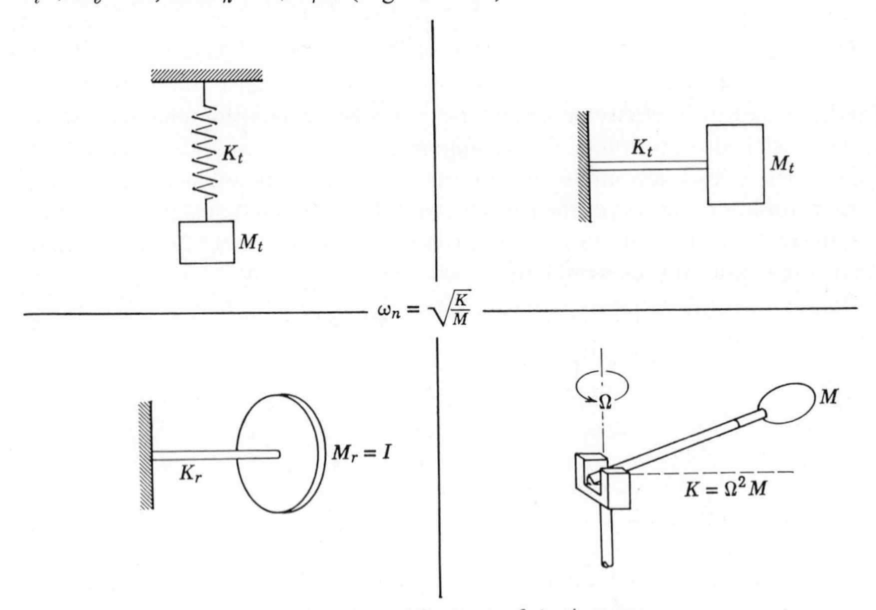

Figure 11.14 Natural frequency of simple systems.

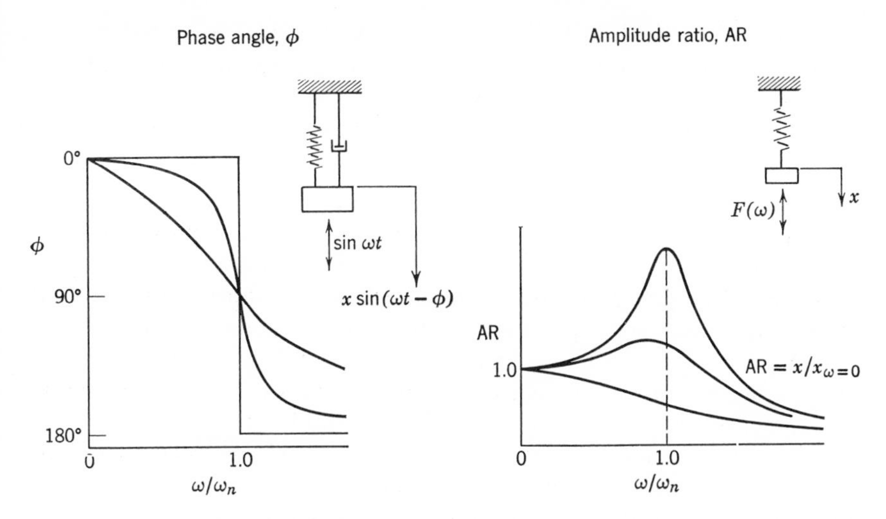

Figure 11.15 Phase angle and amplitude ratio versus frequency ratio for various amounts of damping.

Another concept of vibration that is most important is that of resonance and resonant amplification of a forcing function (an oscillatory load in the normal case; $F = F_{0_j} \sin j\omega t$). At resonance (neglecting damping), since the inertial and elastic loads exactly cancel independent of amplitude, the application of any load whatsoever (except at a node point, if any) at the resonant frequency will be unbalanced except by continuing increases in amplitude of the motion without limit. Practically, except for self-excited systems, there is always a finite amount of damping present which will limit the amplitude to some finite value, depending on the amount of damping present. This phenomenon is referred to as resonant amplification. As noted in Figure 11.15, some degree of finite amplification or attenuation of the internal loads and stresses in the vibrating system occurs for all forcing frequencies (except 0) as compared to quasi-static conditions. Other vibration concepts, which it is helpful to consider, especially in regard to helicopter rotor blades, are those of multiple degrees of freedom, mode shapes, nodes, and coupling. Degrees of freedom may be defined as the total number of possible simultaneous vibratory motions of all the mass elements of a system within the acting constraints and elastic restraints of the system. A continuous free beam of any finite length will thus have theoretically unlimited multiple degrees of freedom in two directions of bending and in torsion. The same is true of a rotor blade. There will be as many mode shapes and corresponding natural frequencies as there are degrees of freedom. Fortunately, only the lowest few natural frequencies and their mode shapes are of concern because:

1. They require less energy to excite to a given amplitude and are usually more lightly damped aerodynamically.
2. The forcing loads decrease markedly with each higher harmonic of the rotor frequency.

Each higher mode shape (or natural frequency) has one additional node point (i.e., points having no vibratory motion regardless of amplitude for the particular motion). A further important consideration is that no response whatsoever in a given mode of vibration will occur for a forcing load (even at the natural frequency of that mode) applied to any node point in that mode. Coupling of vibratory modes occurs where the vibratory forces (mass or elastic) of one vibrating system are simultaneously involved in another vibrating system. For instance, coupled flapwise bending and torsion may occur due to the mass and spring elements not lying in a straight line; also, coupled chordwise and flapwise bending may exist due to blade twist or due to misalignment of the retention straps or rotor hinges.

11.8.1 Rotor System Resonances

From the above discussion it is readily understood why the helicopter manufacturer employs every effort to avoid rotor blade resonance with at least the lower harmonics of the rotor frequency. In practice, the designer calculates at least the first three flapwise, the first two chordwise, and the first two torsion modes for an articulated rotor system. Elastic and mass centers are determined and chordwise mass placement set either for negligible or favorable coupling. Particularly to be avoided, if at all possible, are blade resonant frequencies at $NP \pm 1$, because they tend to be transmitted directly on through to the airframe (at nonrotating NP frequency). An additional consideration relates to the difference in vibratory characteristics imposed by physical hinges in the rotor retention area. Ideally, a hinge becomes a forced node point for all modes of any vibratory system which include this particular hinge motion. Thus, the rigid rotor will have an entirely different set of vibratory mode shapes and natural frequencies to avoid than does the articulated system. Another important effect on rotor blade natural frequencies and mode shapes is that of rpm in the form of centrifugal stiffening; for instance, even a chain (i.e., no bending stiffness whatsoever) will exhibit a first nonrigid-body frequency of approximately $2.5P$.

11.8.2 Other Systems

The airframe and power transmission system may also be susceptible to resonant amplification of rotor induced loads, usually at NP frequency. The system natural frequencies are calculated and then checked experimentally. Stiffness, mass, or damping changes are incorporated where necessary to avoid resonance with known forcing functions. Occasionally, because of system complexity, so many natural frequencies may exist at or near the exciting frequencies that operation at a resonant frequency may be unavoidable at some rotor rpms. This problem may be solved by adding operating restrictions or by suitable detuning or damping methods.

11.9 FATIGUE TESTING AND SERVICE LIFE SUBSTANTIATION

As mentioned in Section 12.6 the determination of the actual fatigue strength characteristics of the structure is an important step in the substantiation of the safe service life of each fatigue loaded helicopter primary structural component. Reference [4] presents several acceptable methods for obtaining the fatigue properties of the structure. Also, acceptable methods of utilizing the load spectra information in combination with the fatigue strength information in order to arrive at engineering estimates substantiating the safe service lives of these helicopter components are presented. These approaches, though by no means perfect, provide guidance material for reasonable and conservative engineering approaches to the problem.

The actual fatigue strength characteristics of the structure may be determined by (a) analytical methods, (b) testing methods, including *S-N* Curves, spectrum tests or major system tests, (c) fail-safe evaluation, or (d) combined fail-safe and safe-life evaluation.

Analytical methods are acceptable only where allowance is made for the effects of stress concentrations and other significant factors and where an additional factor of safety of 3 is included. Fatigue test data for specimens or parts with similar stress concentrations are used to establish the basic fatigue properties of the particular structure see also [15], [18], and [19]. Considerable engineering judgement must be used in the application of this data to actual structures. For this reason the analytical approach must always be used with caution, and the fatigue testing of actual components or the critical portions thereof is generally preferred. The stress concentration factor used in the analysis must be adequate to allow for surface conditions, fabrication variables, fretting, size, and discontinuity effects, as well as, the stress concentrations around bolts and rivets, threads, fillets and other notches. A technique based on the Goodman diagram may be used to reduce the allowable oscillatory endurance limit stresses based on completely reversed loading in order to account for the effect of steady stress. This is done by reducing the oscillatory endurance limit in a manner directly proportional to the ratio of the actual steady stress to the material yield stress. The safety factor is then applied in order to arrive at an "operating boundary" stress. If all stresses from the flight load spectrum are less than this value, no fatigue testing is necessary and the component is viewed as having "unlimited" life. Wherever possible, in lieu of this purely analytical approach, the "operating boundary" stress is based on actual fatigue tests and service experience.

In the determination of safe service life, we are primarily concerned with oscillatory or vibratory loadings on structural components, since it is this type of repeated loading that causes fatigue damage and ultimately a failure. However, a complete engineering approach must also include the once-per-maneuver variation in the mean stresses as well as the ground to air cycle effect on mean stresses. In the consideration of fatigue damage, I will now review a number of fundamental ideas or concepts regarding fatigue loadings, which would be of little concern in the consideration of static or infrequently applied loadings.

First, we may think of the endurance limit as the maximum cyclic loading that the structure will withstand for an indefinite period of time without detectable damage. Consequently, in our consideration of the instrumented flight loads data, we are always primarily concerned with those cyclic stresses at or above the endurance limit. Second, notches, discontinuities, corrosion pits and other stress concentrations which would be of little concern in an infrequently loaded structure because of alleviation due to yielding of the metal, become of the utmost importance in determining the fatigue life of the structure because the fatigue stresses are generally well below the metal yield point. Thus, in many respects, a structure subjected to fatigue loading behaves in a manner similar to a brittle material under static loading (like glass). Third, fretting or wear damage to rubbing surfaces has the effect of increasing the stress concentration factor versus the number of cycles and usually increasing the cyclic stresses as well. Fourth, safe life prediction must also allow for the statistical scatter which is inherent in fatigue test results for any given component subjected to a given cyclic loading. Fifth, notch sensitivity must be considered in the fatigue evaluation of a structure, since some materials and some types of structures will exhibit far more loss of fatigue strength due to a given theoretical stress concentration than others. Thus low-strength materials with high ductility and toughness generally have the lowest notch sensitivity, and the ultra high strength materials with low ductility, and especially at high steady stress levels, usually exhibit very high notch sensitivities. Finally, the effect of the steady or mean stress on the fatigue life must be accounted for. Analytically, a Goodman diagram is used.

To summarize, the primary purpose of both bench testing and fatigue analysis is "to provide a rational prediction of the safe fatigue service life of the helicopter and its components." From the engineer's standpoint, this prediction is based on the following factors:

1. Knowledge of the maneuvers and associated stresses, particularly the cyclic stresses, expected to be encountered in normal operation.
2. Knowledge of the frequency of occurrence of these maneuvers and the associated stresses.
3. Knowledge of the fatigue strength of the structure, including all components.

Further discussion will be concerned only with the latter. Generally, the main components of a helicopter which must be substantiated to the FAA with respect to safe service life are the following:

1. Main and auxiliary rotor systems, including blades, hubs, and retentions.
2. Power transmission systems, including rotor shafts, drive shafts, couplings, and gearboxes.
3. Rotor control systems.
4. Other parts which usually also require attention are: pylons and support structure, tail surfaces, tail booms, and landing gear.

Substantiation of fatigue service life to the FAA may be accomplished either by analysis or by fatigue testing (or by a combination of both). Unquestionably, the former method of substantiation is by far the quickest and cheapest. Because of the additional unknowns to be allowed for, however, such as prediction of stress concentration and extrapolation from specimen fatigue data, plus the normal errors involved between the theoretical and actual stresses for a given loading, as well as statistical scatter, the FAA imposes an arbitrary additional safety factor of 3 when using this method. Therefore only those components exhibiting cyclic loads far below the nominal endurance limits can be substantiated by this method. Such components may include control rods and levers, power transmission shafting, landing gear, tail boom, and rotor pylons. Also considerably greater care in the detail stress analysis is required for the prediction of fatigue stresses than is normally considered justified for the prediction of static strength, since no stress relief due to local yielding can be counted on.

11.9.1 Substantiation of Fatigue Service Life by Fatigue Testing

The following three methods of testing are considered acceptable by the FAA:

1. Laboratory testing of structural components or portions thereof under simulated steady plus cyclic loadings.
2. Testing of a complete rotor system on a whirl tower.
3. Actual flight demonstration.

The latter method is normally avoided because of the safety hazard to the pilot and the helicopter and because it is usually the most costly method. The whirl tower test is sometimes used but is again more costly and more difficult to control than laboratory testing. Laboratory testing is generally preferred because of lower cost, improved safety, and better engineering control over the test conditions and loads than for any of the other methods. Further discussion of fatigue testing will therefore be limited to laboratory type testing. The FAA specifically recognizes three general types of laboratory fatigue testing.

Establishment of S-N Curves. In this type of testing the knee of the *S-N* curve is generally found, indicating the endurance limit. Endurance limit, for test purposes, is defined by the FAA as 10^7 cycles for steel and 5×10^7 cycles for aluminum alloys. Scatter of the test data is allowed for by reducing from an average curve through a minimum of four test points by a factor depending on the number of specimens tested and the observed variability in results. Steady stresses may be allowed for either by running all testing at the highest measured steady stress or by Goodman diagram extrapolation between two *S-N* curves determined at each of two different steady stress levels. Coaxing to show a slightly increased endurance limit is always avoided. The prime advantage of the *S-N* type of testing is that the testing can be run to completion prior to completion of the instrumental flight loads program [5].

Cyclic Unit Testing or Load Spectrum Testing. In this type of testing a spectrum including a proportional amount of time at each stress level encountered in the instrumental flight loads test program is repeated, until a failure is achieved. Again, at least four specimens should be tested, and the lowest life of these four is used for service life prediction with a rational further reduction to allow for scatter. This method of testing requires the random application of the high and the low cyclic stresses and in sufficiently small repeated spectrum blocks to realistically represent the applied service loads. This method of testing is rather infrequently used because it requires a more costly test setup and an elaborate control of the loading program. Also, this type of test cannot usually be run to completion until after the flight loads program has been completed and the data reduced. The availability of automatically programmed and monitored fatigue testing equipment, however, is currently making spectrum testing more practicable. Adjustment of test results to new or changed flight loads data must still be handled by cumulative damage methods, however [6].

Modified Test Procedure. This is a combination of the two types of fatigue testing in which the endurance limit for the structure is first determined, and only those loads that are above this established endurance limit are included in the final spectrum testing. If the entire load spectrum for all maneuvers is below the endurance limit after due allowance for scatter, no further testing is required. This method also has the advantage that the establishment of the endurance limit for a structure can be accomplished before the flight-loads program. Usually, only a few of the more severe loading conditions would thus have to be included in the final spectrum testing.

11.9.2 Design of the Fatigue Test Setup

A few principles for designing the most compact, economical, and easily controlled fatigue test setups, Figure 11.16, will now be discussed. The single most important principle utilized wherever possible in the design of fatigue test setups, is that of cyclic load amplification due to operation at the natural frequency of vibration. In this type of system, the actuating load input is ideally only that required to overcome the damping in the system. For a lowly damped system then, the actuating loads for the test may thus be reduced by the order of 50 to 1. Also, power requirements are greatly reduced, and actuator component loads are only a small fraction of the load felt by the test specimen. Thus, the life of mechanical test equipment components is

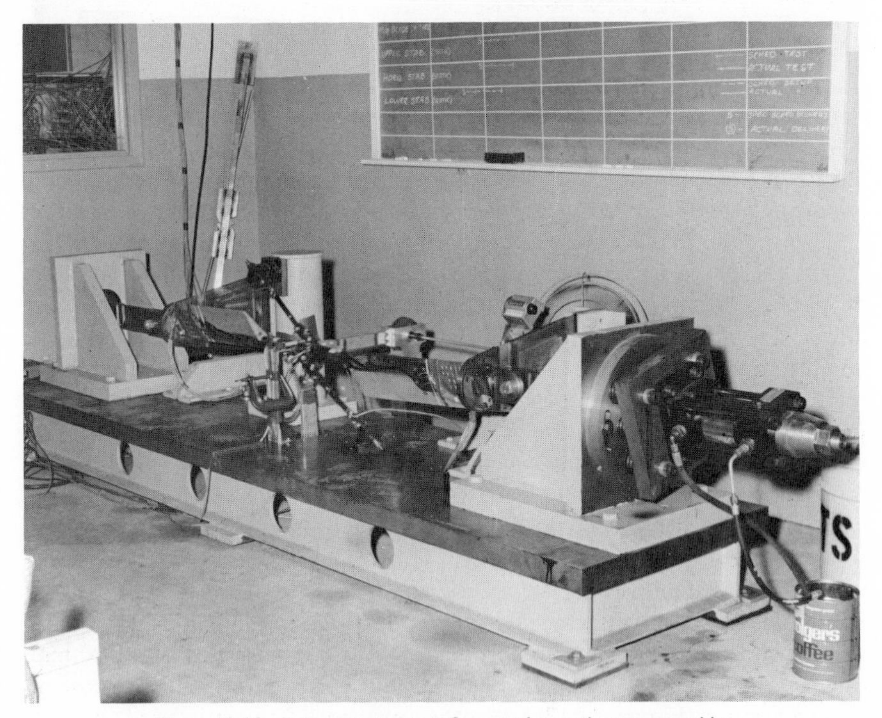

Figure 11.16 Fatigue test set-up for complete tail rotor assembly.

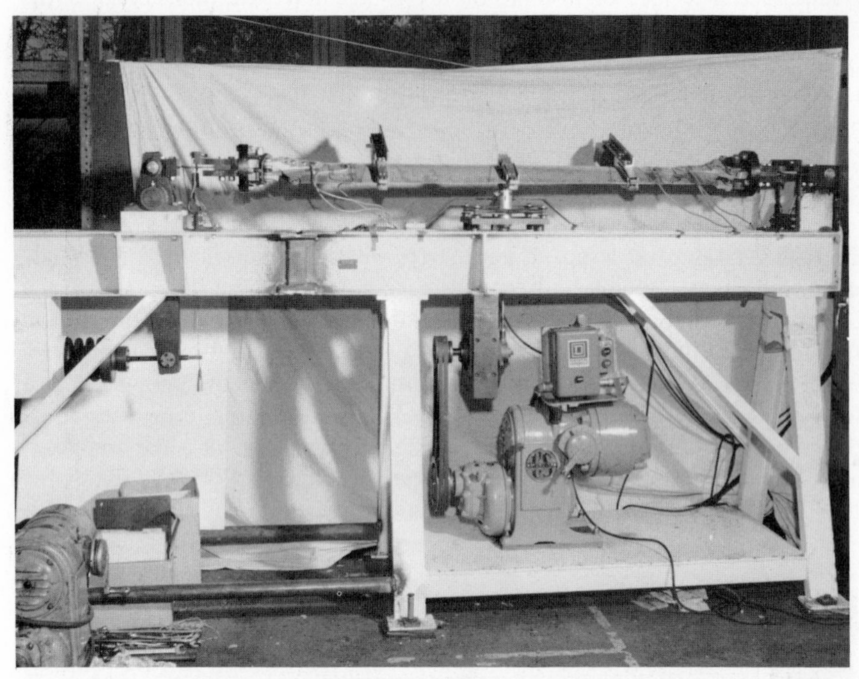

Figure 11.17 Back-to-back or duplicate fatigue testing of main rotor blade root and attachments.

increased tremendously, and a test of this type, once initially checked out, is normally extremely drift-free and trouble-free, and requiring very little monitoring or maintenance attention. Another principle widely used in fatigue test setups is the strap or the plate-type flexure, which takes the place of mechanical hinge components, such as bearings. These flexures are designed to operate below their endurance limit at all the test loads and, once designed and checked out, are likewise trouble-free in operation and require no periodic maintenance. Flexures are inherently low in damping and thus permit the high amplification of loads which is so desirable in resonant vibratory type fatigue testing. Another important principle in the design of fatigue test setups is that of completely locked-in steady loads for the system. This is accomplished by preloading springs or straps and is sometimes referred to as a "boot-strap" type of loading. This principle is important in that it permits the test apparatus to be essentially a self-contained unit with a minimum of external attachments and supports which may introduce damping or drift during testing. Back-to-back, or duplicate testing of more than one specimen at a time may also be utilized to reduce the time and cost of fatigue testing (Figure 11.17). Another principle frequently used in both resonant type and forced type fatigue testing is that of coupling. This consists of using the primary load input into a specimen, for instance, flapwise bending, to simultaneously induce one or more other types of loading, such as torsion or chordwise bending. For testing purposes, coupling can be achieved through system constraints, such as additional supports to ground or through special alignments of hinges of flexures, or it may be achieved by the relative locations of the spring and mass variables of the vibrating system (for a resonant type system). Additional springs and masses are often incorporated into a resonant type system in order to achieve the desired load distribution, to provide dynamic coupling, to produce additional loadings, and to vary the operating frequency (resonance) of the system (see Figure 11.18). Types of actuators for the application of cyclic loads which have been used in the past for fatigue testing are: electrodynamic, electro-hydraulic, spring and crank, and rotating eccentric mass. The rotating eccentric mass is not often used because the applied force varies not only with the amplitude of the eccentricity but also with the square of the frequency. Most frequently

Figure 11.18 Fatigue test of main rotor hub and blade retention system utilizing dynamic coupling to simulate complex loads and motion.

used for low frequency fatigue test actuation is the spring and crank because the applied force varies only with amplitude which can be readily controlled by a variable crank arm. The spring and crank is used instead of a rigid connecting rod and crank in order to reduce the actuator forces and power requirements during starting. Electrodynamic and electro-hydraulic actuators are used for the higher test frequencies, but also require considerably higher capital investment. The spring and crank actuator is simple, inexpensive, and easily adapted to a variety of test requirements where constant amplitude loading only is desired. Test instrumentation generally provides continuous monitoring of the loads and stresses and the number of cycles applied. The more important fatigue tests are set up so as to automatically shut themselves off in case the amplitude of motion varies beyond permissible limits, such as might happen during failure of the specimen. Also crack detection wires may be used to achieve early failure detection. Low frequency-of-occurrence loadings (e.g., once per flight, such as start-stop cycle, on once per maneuver) are usually applied by automatic cycling hydraulic or electro-hydraulic actuators.

11.9.3 Service Life Determination Including a Typical Problem

From the flight loads information used in conjunction with the fatigue test or analysis information, the allowable safe service life of a given helicopter component is finally determined, and is included in the FAA Type Data Sheet and in the Maintenance Manual. If all cyclic loads lie below the test endurance limit, including allowance for scatter, then an indefinite service life is assigned to that component. The same is true where an analytical substantiation, including the safety factor of 3 vs. endurance limit, is used. Where finite or limited life is indicated, however, either the cumulative damage hypothesis in combination with the *S-N* test data reduced by an appropriate scatter factor may be used in combination with the measured flight loads spectrum to arrive at an estimated fatigue life, or, the cyclic unit or spectrum testing method is used wherein the safe service life to be permitted would be based on the worst of at least four specimens tested, as again reduced by an appropriate reduction factor (in the life direction).

Analytical substantiation may also be performed using the cumulative damage hypothesis plus the safety factor of 3 (in some cases the analytical safety factor is permitted to vary to 1.5 at 1000 cycles). The FAA, however, is generally less receptive to analytical substantiation of fatigue loaded components than to fatigue test substantiation. Should the allowable service life of the component thus arrived at appear unduly low, one or more of the following may be used to improve the fatigue service life: extreme clamping preload of mechanical joints to prevent fretting, peening, or rolling the metal surface in critical areas to induce residual compressive stress, incorporation of increased fillet radii or other reductions in stress concentration, use of improved materials, or use of adhesive bonding to eliminate mechanical or welded attachments. Also, there is a growing appreciation of the improved safety derived from "fail-safe" components, a good example of which is the

PROBLEM

Allowable *S-N* fatigue test data (reduced):

1. ±1400 inch lbs — 260,000 cycles
2. ±900 inch lbs — 2,000,000 cycles
3. ±625 inch lbs — 14,000,000 cycles
4. ±500 inch lbs — 50,000,000 cycles (endurance limit)

TABLE 11.1

Condition	Rotor (rpm)	Percent Occurrence	Flight Measured Cyclic Load (± in. lb)
Power on right turns (2700)[a]	450	0.03	763
Power on left turns (2700)[a]	450	0.03	748
Power on right turns (2500)[a]	420	2.97	724
Power on left turns (2500)[a]	420	2.97	795
Power on pull-ups (2700)[a]	450	0.04	701
Power on pull-ups (2500)[a]	420	0.46	682
Power on partial power descent	414	3.00	752
Power on approach (2500)[a]	420	1.00	690
Power on approach (2700)[a]	450	2.00	708
Autorot. pull-up (415)	426	1.90	536
Autorot. autorot. landing	465	2.50	633
Special conditions Simulated power failure	420	0.10	937
Special conditions Altitude simulated by high-speed turns	420	0.05	641

[a] Engine rpm

(All other loading conditions exhibited cyclic stresses below the endurance limit.)

(For simplicity, neglect ground-air cycles and once per maneuver cycles, also assume no cyclic stress variation within each maneuver.)

Using the methods of rotor service life determination mentioned above and the reduced *S-N* fatigue test failure points listed above [3, 3a], calculate the allowable safe service life for the blade loading spectrum shown in Table 11.1. Assume that all cyclic loads are applied at one per rev of the rotor. The *S-N* test data presented was all obtained using the highest steady stress encountered in operation.

laminated strap type rotor retention [20]. A "fail-safe" structural component is one which still retains a large percentage of both its original static strength and its life to first readily detectable fatigue damage, after the occurrence of such detectable fatigue damage. Thus a rotor retention system of 20 strap laminations would obviously retain 75% of its original strength after the failure of five of the laminations.

11.9.4 Fail-Safe Components

A fail-safe structural component may be defined as one wherein the fatigue damage can with high assurance be detected prior to any dangerous loss of strength of the component. This concept [21], while only recently recognized in FAA requirements for helicopters, has nevertheless had defacto recognition by both industry and FAA engineers at the working level for many years, and has been recognized in both FAA and military requirements for fixed wing aircraft for many years. Both industry and FAA appreciation and understanding of the safety advantages of the fail-safe structural approach in regards to fatigue loading (as is the main concern of helicopters) lagged that in regard to quasistatic loadings (as of main concern to fixed wing aircraft) by many years, because of the more recent state of the art development in regard to fatigue loading as compared to quasistatic loading. I would categorically state without fear of contradiction, however, that the application of fail-safe structural engineering principles are even more important in respect to fatigue loaded structure of the helicopter than it has been in respect to the quasistatically loaded structure of the fixed wing aircraft. The gains to the helicopter operator and to the public due to the implementation of the fail-safe concept into the important fatigue loaded helicopter structural components are:

1. Elimination of the safety hazard of possible, though improbable, sudden catastrophic failure of major components without warning in flight.
2. Elimination of the uneconomic early retirement of many safe and reliable components in order to remove the very few unsafe components.
3. Improved inspectability so that the hazards of unusually severe operating conditions or maintenance neglect will not lead to catastrophic failure in flight.
4. Reduce the extreme level of skill required for maintenance and inspection by the operator in order to assure continued safe airworthiness in spite of human errors which cannot be completely eliminated—either in manufacture or in operation.

The fail-safe structural component by its redundancy, its inspectability, and its reserve strength will provide both adequate residual static strength and adequate residual fatigue life after any detectable initial fatigue failure, if one should ever occur in service for any reason whatsoever.

Any initial application of the fail-safe structural concept to helicopter components might involve a combination of replacement time (safe-life) and fail-safe evaluation. Thus it may be possible to safely extend the replacement time of the safe-life components which exhibit limited fail-safe capability by using a combination of the safe-life and the fail-safe concepts. In this case the replacement time may then be based upon the combined probability of not initiating a fatigue crack before the replacement time and the probability that the crack, if initiated, will be detected prior to catastrophic failure or loss of the limit load capability of the structure. The latter involves probability of detection, including effectiveness of inspection, the inspection time interval, and the residual fatigue life after a detectable partial failure. The end result of this combined approach would be to permit the use of a less conservative reduction factor in assigning the safe retirement life for the structure, while retaining an equivalent over-all safety level.

A complete fail-safe approach to the fatigue life problem of a primary structural component would involve no fixed assignment of safe-life, but would concentrate on ensuring that should fatigue cracks initiate, the remaining structure would withstand, with high assurance, service loads without danger of complete failure until the fatigue cracks or other damage was detected. Any evaluation of this type includes establishing which components are indeed fail-safe, definition of loading conditions and extent of damage which the structure needs to withstand, conducting the structural tests and analysis to substantiate that the fail-safe objective has been achieved, and establish the required inspection frequency and procedures to assure detection of the fatigue damage while it remains at a safe level. Design features used in attaining fail-safe structures are:

1. Selection of materials and stress levels that provide a controlled slow rate of crack propagation combined with high residual strength after initiation of cracks.
2. Design to permit detection of cracks, including the use of crack detection systems, in all critical structural elements before the cracks can become dangerous or result in appreciable strength loss, and to permit replacement or repair.
3. Use of multipath construction and the provision of crack stoppers to limit the growth of cracks.
4. Use of composite duplicate structures so that a fatigue crack or failure occurring in one-half of the composite member will be confined to that half, and the remaining structure will still possess appreciable load-carrying ability.

5. Use of backup structure wherein one member carries all the load, with a second member available and capable of assuming the extra load if the primary member fails.

Those points within the structure wherein fatigue failure may initiate must be identified and substantiated during the Type Certification of the helicopter. The extent of the partial failures must be such as to be obvious during the specified inspections. Such partial failure of a fail-safe structure would usually involve a complete failure of a principal element, although it may involve failure of more than one element or only a partial failure of a single element, depending on the rate of crack propagation, the ease of detection by inspection, and the desired inspection interval. Initiation of fatigue damage in inaccessible or uninspectable areas is very undesirable and avoided through design for a fail-safe structure. Typical examples of such partial failure due to fatigue damage may be as follows:

1. Cracks emanating from the edge of structural openings or cutouts which can be readily detected by visual inspection of the area.
2. A circumferential or longitudinal skin crack in the basic fuselage structure of such a length that it can be readily detected by visual inspection of the surface area.
3. Complete severance of interior frame elements or stiffeners in addition to a visually detectable crack in the adjacent skin.
4. Failure of one element where dual construction is utilized in components.
5. Failure of primary attachments, including control hinges and fittings.

Determination of the probable failure locations should be by test, analyses, or preferably both. This determination is based upon stress levels, strain surveys, fatigue analyses, fatigue tests, points of high stress concentrations or discontinuities within a structure, detail design, processing or fabrication considerations, service experience, and, last but not least, sound engineering judgement. The detection of fatigue cracks, or damage, prior to their becoming dangerous is the ultimate consideration in assuring the safety of a fail-safe structure or component. Therefore, the manufacturer and the FAA must provide the helicopter operator sufficient guidance information to assure that any fatigue damage will most assuredly be detected while at an early and safe stage.

11.9.5 Future Fatigue Testing Trends

The present trend is toward more automation, both in recording the flight data, in reducing and counting it, and in applying the spectrum data to the control of the fatigue testing. The reason for this trend is primarily to reduce the overall development program time required for a new helicopter design and also to eliminate the unavoidable human errors; not just to save testing manpower or cost "per se." Taped flight loads have already been obtained in several cases for actual military helicopter operations. Also, complete helicopter airframe fatigue tests have been experimentally controlled by tape inputs. Automatic stress and load spectra counting devices have also been used [17]. An automated spectrum type fatigue test has the inherent advantage of bypassing the entire controversy of the applicability cumulative damage theory [1], [16], and [17]. Problems yet to be solved in automated testing are (a) inability to apply judgement and to detect faulty data, (b) inability to detect the sources of the more severe fatigue loadings and to thus suggest the desirability of certain operating restrictions; and (c) inability to select or check the degree of maneuver severities before using the data without extreme complication of the system. Thus, it looks as though the engineer will still have to be around for a while, in spite of automation. The current trend toward more fail-safe structure also adds the following testing considerations: (a) residual static strength after detectable partial failures or cracks; (b) detectability of fatigue damage by inspection, etc.; (c) crack propagation rates for determining the inspection time intervals, and (d) the initial safe-life for a fail-safe structure.

11.10 HELICOPTER SERVICE PROBLEMS

Lack of maintenance. A flapping hinge bearing which, due to poor maintenance practices, deteriorated in its function to such an extent that blade articulation was impaired, resulted in greatly accelerated fatigue damage and leading to loss of the blade in flight. Investigation showed the following maintenance history: (a) wrong lubricant had been used at one time; (b) grease purging channel subsequently blocked; (c) only partial replacement of the damaged bearings which had been caused by the inability to purge the lubricant; (d) continued flight operation until the accident without correction of the observed lubricant blockage.

Out-of-track or out-of-balance rotors create unnecessary vibratory forces which are transferred to the airframe, the controls, and the crew and which create fatigue problems in each case. In one case, cracking of the main rotor support member occurred, and in another, cracking of the vertical and horizontal tail surfaces occurred. Fortunately, the affected airframe structure in these cases was sufficiently "fail-safe" that no serious consequences ensued.

Failure to correct minor problems can cause major problems due to fatigue. For instance, minor dents, nicks, or scratches, possibly aggravated by corrosion pits, if not corrected as soon as they occur on critical components such as rotor blades, can seriously reduce the service fatigue life. Smoothing out allowable defects and replacement of the corrosion protection is essential to safe maintenance. Critical fatigue components accidentally damaged beyond the repairable limits must be replaced.

With the increasing trend toward "fail-safe" structures wherever possible for fatigue applications, more and more emphasis will be placed upon inspection to find the occasional low life parts which we must learn to expect statistically. As noted in [2], as production volume and number of components in service increases the less practicable it becomes to rely solely on the "safe-life" concept as covered by existing FAA regulations [3] and [4]. With the present lack of emphasis on inspectability or on "in-service" inspection, we are missing an opportunity for considerably improving the safety record in regard to fatigue failures of helicopter components.

Abuse. Helicopters may be occasionally operated beyond restricted limits pertaining to rpm, gross weight, forward speed, blade balance, and blade track. For instance, repeated premature engine coupling shaft failures which occurred on one type of helicopter being exported to a certain foreign country were found to be due to a pilot custom of racing the engine, declutched, well beyond the operating restrictions every time it was started up. Thus, a lowly damped vibration mode was excited, resulting in an extreme fatigue loading, and bearing no relation whatsoever to normal flight operation with the clutch engaged.

Another source of premature fatigue failures are physical damage in the form of dents, nicks, and scratches due to dropping tools, hanger collisions, or careless handling or shipping. Dents cause local bending stresses which superimpose upon the normal fatigue stresses, thus shortening the life to failure. Nicks or gouges may leave severely deformed metal, containing possible microscopic cracks and severe residual stresses which are also very detrimental to the fatigue life.

Chemical contamination from crop spraying or dusting materials may cause stress corrosion cracking of certain structural materials. Some aluminum alloys (such as 7075 *T*6), high strength steels (such as for springs), and stainless steels (such as 17-7 *PH* or 440 *C*) may also be affected. Precautions in the form of special heat treatment or of paints or platings may be necessary or even a change in the structural material.

Unusually severe corrosion environment may also cause spontaneous cracking of structural metals, and surface pitting and galvanic corrosion can act as abnormal stress raisers on critical structural parts. Steel or aluminum alloys when exposed to severe corrosion conditions need special maintenance treatment such as frequent cleaning and waxing or painting (similar to the trim on your car).

Another potentially serious form of abuse is the re-use of wrecked and damaged components. Welding of heat-treated parts, riveting patches or splices in critical areas, and the unknown residual stresses or defects are but a few of the dangers from this practice.

Neglect. Improper repairs and improper assembly can have major effects upon fatigue life. You need only to think of the mechanic who doesn't know whether his torque-wrench reads in ft-lbs or inch-lbs, or who leaves out a key spacer in a bearing installation because he didn't take the time to read the maintenance manual. Riveted patches applied to an originally smooth structure are also very detrimental to fatigue life (as noted above).

Overloading and exceeding the placard speeds both tend to increase the rotor blade fatigue loads and may easily shorten the fatigue life by as much as one-half. Incompetent piloting is another factor affecting component fatigue life. Carelessness in regard to correcting observed abnormal vibrations or engine roughness will have adverse effects. Abrupt maneuvers, frequently executed, will reduce fatigue lives of many components.

Failure to comply with servicing recommendations of the manufacturer can reduce the life of many components. Exposure to weathering, erosion, wind, sand, etc., cause more rapid deterioration of the structure. Adequate mooring and hangaring are important in achieving the desired service lives.

Rough Handling. The effects of dents and dinges on the fatigue loaded components has already been pointed out. Parts are also often damaged by dropping them during storage or transportation. Even laying an aluminum alloy part carefully over the steel edge of a workbench may cause detrimental scratches and gouges.

Light gauge parts such as rotor blade trailing edges, and horizontal or vertical tail surfaces may easily be damaged while moving the helicopter on the ground or while mooring or hangaring it. It is impossible to build a helicopter with the ruggedness of an automobile and still have it fly. Dents at the trailing or leading edges of these surfaces may easily lead to a fatigue crack which eventually propagates through critical primary structure.

Recently, an incident of rotor blade cracking was traced to an area damaged by electrical sparks, which had left essentially small cast nuggets to act as metallurgical notches in an otherwise defect free surface.

Another form of rough handling damage occurs due to forcing the assembly of mismatched or deformed parts. Sometimes this may occur because of wreck damage or due to abnormal manufacturing tolerances between original and replacement parts. It may also occur because of improper support of the mating parts or of the helicopter. The damaged surfaces as well as the associated residual stresses may occur in an area already critical for fatigue, thus reducing the life.

Let us remember that no matter how well the helicopter is designed and built it will not provide the expected service life without reasonable care during operation, be it piloting, maintenance, inspection, or environment.

REFERENCES

[1] "Symposium of Fatigue Tests of Aircraft Structures: Low Cycle, Full Scale and Helicopters," ASTM, Special Technical Publication No. 338, 1963.

[2] H. T. Jensen, Reliability Concepts, Fatigue Loaded Helicopter Structures, A.H.S. 18th Annual Forum, May 1962.

[3] F.A.R. Part 27, Airworthiness Standards: Normal Category Rotorcraft.

[4] F.A.A. Advisory Circular, Fatigue Evaluation of Rotorcraft Structure.

[5] Lloyd Kaechele, "Probability and Scatter in Cumulative Fatigue Damage," Memorandum RM-3688-PR, December 1963, The Rand Corp.

[6] Lloyd Kaechele, "Review and Analysis of Cumulative-Fatigue-Damage Theories, Memorandum RM-3650-PR, August 1963, The Rand Corp.

[7] *CAL/TRECOM Symposium Proceedings*, Vol. I, *Dynamic Load Problems Associated With Helicopter and V/Stol Aircraft*, June 26–28, Buffalo, N.Y.

[8] AHS, Proceedings of 19th Annual Forum, May 1963, Washington, D.C.

[9] Gessow and Meyers, *Aerodynamics of the Helicopter*, Macmillan, New York, 1952.

[10] den Hartog, *Mechanical Vibrations*, McGraw-Hill, New York, 1944.

[11] "Preliminary Flight Test Data—*XH*-51*A* Rigid Rotor High Speed Flight Program," U.S. Army, TRECOM Interim Report No. 4, 7, and 9.

[12] J. F. Ward and R. J. Huston, AHS, *Proceedings of the* 20*th Annual National Forum, May 1962, A Summary of Hingeless-Rotor Research at NASA-Langley*, NASA.

[13] "Stress Determination and Fatigue Life of Generated Bevel Gears," Gleason Works Rochester, N.Y., 1960.

[14] Wells Coleman, SAE Preprint No. 627, "An Improved Method for Estimating the Fatigue Life of Bevel Gears and Hypoid Gears," Gleason Works, Rochester, N.Y.

[15] Aircraft Industries Association, "Aircraft Fatigue Handbook," ARTC Aircraft Structural Fatigue Panel, 1957.

[16] *Proceedings of the WADC Symposium, Fatigue of Aircraft Structures*, 1959.

[17] "Investigation of the Representation of Aircraft Service Loadings in Fatigue Tests," Technical Note No. ASD-TR-61-435, July 1962.

[18] Grover, Gordon, and Jackson, "The Fatigue of Metals and Structures," Technical Note No. NAVWEPS 00-25-634, revised June 1, 1960.

[19] Sines and Waisman, *Metal Fatigue*, McGraw-Hill, New York, 1959.

[20] L. G. Hohnson, *The Statistical Treatment of Fatigue Experiments*, Elsevier, New York, 1964.

[21] H. G. Smith, *Fail-Safe Structural Features of the Hughes OH-6A*, AHS 22nd National Forum, 1966.

Appendix

I. In any given flight condition each main rotor blade is exposed to essentially the same dynamic and aerodynamic variations at identical azimuth positions around the rotor disc, leading to identical (except for phase shift due to blade relative positions in the rotor) periodic functions repeating at the rotational frequency, $\Omega t (= \dot{\Psi} t) = 2\pi$. This same reasoning applies also to the tail rotor. Thus all helicopter rotor loads and motions are normally expressible in the form of a Fourier series:

$$\begin{aligned} f(\Psi) = A_0 &+ A_1 \sin \Psi + A_2 \cos \Psi \\ &+ A_3 \sin 2\Psi + A_4 \cos 2\Psi + \cdots \\ &+ A_{n-1} \sin \frac{n}{2} \Psi + A_n \cos \frac{n}{2} \Psi. \end{aligned}$$

II. Example of harmonic blade loading due to helicopter forward speed, V. Velocity at any blade element, $U = \Omega r + V \sin \Psi$. Aerodynamic forces at any blade element $F = CU^2$, or

$$\frac{F}{C} = \Omega^2 r^2 + 2\Omega r V \sin \Psi + V^2 \sin^2 \Psi,$$

or

$$\begin{aligned} \frac{F}{C} &= \Omega^2 r^2 - 2\Omega r V \sin \Psi + V^2(\tfrac{1}{2} - \tfrac{1}{2} \cos 2\Psi) \\ &= \Omega^2 r^2 + 2\Omega r V \sin \Psi + \frac{V^2}{2} - \frac{V^2}{2} \cos 2\Psi. \end{aligned}$$

Thus even in this simplified illustration steady plus first and second harmonic blade loads are indicated. Also, even higher harmonic loads can be shown to exist because of nonuniform inflow, as well as higher harmonic flapping, feathering, and lag motions.

III. Coriolis Forces

These are mass, or inertia, forces resulting from accelerations arising from blade momentum fluctuations due to the flapping motion. As an example, the simple case of a point mass, m, at distance, r, from the center of rotation is analyzed:

$$\begin{aligned} \dot{\Psi} &= \text{angular velocity of rotor,} \\ \beta &= \text{flapping angle,} \\ &= \beta_0 + \beta_1 \sin \dot{\Psi} t, \\ \dot{\beta} &= \beta_1 \dot{\Psi} \cos \dot{\Psi} t, \\ \dot{r} &= r\beta\dot{\beta}. \end{aligned}$$

Accounting for momentum changes in the tangential direction only of both the $mr\dot{\Psi}$ and $m\dot{r}$ (the two in-plane) momentum vectors,

$$\text{(1)} \qquad \overrightarrow{\frac{d(mr\dot{\Psi})}{dt}} = \overrightarrow{m\dot{\Psi}\dot{r}},$$

$$\text{(2)} \qquad \overrightarrow{\frac{d(m\dot{r})}{dt}} = \overrightarrow{m\dot{r}\dot{\Psi}},$$

$$\text{Total (1) + (2)} = 2m\dot{r}\dot{\Psi} = 2mr\beta\dot{\beta}\dot{\Psi}.$$

or

$$\text{tangential force} = 2mr\dot{\Psi}(\beta_0 + \beta_1 \sin \dot{\Psi} t)(\beta_1 \dot{\Psi} \cos \dot{\Psi} t),$$

$$\text{tangential force} = 2mr\dot{\Psi}[\beta_0\beta_1\dot{\Psi} \cos \dot{\Psi} t + \beta_1{}^2\dot{\Psi} \sin \dot{\Psi} t \cos \dot{\Psi} t].$$

$\therefore$ maximum coriolis forces are

$$\text{1st harmonic} = 2mr\dot{\Psi}^2\beta_0\beta_1,$$

$$\text{2nd harmonic} = mr\dot{\Psi}^2\beta_1{}^2.$$

Additional note: Coriolis forces can never combine in such a manner as to produce net steady in-plane loads transmitted from the rotor to the airframe.

IV. Loads Due to Pitching or Rolling Motion of the Helicopter

These loads are primarily of importance for an ideal rigid rotor; for example, because of pitching (rolling) acceleration, $\ddot{\theta}$

(1) rotor C.G. acceleration (in-plane) is due to rotor height above helicopter C.G., Z_R

Hub (in-plane) force $= m_R Z_R \ddot{\theta}$.

(2) hub moments due to momentum changes;
  (a) due to the $(I_R/2)\dot{\theta}$ change

$$\overrightarrow{M_y} = \overrightarrow{\frac{d[(I_R/2)\dot{\theta}]}{dt}} = \overrightarrow{\frac{I_R}{2}\ddot{\theta}}.$$

  (b) due to the $I_R\dot{\Psi}$ change

$$M_x = \overrightarrow{\frac{d(I_R\dot{\Psi})}{dt}} = \overrightarrow{I_R\dot{\Psi}\,\dot{\theta}}.$$

V. Effect of Rotating vs. Nonrotating Frame of Reference on Harmonic Loads

Any harmonic load (or moment) vector will split half each to the next higher and lower harmonics when it is transmitted from a rotating to a nonrotating structure (or frame of reference) and vice versa. Thus, as an example,

$$\text{rotating hub force} = P_0 \sin n\dot{\Psi}t$$

Then

$$\text{nonrotating hub force} = (P_0 \sin n\dot{\Psi}t) \sin \dot{\Psi}t$$

$$= \frac{P_0}{2}\cos(n-1) - \dot{\Psi}t\,\frac{P_0}{2}\cos(n+1)\dot{\Psi}t.$$

VI. Compensation of Harmonic Rotor Loads Due to Multiple Blades (to Nonrotating Reference)

(1) 2 blades—rotating:

$$F'_{b1} = P_0 \sin \dot{\Psi}t + P_1 \sin 2\dot{\Psi}t$$

$$F'_{b2} = P_0 \sin (\dot{\Psi}t + 180\ ) + P_1 \sin 2(\dot{\Psi}t + 180^0).$$

nonrotating:

$$F_{b_1} = \frac{-P_0}{2}\cos 2\dot{\Psi}t + \frac{P_0}{2}\cos\dot{\Psi}t - \frac{P_0}{2}\cos 3\dot{\Psi}t$$

$$F_{b_2} = \frac{-P_0}{2}\cos 2\dot{\Psi}t - \frac{P_0}{2}\cos\dot{\Psi}t + \frac{P_0}{2}\cos 3\dot{\Psi}t$$

Total: $F_b = -P_0 \cos 2\dot{\Psi}t.$

(2) 3 blades—perform as a class problem.

(3) In any rotor system a steady rotating vector will give rise to a first harmonic nonrotating vector.

(4) The rotor harmonic loads which are of major concern as fatigue loads on the (nonrotating) airframe are those transmitted at $n\Omega$, i.e., those due to $n - 1$ and $n + 1$ rotor harmonics (where n = no. blades and Ω = rotational frequency).

Index